A Series In

The History of Modern Physics, 1800-1950

The History of Modern Physics, 1800-1950

TITLES IN SERIES

VOLUME I
Alsos
by Samuel A. Goudsmit

VOLUME II
Project Y: The Los Alamos Story
Part I: *Toward Trinity*
by David Hawkins.
Part II: *Beyond Trinity*
by Edith C. Truslow and Ralph Carlisle Smith

VOLUME III
*American Physics in Transition: A History of Conceptual
Change in the Late Nineteenth Century*
by Albert E. Moyer

VOLUME IV
*The Question of the Atom: From the Karlsruhe
Congress to the Solvay Conference, 1860-1911*
by Mary Jo Nye

INTRODUCTORY NOTE

The Tomash series in the History of Modern Physics offers the opportunity to follow the evolution of physics from its classical period in the nineteenth century when it emerged as a distinct discipline, through the early decades of the twentieth century when its modern roots were established, into the middle years of this century when physicists continued to develop extraordinary theories and techniques. The one hundred and fifty years covered by the series, 1800 to 1950, were crucial to all mankind not only because profound evolutionary advances occurred but also because some of these led to such applications as the release of nuclear energy. Our primary intent has been to choose a collection of historically important literature which would make this most significant period readily accessible.

We believe that the history of physics is more than just the narrative of the development of theoretical concepts and experimental results: it is also about the physicists individually and as a group—how they pursued their separate tasks, their means of support and avenues of communication, and how they interacted with other elements of their contemporary society. To express these interwoven themes we have identified and selected four types of works: reprints of "classics" no longer readily available; original monographs and works of primary scholarship, some previously only privately circulated, which warrant wider distribution; anthologies of important articles here collected in one place; and dissertations, recently written, revised, and enhanced. Each book is prefaced by an introductory essay written by an acknowledged scholar, which, by placing the material in its historical context, makes the volume more valuable as a reference work.

The books in the series are all noteworthy additions to the literature of the history of physics. They have been selected for their merit, distinction, and uniqueness. We believe that they will be of interest not only to the advanced scholar in the history of physics, but to a much broader, less specialized group of readers who may wish to understand a science that has become a central force in society and an integral part of our twentieth-century culture. Taken in its entirety, the series will bring to the reader a comprehensive picture of this major discipline not readily achieved in any one work. Taken individually, the works selected will surely be enjoyed and valued in themselves.

A Series In

The History of Modern Physics, 1800-1950

VOLUME II

Project Y: The Los Alamos Story

Part I. *Toward Trinity*
Part II. *Beyond Trinity*

PUBLISHER'S NOTE

This book is an edited version of the LAMS-2532 report written in 1946 and 1947, originally titled, *Manhattan District History: Project Y, The Los Alamos Project.* Editors at the Laboratory in Los Alamos have added material declassified since the original report was issued, and have made minor alterations to the text in the interest of clarity and readability. In no way has the original factual material been changed.

We are grateful to the Los Alamos National Laboratory for the photographs that appear with the text, and to its staff for the considerable effort expended in editing and typesetting the volume. Special thanks are due also to Robert D. Krohn who has so ably coordinated this joint effort.

Project Y: The Los Alamos Story

PART I

Toward Trinity

BY DAVID HAWKINS
With a New Introduction by the Author

PART II

Beyond Trinity

BY EDITH C. TRUSLOW
AND RALPH CARLISLE SMITH

Tomash Publishers

LOS ANGELES / SAN FRANCISCO

Library of Congress Cataloging in Publication Data

Manhattan District history, Project Y, the Los Alamos Project.
    Project Y, the Los Alamos Project.

(History of modern physics, 1800-1950; v. 2)
    Reprint. Originally published: Manhattan District History, Project Y, the Los Alamos Project. Los Alamos: Los Alamos Scientific Laboratory, University of California, 1961. (LAMS; 2532) With new intro.
    Includes indexes.
    Contents: pt. 1. Toward trinity / by David Hawkins — pt. 2 Beyond trinity / by Edith C. Truslow and Ralph Carlisle Smith.
    1. Manhattan Project—History. 2. Los Alamos Scientific Laboratory—History. I. Hawkins, David, 1913-   . II. Truslow, Edith C. III. Smith, Ralph Carlisle, 1910-   . IV. Los Alamos Scientific Laboratory. V. Series. VI. Series: LAMS (Los Alamos Scientific Laboratory); 2532.
QC773.3.U5M25 1982   623.4'5119   82-50751
ISBN 0-938228-08-0

Preface

THIS VOLUME tells an important part of the story of one of the greatest scientific achievements in history—the story of the founding of the Laboratory at Los Alamos through the successful completion of its secret mission to create the first atomic bomb. The celebration of the fortieth anniversary of the establishment of the Laboratory provides an appropriate occasion to reprint this important and significant historical report, compiled in 1947, from internal documents. Few individuals, even those who are knowledgeable about that early era, will read this work without feeling a sense of profound discovery at the importance of those daily events which led to the most dramatic technological development of this century.

The United States and the free world in general, owe a great debt to those who, prior to World War II, saw that science and technology could be vital for the preservation of freedom. It was such a group that assembled at Los Alamos. Science and technology have long shaped the outcome of international conflicts, domestic policies, and society in general. But World War II created two important watersheds. First, the magnitude of developments from the world of science was far greater than those in any previous conflict. Second, the public recognition of the positive results of scientific developments was extensive, deep, and profound. Those results have created the problem of how science and society can best survive and prosper together. We constantly struggle with this issue, sometimes successfully, sometimes with less serenity about the outcome.

This account covers three critical periods: the initial formation of the Laboratory, the successful completion of the monumental task assigned to Los Alamos, and the sudden changes that occurred in the sixteen-month period after the abrupt end of hostilities brought about by the technological developments produced at Los Alamos. Each of these periods created tremendous strains for those who guided the activities at Los Alamos during these years. Each period was successfully managed by the hand of a wise leader supported by brilliant and dedicated workers.

This history does not pretend to provide the entire record of the activities at Los Alamos, nor is it couched in the rich language appearing in some of the histories that have been compiled since the publication of this volume. Time did not allow such polish. But this history has an authenticity, readily apparent to the reader, that flows directly from its foundation resting in the extensive documentation provided by the reports of the technical divisions and from other material written at Los Alamos.

The account of this successful venture provides stimulating reading. Those who examine this volume will agree that the men and women who made this massive undertaking into a successful reality rendered a profound public service to their country.

Los Alamos today continues in the tradition which began with the formation of the Manhattan Engineer District Project. The current work of the Laboratory, though much broader in scope than during the war, continues the effort of exploring the frontiers of science to search for concepts and applications that will be of use in meeting the important needs of national defense and energy security. The Laboratory has always been a place that is alive, and that has meant new growth and development. Los Alamos is now many times larger than it was during the Manhattan Project. More than forty percent of our programs fall in nonweapons activities such as energy technology development, life science research, magnetic fusion energy, and environmental assessment. But our major activity remains weapons research and development, both nuclear and nonnuclear.

We build upon the fine foundation laid by the people of vision who, forty years ago, knew that concentrated scientific work would be of great value in solving important national needs. That tradition and that vision guide our path today.

DONALD M. KERR, *Director*
Los Alamos National Laboratory
Los Alamos, New Mexico
November, 1982

Introduction

THE PRESENT HISTORY, Volume I, was first produced in 1947 as a manuscript of the Los Alamos Laboratory, Manhattan District, U.S. Corps of Engineers. As its author, I had the privilege of deciding upon its security classification—*Secret* or *Top Secret*. It had been argued, by our Intelligence Officer, that its synoptic and qualitative treatment would make it attractive reading to a foreign power: he recommended *Top Secret*. I opted for *Secret*; the book did not tell a reader how to make a bomb. It was a history, not a recipe book. The recipe book, by contrast, would be a ton or two of documents and drawings.

In any case, the half-life of secrets, even top-secrets, is short. The book and its sequel were declassified in 1961, with only fairly minor deletions. Norris Bradbury, who, at that time, was the second Director of the Laboratory, sent me a copy of the report. With it, he enclosed this letter:

> Well, here is the famous history after all these years with really rather few omissions on account of classification! Although nearly sixteen years have passed, I expect the words will seem very familiar to you and will bring back additional vivid personal recollections. We thought you would enjoy having copies of these two volumes for your own personal library.

The volumes as thus edited were published by the Office of Technical Services, U.S. Department of Commerce. Before that time, they had remained on the shelves of the library at Los Alamos. If my story was read in that interval, it was read as a technical summary; I doubt that the motive would often have been one of historical curiosity. To use Robert Oppenheimer's expression, voiced in the wider military-political context of that period, it was "too soon for history, too late for advocacy."

During those years, I was, myself, so to speak, declassified, and had no access to the book or to our erstwhile files. I well remember my first rereading of it after declassification, early in 1962, and some of the reactions it provoked. The first was a strong revival of memory, though not strong enough to spot readily those deletions referred to above. The second reaction was a kind of disappointment, a sense that of such extraordinary people, events, and developments I had produced a record so limited by the style and restraint of an official military history. Could I not have emulated, say, the style of Sir Thomas Mallory, or that of a Bernal Diaz in his chronicle of the conquest of Mexico?

Actually, the charge I had been given, to write this history, was, in fact, more restrictive. In early 1945, Robert Oppenheimer offered me the chance to produce a technical, administrative, and policy-making history of the wartime Laboratory, and I had accepted with great pleasure, though not without misgivings as to my competence: I had no historiographic training.

I arrived at Los Alamos in early May, 1943, some six weeks after the formation of the Laboratory, having hastily left a job in Berkeley as Instructor in Philosophy. I was a partial and junior replacement for my friend and department chairman, William R. Dennes (see below, p. 33), who had earlier come to Los Alamos but decided not to stay. As a youthful amateur of physics and mathematics, I soon learned something of the underlying commitment and substance of this extraordinary project, and of the style of those who worked at it. In its early months, the Laboratory was distinguished by an almost total lack of professional administrative officers. As an administrative assistant to Robert Oppenheimer, I was therefore given many odd jobs and was able, as a result, to develop a fairly coherent picture of things as they evolved in the Laboratory with regard to our relations with the U.S. Engineers and with the civilian community. I came to know most of the scientific staff personally, and in the interest of getting the young ones deferred from the draft, which was one of my jobs, it was necessary to learn about their professional backgrounds. The young ones were, in fact, most of the staff (see Graph 1, p. 483).

When the time came to write this story, I surely did not know enough to deal with all of the technical problems and policy questions, but at least I knew enough to probe further, to ask some of the right questions of some of the right people. It was an odd experience, after the rather intense activism of the first year and a half, to become the observer and chronicler of things still in the state of being born. My very competent assistant, Emily Morrison, and I even played a bit at the game of drafting an account of things which had not yet happened.

I was aware, at the time, of this shift in perspective, and commented on it in the Preface. Soon I began to see how the various divisions of the work might flow together to a predictable end. Where those involved in the work mainly saw uncertainties and problems, I began, too easily, to smooth these over, seeing inevitable outcomes. So I think I could indeed have tried harder to encapsulate the life, its personalities, presuppositions, and tensions. I do not now think this failure was a serious distortion.

General Groves remarked to Oppenheimer, with a touch of official humor, that he might as well get someone to do the history who was a loyal supporter. Indeed, I was that, like most others of us in that place and time. When he first read my manuscript after the war, Oppenheimer remarked, rather darkly, that I was surely adept at "walking on eggs." He was given to dark sayings, and would carry the discussion no further. He must have meant that I had avoided or touched too lightly upon some of the conflicts and controversies of which he had, himself, been keenly aware. I did not take the remark to be altogether a compliment.

I had, of course, been aware of some of those policy conflicts, and had found them hard to dig into. Documentation was scarce or absent, personal recollections were cautious and polite, and in 1945, the prevailing mood was hardly one of reminiscence. In the end, I must say I did not find those crises and controversies very significant. Oppenheimer was indeed a superb leader of that wartime weaponeering enterprise. In policy matters and planning he was often able to achieve a consensus. He did so through a remarkable art of analysis and reformulation, in which initial disagreements fell into place as components of a consistent pattern. But his manner, usually responsive and gentle, did not always conceal the darts which a mind so quick could sometimes impulsively fire. On balance, I am now—thirty-five years later—less critical of the style and omissions of the account as written. It pays central attention to the one overwhelmingly central figure; it is an early chapter in the biography of the bomb. The reader who wishes to find some compensation for these omissions may wish to read *No High Ground* by Fletcher Knebel and Charles W. Bailey III (New York: Harper & Row, 1960) an excellent work about the final developments of the year 1945. For a closer look at Oppenheimer's life, Alice Kimball Smith and Charles Weiner have edited *Robert Oppenheimer, Letters and Recollections* (Cambridge, Mass.: Harvard University Press, 1980) which is highly recommended.

There were important wider issues affecting the wartime history of Los Alamos not dealt with in the present book. For the treatment of some of these, there is an excellent official history of the entire Manhattan Project, *The New World, 1939/1946: A History of the United States Atomic Energy Commission, V. I,* by Richard G. Hewlett and Oscar E. Anderson (University Park, Pa.: Pennsylvania State University Press, 1962). A still broader context—that of the political-military history of World War II—as this affected and was affected by the bomb project, is not yet fully explored. A fascinating interpretative history, based in part on important materials recently made available, is *A World Destroyed* by Martin J. Sherwin (New York: Knopf, 1975). This is a work of excellent scholarship, and one which certainly has caused me to modify some of my earlier beliefs—in particular, those which I and others gained, I think, from Oppenheimer—about President Roosevelt's perceptions and plans.

All or most of such matters, of course, lay beyond my purview. But there was one kind of development I must mention, a subjective one, which took place in parallel with the technical work of weapon development, and about which I surely knew at first hand. Everyone's life was being changed, changed radically I think, and irreversibly. Many of us were aware of those changes at the time, though I think even the most reflective of us were inadequately aware of them. I could not then have generalized about the nature of these changes, and I am not much more able to do so in retrospect. We all did know we were involved in something which would alter the nature of the world. We understood less, perhaps, the reflex effect upon ourselves.

By mid-1945, there were some 400 scientific staff members, including senior administrators. Through weekly meetings of the Coordinating Council—group leaders and above—everyone had access to everything going on, including decisions most recently taken. Through the weekly colloquium, staff members were invited to hear reviews and reports of research. At Los Alamos there was no internal compartmentalization above the staff member level. This reflected an organizational principle for which Robert Oppenheimer had fought hard, against all "need-to-know" habits of secret research and development. The policy called "need-to-know" was euphemistically misnamed; it should have been called "need-not-to-know." Its aim was to extend the half-life of military secrets. But, unfortunately, A's need to know the fruits of B's research is sometimes evident neither to A nor to B separately, nor to the official who restricts their communication. Robert Oppenheimer's hard-won battle to escape from internal compartmentalization at Los Alamos was a major factor of its extraordinary wartime morale. His victory was matched, however, by that of General Groves, who insisted on the extreme isolation of Los Alamos from the rest of the Manhattan Project.

I suppose it is possible that some of the scientific staff simply took it as a technically exciting job, a pleasant wartime retreat among the hills of a once very violent (but very ancient) volcano, and thought little more about the fact that they were personally involved in a development which would, in one way or another, alter the whole pattern of world affairs. But from retrospective evidence, a good majority was touched profoundly.

A part of this reflex effect lay simply in the transformation of academic physicists, chemists, and mathematicians into creators of a radically new weaponry. More consciously in some cases, less in others, they were deeply affected by moral-political concerns about the consequences, yet at the same time, flattered into self-consciousness by a sense of its overwhelming significance. I do not imply uniformity in the way these contrary pressures changed us; however powerfully human beings are affected by novelty they, in the end, try to reestablish some continuity with the past. In such a humanly diverse group as we were then, the changes I have seen or inferred, the radical changes, have also been diverse.

But of course the similarities were striking. First of all, these youthful scientists had, or could acquire, far more easily than most, a ready intuition—all too rare in the rest of the world, as we were to learn—of what it would mean to see something in the order of a million-fold increase in energy release per kilogram of unstable matter, from atomic to nuclear exergonic chemistry. This readiness to understand came from an education in physics which can bring with it a deep appreciation of the significance of the powers of ten, of the ways in which the whole face of nature changes with any large change of scale. Second, these same scientists had advanced notice, over two or more years during which they lived, almost alone, with a virtual reality of which the rest of the world knew nothing; they had some chance to explore imaginatively its future ramifications and confront its uncertainties. For both of these reasons, whether consciously in prospect or afterward in later life, they share still, after a third of a century, a sense of being marked by an irreversible commitment.

Robert Oppenheimer once dramatized this commitment by saying that scientists "had known sin." His statement angered some who believed their own involvement in the weaponeering had been morally justified and who took him to be voicing some sort of invitation to repentance. That was not what he meant; they did not understand him. He was not speaking the language of established morality but that of religion or philosophical ethics, of the Garden of Eden, of lost innocence. Freeman Dyson made the same observation in Faustian terms when he spoke of a compact with the Devil. These were both ways of saying that any major deliberated choice we make, one which breaks new pathways, creates a compact: we are held irrevocably to its consequences. The knowledge of sin, the compact with the serpent, is nothing other than a metaphor to describe the fullest and deepest awareness we have of what it means to live fully with the processes and consequences of personal choice.

One clear manifestation of this as expressed by scientists throughout the Manhattan Project was the burst of political activity which followed the end of the war with Japan. The indispensable record and analysis of that early period is told in Alice Kimball Smith's work, *A Peril and a Hope: The Scientists' Movement in America 1945-47* (Chicago: University of Chicago Press, 1965). It is my interpretation—which I believe Alice Smith's work will sustain—that the burst of activity was not all "political" in a narrow or partisan sense. In fact, no political party here or abroad, left or right, did, in that time, give evidence of adequately understanding the newly created issues which were defined around the possibilities of nuclear war. What we begin to see was a long series of improvisations, each driven by those previous to it, aimed to deter nuclear war by steadily increasing the number and power of its instruments. The result has, in fact, been a kind of metastable equilibrium, with the long-range issues of military nationalism even farther from solution than they seemed to have been in 1946. There has been no peace even though nuclear weapons have not been used. But even though they have not been used, their potential has grown more devastating, and a sensitive and trembling needle on the scale of probabilities goes dangerously up in every major international crisis. The kinds of political proposals which arose from discussion among the Los Alamos and other Manhattan Project scientists, and which centered around the possibility of denationalizing or internationalizing the whole of nuclear weapon control, research, and technology, implied a clear anticipation of all this. If those schemes seemed or still seem farfetched, one must come back to the fact that the weapons themselves were much farther fetched, and that no stable equilibrium to prevent this use appears yet at hand. What Harry Truman described, in 1945, as ''a new force too revolutionary to consider within the framework of old ideas'' has not yet sufficiently forced a recognition of any framework of ideas within which it *can* be adequately considered. Ideas cannot be forced. Those early proposals of the Manhattan scientists, the sense of which Truman's first message to Congress partially and transiently reflected, may or may not have defined a then practicable national policy, a realistic solution. They did, at any rate, define what Leo Szilard once called a *mathematical* solution. A mathematical solution, he said, is one which, even though seen as impractical, adequately *defines* the problem.

I have spoken briefly of the history of this book, of its background and context, and of some things omitted on the subjective side, which a less circumscribed account would have treated at length. I suggested a defense of these omissions: the book is, primarily, a biography of the bomb. It is the central character. But this is only one chapter, a first. To continue the emphasis on this central character, I wish to suggest the need for a sequel: an objective account of all nuclear weapon developments since that early time. Without such a sequel the present story has only a diminishing interest. This later part of the biography would cover three decades rather than three years. In comparison with these more recent achievements, those of wartime Los Alamos seem small: yet the latter have provided the nucleus and set the pattern for all that has happened since.

In suggesting such a sequel, let me first step back to some prenatal times, back to the discovery of fission surely, and even farther. The fact that the whole story of fission power and of the nuclear weapons depends absolutely on a rather trivial detail of terrestrial history is not always appreciated. At the end of the periodic table—the classical table, before the transuranics—lies uranium, famous first at the end of the nineteenth century for its newly-found activity, and then for its generation of radium. The properties of uranium exhibited initial evidence of the character and magnitude of possible inner energy changes occurring in some of the atoms. Though widely distributed in the earth's mass, uranium is scarce indeed in concentrations high enough to allow its extraction for dollars to the pound. A kilogram of this element, almost all of it U238, contains about seven grams of U235, the scarcest of the scarce. Without that miniscule amount of U235, there would have been nothing for Fermi's pile to work on, nor indeed, any other; nothing to enrich, and no re-creation of plutonium, otherwise long gone from the planet's early history. Thus, no Oak Ridge and no Hanford. Without fast fission of these isotopes, in turn, no thermonuclear weapons. There may, someday, or somewhere else, be alternative paths to the large-scale release of the nuclear energies trapped in the heaviest elements or the lightest. That would be for another planet, another time-track. In our own history, at least, this very minor accidental isotopic fact has provided the small fiber from which all the fabric of the nuclear era has been spun and woven.

Against this background of so minor a cosmic-terrestrial detail, one should then more keenly appreciate the story of Oak Ridge and Hanford, with their massive investments in uranium separation and plutonium production. For a sense of scale, one should notice the momentous arrival, at Los Alamos, of the first one hundred grams of highly enriched uranium (in bulk; a few milliliters) in June, 1944. The first plutonium (not yet from Hanford) had come in milligram amounts, a few months earlier (cf. p. 79). These were the minute forerunners of a man-made transformation which nature had barely allowed us to begin, but which, thus allowed, grew rapidly. Graph 12 (p. 494) shows the earliest exponential part of a logistic curve which would only have begun to level off far later, when grams and kilograms had been replaced by tons.

Indeed, every other measure would show the same pattern of growth. It took three years to produce the first fission explosions, the Trinity test, and the bombings of the two Japanese cities. Since then, in only ten times that interval, there have been hundreds of tests with a total energy yield of many thousand fold. By such measures, the uses of nuclear fuel for power production have, to date, been minor.

Another graph would show costs. There was a figure for the direct cost of the Manhattan Project as a whole. The sum was two thousand million dollars, of which the Los Alamos budget was only a small fraction, perhaps two to five percent. To account for all the later activities which were originally concentrated in Los Alamos would be to survey many diverse installations, all of them scaled up in kind and quantity from the original. The two thousand million figure translates roughly into two hundred thousand man-years. (The *direct* labor cost rose to about one hundred twenty thousand employed.) For any reasonable multiplier, the later growth became a not negligible fraction of the capital and—mostly skilled—labor force, and implies a radically new industry increasingly enmeshed in the whole economy.

Yet that wartime investment was, itself, a major scaling up of any previous military commitment to research and development. It already begins to confront the historian with all the problematic aspects of very large-scale governmental enterprises. This confrontation brings me back to the wartime growth of the nuclear enterprise. There was an in-joke, current at the end of the war, that General Groves had been picked to run the Project because in earlier assignments he had managed to spend more money than had any previous general. He had been Deputy Chief of Construction for the Army, and *inter alia,* had been in charge of the construction of the Pentagon building. Groves was, in fact, an extremely capable administrator who carried out an immensely complicated wartime assignment. His memoir has the—unfortunately—unoriginal title, *Now It Can Be Told* (1962; reprinted New York: Da Capo, 1975). It is a book which, for many historian's reasons, should be carefully studied: for matters of historical fact, of course, as perceived by the Project's chief and powerful executive; but especially for the self-revealing openness with which Groves presupposes in his readers, his own system of beliefs and perspectives. Groves did not appear to understand that the project would change the world. The scientists did. Yet it was he, and not they, who in the end, created this military armamentarium and who gave it life and direction within an institutional framework which in the end it would radically destabilize. Groves' memoir gives a powerful impression of the build up of an institutional momentum which, though generated and directed from small beginnings and highly controlled, grew great enough to almost limit or even predetermine many later decisions. Groves was the Project's major creator and yet he well-nigh appears, in the end, as its creature.

With that observation I come back to what I have called the biography of the bomb. If the institutional momentum of this extraordinary military development—which annihilates all historical, Clausewitzian conceptions of warfare—is not wholly beyond intelligent control, then we have time to go back to what Truman should have meant when he said that the release of nuclear energy had created a force too revolutionary to be considered within the framework of old ideas. The ideas which have, in fact, dominated military matters during the intervening third of this century are unhappily not new; they are very old ideas, and all too familiar. They need to be reexamined with just that question about the nature of the new force and any new framework needed to control it. This is the final and most difficult part of the biography of the bomb to understand and to write. It has been too long neglected. Such a history cannot, of itself, create a new framework of international policy, but it can powerfully criticize and discipline our search for such a framework, and contribute to its definition. It is not too soon for that kind of history; nor too late for the advocacy which it can help create and support.

DAVID HAWKINS
Boulder, Colorado
September, 1982

Contents

ORIGINAL PREFACE TO PART I

Project Y, the Los Alamos Project, was part of the Development of Substitute Materials (DSM) project devoted to developing the atomic bomb. This branch of the DSM, created early in 1943, was the center of bomb development and production, as distinguished from development and production of nuclear explosive material.

The history of the DSM project is noteworthy, not only because of the remarkable achievements and potentialities of nuclear technology, but also because of the wartime character and motivation of their initial development. Because of its large social cost, a scrupulous accounting of the entire venture was required. Project Y occupied a crucial position; the success of the entire DSM undertaking depended upon its success.

The nature of Project Y required a careful accounting of its technical, administrative, and policy-making activities. This document is a record, not an interpretation of events. However, it has not been forgotten that these events took place within a wider context, the evolution of organized scientific research and world technology.

Another limitation inherent in an official record is omission of many subjective factors. The success of so complex and uncertain a venture as Project Y depended upon its ability to convey explicit and publicly accountable information. But this ability depended, in turn, upon an accumulation of experience and skill in technical and human affairs inseparably associated with the qualities, and even the vagaries, of personality. What appears in retrospect as a natural unfolding of possibilities acquired this appearance only through the interaction and, on occasion, the clash, of opinion in an atmosphere dominated by the problematical and the uncertain. The omission of the subjective is inevitable in an account that must be based upon objective evidence.

However, these omissions do not distort the picture seriously, as they would if important occurrences and tendencies were not objectively justified. The pattern of development is so largely rational as to be a tribute to the unity of purpose of all concerned: administrators and scientists, civilian and military. Much of the credit must be given to the Director, J. Robert Oppenheimer, for his general leadership and, especially because he understood the need for unity and sought in every way to foster it. Members of the Los Alamos Laboratory who provided assistance were J. A. Ackerman, S. K. Allison, E. Anderson, K. T. Bainbridge, C. L. Critchfield, Priscilla Duffield, A. C. Graves, L. H. Hempelmann, A. U. Henshey, H. I. Miller, Emily Morrison, Philip Morrison, N. H. Ramsey, Frederick Reines, Ralph Carlisle Smith, and R. F. Taschek. Especially helpful were Emily Morrison and Priscilla Duffield in their ingenious research into the records of an organization that was frequently too busy to be concerned with posterity. Mrs. Morrison has prepared the graphical material, has drafted several of the chapters, and has given invaluable general assistance. Finally it must be made clear that all errors of fact in this record are the sole responsibility of the author.

DAVID HAWKINS
6 August 1946

PART I

Inception Through Mid-August 1945

by David Hawkins

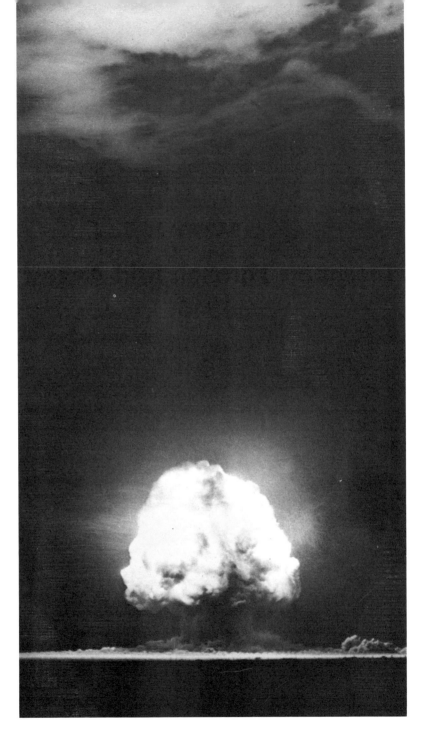

The detonation of the world's first nuclear device took place at 5:29 a.m. on July 16, 1945. "Trinity" was the code name for the test.

CHAPTER 1

Introduction

OBJECTIVE AND ORGANIZATION

Reasons for the New Project

Early in the Development of Substitute Materials (DSM) project, the most urgent requirement was for large-scale production of nuclear explosives. There could be no atomic bomb without usable amounts of fissionable materials. Separation of ^{235}U and production of ^{239}Pu both presented major scientific and industrial problems, so there was little need or time for detailed theoretical or experimental work on the mechanism of the nuclear explosion. Such work had not progressed far beyond showing the probable feasibility and effectiveness of the fission bomb as a war weapon. However, by mid-1942, the scientific and engineering problems of developing such a weapon called for early, intensive effort. Overall responsibility for the physics of bomb development had been given to the Metallurgical Laboratory of the University of Chicago. However, that organization was geared to develop the slow-neutron chain reaction as a source of plutonium. Work on fast-neutron chain reactions for bomb development had been largely subcontracted.

A more concerted program of bomb development began with the June 1942 appointment of J. Robert Oppenheimer, from the University of California, as Director of the work. Although associated with the Metallurgical Laboratory, Oppenheimer worked at the University of California with a small group of theoretical physicists. In coordinating the experimental work on fast-neutron physics, he was assisted by J. H. Manley of the Metallurgical Laboratory, and later by E. M. McMillan, who joined Oppenheimer's group in Berkeley.

Late in June 1942, Oppenheimer, J. H. Van Vleck, R. Serber, E. Teller, E. J. Konopinski, S. P. Frankel, H. A. Bethe, E. C. Nelson, and F. Bloch met at Berkeley to discuss the theory of the bomb and to plan future work. Much of the discussion was devoted to a new type of explosive

3

reaction, a thermonuclear reaction in deuterium that Teller had considered. The theory of the shock waves produced in a chain reaction explosion was discussed in light of work by Bethe and Van Vleck. The damage expected in terms of energy release was discussed qualitatively by scaling up from small explosions and comparing them with such disasters as the Halifax explosion. Completed theoretical and experimental work also was reviewed thoroughly. Rough but qualitatively reliable data were available from work done under Metallurgical Laboratory contracts. Relevant information had been obtained from British work by Peierls, Fuchs, Davison, and Dirac. British theoretical results also were available. Although much of the discussion was not along what subsequently became the main line of development, it helped to clarify basic ideas, define basic problems, and indicate that development of the fission bomb would require a major scientific and technical effort.

Subsequently, several conferences were held with experimentalists in Chicago. The University of Chicago had already let several subcontracts to investigate nuclear properties relevant to bomb design. A "loose" organization was formed among Rice Institute; the Carnegie Institution of Washington's Department of Terrestrial Magnetism; the Universities of Wisconsin and Minnesota; Purdue, Stanford, and Cornell Universities; and the Universities of Chicago and California.

By October 1942, the difficulties involved necessitated formation of a new organization. Even initial provision of nuclear specifications for the bomb was seriously hampered by the lack of an organization united in one locality. Clearly, without such an organization the ordnance work would be impossible.

Location

The Project Y site, selected in November 1942, was the Los Alamos Ranch School, on an isolated mesa of the Pajarito Plateau about 40 miles northwest of Santa Fe, New Mexico. The reasons for selecting such a site typify the character of the project. First, a large proving ground with a climate suitable for outdoor work in winter would be needed. Second, the site must be remote from both seacoasts and the possibility (at that time not negligible) of attack. Other locations that satisfied these requirements would have been more accessible, but inaccessibility would not create serious problems for a small project such as this was intended to be. Its subsequent growth was unforeseen. In accord with prevalent military

4

security policy, inaccessibility was a deciding factor in favor of this location.

During 1942, steps were taken to transfer the entire DSM project from the Office of Scientific Research and Development (OSRD) to the Manhattan District. The highest secrecy had to be maintained throughout the entire program, because the new subproject was its most secret part. The need for unusual isolation was met, in part, by the site's inaccessiblity. Access from populated areas was difficult, except along certain roads and canyons, owing to the line of cliffs forming the eastern edge of the Pajarito Plateau. Also, the geographically enforced isolation of project personnel would minimize the possibility that secret information might diffuse outward through social and professional channels.

Choice of a site was not the responsibility of the project director or his staff. Their views, nevertheless, strengthened the arguments for isolation in the minds of the military authorities. The task confronting the project was not ordinary development and engineering but intensive, highly organized research in a discipline only schematically explored. Physicists, chemists, metallurgists, and engineers collaborated to solve difficult problems, many of which could not even be anticipated until the work was well under way. The need for collaboration was emphasized by the fact that the bomb had to be ready for production as soon as usable quantities of nuclear explosive became available. Success would require the highest integration and, therefore, decentralization and mutual confidence. To this end, free communication within the Laboratory was indispensable.

In contrast with the requirements stated by the scientific staff, the normal military procedure for protecting secret information was by subdivision. Each individual would have access only to information immediately relevant to his work. This conflict of scientific and military requirements was, of course, not peculiar to nuclear research. Many of the scientists had been engaged in other war research and were convinced of the evils in obstructing the normal flow of information within a laboratory. They were vigorously opposed to compartmentalization. However, no alternative was acceptable that satisfied military security requirements. Those requirementscould be met by allowing internal freedom and imposing severer external restrictions than would otherwise seem necessary. Adoption of such a policy necessitated an isolated location for the project.

Organization

The Los Alamos site and a large surrounding area were made a military reservation. The fenced and guarded community was made an Army post. The Laboratory, in turn, was built within an inner fenced and guarded "Technical Area." Both military and technical administrations were responsible to Major General L. R. Groves, who had overall executive responsibility. The Commanding Officer, who reported directly to General Groves, was responsible for the conduct of military personnel, maintenance of adequate living conditions, prevention of trespass, and special guarding. Oppenheimer, as Scientific Director, also was responsible to General Groves, whose technical adviser was J. B. Conant.

In addition to his technical responsibilities, the Director was responsible for security policy and administration. This provision was to guarantee that there would be no military control of exchange of information among scientific staff members, but to fix responsibility for maintaining security under these conditions. In security matters, the Director was assisted and advised by a Military Intelligence Officer.

Project Y finances and procurement operations were handled by the University of California as prime contractor. While these operations were largely employing personnel and establishing a procurement office, the University acted under a January 1, 1943, letter of intent from the OSRD. This letter was superseded by a formal contract, W7405-ENG-36, with the Manhattan Engineer District of the War Department, effective April 20, 1943, and retroactive to January 1. This contract, with subsequent supplemental agreements, was the formal basis of the project's operation.

The financial operations of the University of California at Los Alamos were provided for when J. A. D. Muncy was appointed resident Business Officer. Procurement of materials was through a dual organization. In addition to the project's procurement division, the University established a special purchasing office in Los Angeles, primarily for security reasons. It might be possible to determine both the nature and progress of the work from knowledge of its procurement operations. Therefore, goods ordered through the Los Angeles office were received there and transshipped to Los Alamos. The procurement offices at the site were under the direction of D. P. Mitchell, who had long been in charge of laboratory procurement for the Physics Department at Columbia

University, and later for a National Defense Research Council (NDRC) project there.

The responsibilities of the military and contractor organizations, and the scope and purpose of Project Y, were outlined in a February 25, 1943, letter to Oppenheimer from General Groves and Conant (Fig. 1). The letter also covered future organization of the project, which was expected to remain an OSRD-type organization while engaged mainly in nuclear research. Later, during the dangerous work of bomb development and assembly, the operation would be conducted on a military basis, with opportunity for civilian staff members to be commissioned as officers. This reorganization proved unnecessary, because anticipated difficulties in the initial form of organization did not occur.

Personnel, Materials, and Construction

Finding personnel was difficult. Work began when the country's scientific resources already were fully mobilized for other war work; many who were willing to join the project already had made other commitments. The nucleus of the organization came from groups engaged in fast-neutron work under Oppenheimer, and they transferred their work and equipment to Los Alamos. Others were released to come to Los Alamos, in part, through the assistance of Conant as Chairman of the NDRC. The greatest difficult was in obtaining an adequate technical and administrative staff; they also came mainly from groups fully employed in war work. Here, the disadvantages of isolation and restriction weighed heavily, but were largely overcome among the scientific staff by interest in the work and recognition of its importance.

Among those who had worked previously under Oppenheimer were: from the University of California, Robert Serber and E. M. McMillan, as well as E. Segrè, J. W. Kennedy, and their groups; from the University of Minnesota, J. H. Williams and group; from the University of Wisconsin, J. L. McKibben and group; from Stanford University, F. Bloch, H. H. Staub, and group; and from Purdue University, M. G. Holloway and group. Others included: from the Radiation Laboratory of Massachusetts Institute of Technology, R. F. Bacher and H. A. Bethe; from the Metallurgical Laboratory of the University of Chicago, Edward Teller, R. F. Christy, D. K. Froman, A. C. Graves, J. H. Manley, and group; from Princeton University, R. R. Wilson and group. J. E. Mack, and R. P. Feynman; from the University of Rochester, V. F. Weisskopf; from

7

the Bureau of Standards, S. H. Neddermeyer; from the Ballistic Research Laboratory at Aberdeen, D. R. Inglis; from the University of Illinois, D. W. Kerst; from Barnes Hospital, St. Louis, Dr. L. H. Hempelmann; from Memorial Hospital, New York, Dr. J. F. Nolan; from the National Research Council, C. S. Smith; from Westinghouse Research Laboratories, E. U. Condon; from Columbia University, E. A. Long; and from the Carnegie Institution of Washington Geophysics Laboratory, C. L. Critchfield. Many of these men, having accepted temporary wartime assignments in the above-listed institutions, were on leave from other universities.

Procurement of laboratory equipment, machinery, and supplies was difficult and time-consuming. Even a specialized laboratory requires a great variety of materials and equipment, and any laboratory depends upon accumulated stocks and equipment that can be converted to new uses. Although much material was ordered in advance, procurement, being indirect and newly organized, was slow.

Some groups brought their own special equipment. The largest single item was the cyclotron lent by Harvard University. Before coming to Los Alamos, the Princeton group under R. R. Wilson had gone to Harvard to become familiar with the operation of this cyclotron and to disassemble it for shipment. McKibben's group brought two Van de Graaffs (electrostatic generators) from the University of Wisconsin. Manley's group brought the Cockcroft-Walton accelerator (D-D source) from the University of Illinois. The Berkeley group brought chemical and cryogenic equipment. This equipment allowed work at Los Alamos to begin much earlier than otherwise would have been possible.

Oppenheimer, Manley, and McMillan drafted the initial plan of the Laboratory. It provided for about 100 scientific staff members and a somewhat larger number of administrative, technical, and shop employees. The Laboratory included Building T (Fig. 2), an office building for administrative personnel, the theoretical physics group, a library, a classified document vault, conference rooms, a photographic laboratory, and a drafting room. Building U was a general laboratory and Building V was a shop. Buildings W, X, Y, and Z were specialized laboratory buildings for the Van de Graaffs, the cyclotron, the cryogenic laboratory, and the Cockcroft-Walton accelerator, respectively.

Oppenheimer and a few staff members arrived in Santa Fe on March 15, 1943. Until then, the project had been represented locally by J. H. Stevenson, a Santa Fe resident. Because the laboratory buildings and the

8

housing for Project Y and U. S. Army Corps of Engineers personnel were still under construction, the first project office was opened in Santa Fe. For security reasons, it was undesirable to house the staff in Santa Fe hotels, so they took over local guest ranches temporarily and were later transported to the site. Although the project office remained in Santa Fe, J. H. Williams lived at the site as acting site director.

The Laboratory staff and their families faced life at Los Alamos with enthusiasm and idealism. The importance of the work and its excitement contributed to this feeling, as did the possibility of building a vigorous, congenial community.

However, the actualities of the first months were very difficult. Living at the ranches around Santa Fe presented many hardships. Several families, with many young children, were often crowded together with inadequate facilities. Transportation between the ranches and Los Alamos was haphazard despite great efforts to regularize it. The road was poor and there were too few cars, most of which were in poor condition. Workers were often stranded on the road by mechanical breakdown or flat tires. There were no eating facilities at the site, so box lunches were sent from Santa Fe, but it was winter and sandwiches were not viewed with enthusiasm. The car carrying the lunches often broke down. The work day was irregular and short, and night work was impossible.

Until mid-April, telephone conversations between the site and Santa Fe were possible only over a Forest Service line. One could sometimes shout brief instructions, but discussions of any length, even of minor matters, required an 80-mile round trip.

Friction developed between Laboratory members and the Corps of Engineers' staff, mainly because construction was slow. The contractor could not hire enough labor, he had trouble with the building trade unions, and he did not procure or install the basic laboratory equipment rapidly enough. Pressure to accelerate this work was brought through, and therefore partly against, the military organization. Technical supervisors sometimes were forbidden to enter buildings until the Albuquerque District of the U.S. Army Corps of Engineers had accepted them formally. It was impossible to change even shelf placement or the direction of a door; the buildings had to be completed first and accepted as specified in the original drawings.

The initial problems, elementary and often minute, in retrospect were heightened by an administrative arrangement that presupposed close cooperation, without previous acquaintance, between two groups of widely divergent backgrounds and perspectives, the project members and the military organization. Individually and in detail, these early troubles were of little importance; collectively, however, they had some good and some bad effects on the spirit and tone of the emerging project organization.

The project offices were moved to Los Alamos in mid-April as laboratory space and housing became available.

THE APRIL CONFERENCE

Several conferences were held at Los Alamos in late April to acquaint new staff members and some specially invited consultants with the existing state of knowledge and to prepare a concrete research program. The consultants were I. I. Rabi of the Massachusetts Institute of Technology Radiation Laboratory, and S. K. Allison and Enrico Fermi of the University of Chicago Metallurgical Laboratory, who later became heavily involved in Laboratory work. A special reviewing committee appointed by General Groves also took part. This committee consisted of W. K. Lewis, chairman, M.I.T.; E. L. Rose, Director of Research for the Jones and Lamson Machine Co.; J. H. Van Vleck and E. B. Wilson of Harvard University; and R. C. Tolman, Vice Chairman of the NDRC, secretary of the committee.

Immediately before the conferences, Serber presented an indoctrination course. These lectures reflected the state of knowledge at that time and still constitute an adequate statement of the nuclear physics background.

Theoretical Background

Energy Release. The energy release from nuclear fission is about 170 million electron volts (MeV) per nucleus. For ^{235}U this is about 7×10^{17} ergs per gram. The energy released by TNT is about 4×10^{10} ergs per gram. Therefore, about one kilogram of ^{235}U is equivalent to 17 000 tons of TNT.

Chain Reaction. Large-scale release of energy from a mass of fissionable material is made possible by a neutron chain reaction. In fission the nucleus splits into two almost equal parts that emit, on the average, two to three neutrons. Each neutron may, in turn, cause fission

10

of another heavy nucleus. This reaction can go on until stopped by depletion of the fissionable material or by other causes.

The principal isotope in ordinary uranium, ^{238}U, fissions only under the impact of high-energy (about 1-MeV) neutrons. Neutrons from fission have more than this energy initially; many, however, are slowed by collisions to an energy below the fission threshold of ^{238}U. The result is that each neutron is the parent of less than one neutron in the next generation and the reaction is not self-sustaining.

However, ordinary uranium contains about 0.7% ^{235}U. Neutrons of any energy cause this isotope to fission; in fact, slow neutrons are more effective than fast ones. Therefore, a chain reaction is just possible in the normal isotope mixture or "alloy" if a slowing material is added to bring the neutrons down to the velocities at which they cause fission of ^{235}U most effectively. It is this chain reaction that is used in producing plutonium. Surplus neutrons are absorbed by ^{238}U, giving rise to the unstable ^{239}U, which decays to the end product ^{239}Pu by successive emissions of two electrons.

If the percentage of ^{235}U in uranium alloy is increased, a chain reaction with faster neutrons becomes possible. A concentration is therefore reached at which no special slowing or moderating is needed other than that provided by the uranium itself. The fastest possible reaction is obtained from pure ^{235}U.

Critical Size, Tamper, and Efficiency. In the fast chain reaction occurring in metallic ^{235}U or ^{239}Pu, a further limiting factor becomes crucial. In practice, only a fraction of the fission neutrons causes new fissions. The rest leak out through the boundaries of the material. If the fraction leaking out is too large, the reaction fails to sustain itself. If we consider a spherical mass of fissionable material at normal density, the fraction that leaks out decreases with increasing sphere radius until, on the average, the neutron birth rate compensates for the rate at which they escape from the sphere. In a smaller sphere, a chain reaction dies out; in a larger one, it continues and grows exponentially. This limiting radius is called the *critical radius,* and the corresponding mass is called the *critical mass.*

It is suggested intuitively that the critical radius should be of the same order of magnitude as the average distance that neutrons travel between successive fissions. For fast neutrons, this distance, 10 cm, is much greater than that for slow neutrons. Because of the high cost and limited supply of materials, it was essential to reduce the critical size. Surrounding the sphere of active material by a shell of less expensive material would reflect at least some of the escaping neutrons back into the sphere, and thus decrease the critical mass. Early calculations showed that any of several available reflector or "tamper" materials would very substantially reduce the critical mass.

Given a more-than-critical mass of active material, what is the course of the reaction? Once the reaction starts, the rate of fissioning, and thus the energy release, increases exponentially. The energy released heats the material, which begins to expand. As the density of the active material decreases, the path between fissions increases faster than the radius of the expanding mass, so more neutrons escape. Therefore, at some point the system becomes subcritical and the reaction is quenched. The point at which this quenching occurs determines the explosion efficiency; that is, the percentage of active nuclei fissioned.

The time available for an efficient nuclear reaction was extremely short. Release of 1% of the energy would give the nuclear particles a mean velocity of about a million meters per second. The reaction would be quenched by an expansion of centimeters, so the energy would have to be released within hundredths of a microsecond (μs). Because the mean time between fast-neutron fissions is about 0.01 μs and because most of the energy is released in the last few fission generations, a reasonably efficient reaction evidently was just possible.

Cross Sections. Calculating the static and dynamic aspects of the fission bomb was difficult because of the elaborate theory involved and because the calculations depended on nuclear constants that were not, as yet, well measured. Within the system a neutron may be absorbed or scattered, or it may produce fission. The contributions of each process are measured by the corresponding cross sections or effective target areas that the nucleus presents to an impinging neutron. The total cross section is divided into areas that fission, absorb without fission, or scatter, according to the relative probabilities of the three processes. If the scattering is not isotropic, the angular distribution of scattered neutrons must be specified. All of these cross sections, moreover, depend upon the nucleus involved and the energy of the incident neutron. Calculation of critical mass and efficiency depends on all of these cross sections and on the number of neutrons per fission and the density of the material. Clearly, accurate measurements would entail an elaborate program of experimental and theoretical physics to make the best use of information obtained.

12

Tamper Effects. The tamper decreases the critical mass by reflecting neutrons back into the active material and increases the inertia of the system; therefore, the time that it remains supercritical is also increased. These gains are lessened somewhat by the longer time between fissions of neutrons reflected back from the tamper. The time is lengthened by the longer path and by a loss of energy through inelastic scattering in the tamper. Calculations of the effects of tamper material depend on its absorption and scattering cross sections. Serber's early calculations for a ^{238}U tamper gave critical masses of 154 kg of ^{235}U and 5 kg of ^{239}Pu. Both figures are correct to within a reasonable error. This may be regarded partly as good fortune, because many of the assumptions were rough guesses, but it shows the advanced state of basic theory at the time.

Efficiency, Detonation, and Predetonation. Explosive reactions can be set off by relatively minute forces, the requirement generally being enough disturbance to initiate a chain reaction. Chemical explosives can be protected more or less surely from external forces that may initiate a reaction, but a supercritical mass of nuclear explosive cannot be protected from "accidental" detonation. Chain reactions will begin spontaneously more surely than reactions in the most unstable chemical compounds. Cosmic ray neutrons will enter the mass from outside. Others will be generated within it by the spontaneous fissions that occur constantly in uranium and plutonium. Still others will come from nuclear reactions, most importantly from the (α,n) reaction in light-element impurities. The problems this "neutron background" presents are responsible for much of the project's history. Cosmic rays alone will detonate any supercritical mass within a fraction of a second; other unavoidable sources, within a much shorter time.

The only way to detonate a nuclear bomb is to bring it into a supercritical configuration just when it is to be detonated. The assembly speed required depends upon the neutron background. As the parts of the bomb move together, the system passes smoothly from its initial subcritical to its final supercritical state. However, chain reactions may start any time after the critical position is reached. If assembly is slow compared to the chain reaction rate, so that predetonation occurs, the explosion will be over before the bomb has been assembled to give maximum efficiency. Thus the explosion may occur, with widely varying efficiencies, at any time between achievement of the critical and the final supercritical positions. Decreasing the probability of predetonation and

13

consequent inefficiency requires either faster assembly or a lower neutron background.

The above discussion indicates the magnitude of the assembly problem: to initiate reliably a reaction whose entire course runs only a fraction of a microsecond, while providing high-velocity assembly and low neutron background. The principal source of neutron background is the (α,n) reaction in light-element impurities. To lower this background would require strenuous chemical purification.

The most straightforward method proposed for meeting these difficulties was gun assembly. A projectile of active material and tamper, or active material alone, would be shot through or laterally past a target of active material and tamper. For ^{235}U, the necessary chemical purity and assembly velocity could be attained by known methods. Many difficult engineering problems were involved, but they did not seem insuperable. For ^{239}Pu, both the purity and speed required were somewhat beyond the established range. It seemed, however, that they could be provided by heroic means.

High-velocity assembly and reduced neutron background would decrease the probability of predetonation, but they also would decrease the probability of detonation at the desired time. Unless material could be assembled so as to remain in its optimum configuration for a considerable time, "postdetonation" might also lower efficiency or the system might pass through its supercritical state without detonating. Overcoming this difficulty would require development of a strong neutron source that could be turned on at the right moment. There were theoretically feasible schemes for such an initiator, but their practicability was not assured.

Two other proposed assembly methods were to be investigated. One was a self-assembling or autocatalytic method that operated by compression or expulsion of neutron absorbers during the reaction. Calculations showed that this method would require large quantities of material and give only very low efficiencies. The second method was implosion. The proposal was to blow together a subcritical hollow spherical shell of active material and tamper by detonating a surrounding layer of high explosive. Both methods had the disadvantage, in a short-term war program, of lacking previous engineering implementation like that in the field of gun design.

The Deuterium Bomb or "Super." At the time of the April conferences, another proposal already had received considerable thought

14

and discussion. This was to use the fission bomb to initiate a different type of nuclear reaction from that involved in the fissioning of heavy element nuclei. Fissioning, the disruption of nuclei with liberation of energy, is a somewhat anomalous reaction restricted to the heaviest nuclei. Among the lighter elements, the typical exoergic (energy-producing) reaction is the building up of heavier nuclei from light ones. For example, two deuterium (^{2}H) nuclei may combine to form a ^{3}He nucleus and a neutron, or a tritium (^{3}H) nucleus and a proton. The liberated energy goes into kinetic and radiation. If such a reaction occurs in a mass of deuterium, it spreads under conditions like those that control ordinary thermochemical reactions, so it is called thermonuclear. The cross section for a reaction between two deuterium nuclei is very dependent upon their energy. At low energies the probability of a reaction is very slight but as the temperature of the material rises, reaction becomes more probable. Finally, a critical temperature is reached at which the nuclear reactions in the material just compensate for variouus kinds of energy loss, such as heat conduction and radiation. The thermonuclear reaction is more complicated than indicated because of various secondary reactions.

Among available materials, deuterium had the lowest ignition temperature, estimated to be about 35 kilovolts (keV), or about 400 million degrees, but actually somewhat lower. Once ignited, deuterium is about five times as energy productive per unit mass as is ^{235}U. Therefore, 1 kg of deuterium is more cheaply produced in usable form than ^{235}U or ^{239}Pu. The proposed weapon, using a fission bomb as a detonator and deuterium as an explosive, could properly be called an atomic super-bomb. Development of this superbomb was secondary to that of the fission bomb, but its potentialities were so great that research on it could not be neglected completely.

Most careful attention was given to the possibility that a thermonuclear reaction might be initiated in light elements of Earth's atmosphere or crust. The easiest reaction to initiate, if any, was found to be a reaction between nitrogen nuclei in the atmosphere. It was assumed that only the most energetic of several possible reactions would occur, and that the reaction cross sections were at the maximum theoretically possible values. Calculations showed that no matter how high the temperature, energy loss would exceed energy production by a reasonable factor. At an assumed temperature of 3 MeV, the reaction failed to be self-propagating by a factor of 60. This temperature exceeded the calculated initial temperature of the deuterium reaction by a factor of 100 and that of the fission by even a larger factor.

The impossibility of igniting the atmosphere was assured by science and common sense. The essential factors in these calculations, the Coulomb forces of the nucleus, are among the best understood phenomena of modern physics. There remained the philosophic possibility of destroying Earth, associated with the theoretical convertibility of mass into energy. The thermonuclear reaction, the only method known that could cause such a catastrophe, is evidently ruled out. The general stability of matter in the observable universe argues against it. Further knowledge of the nature of the great stellar explosions, novae and supernovae, will help to resolve these questions. In the almost complete absence of real knowledge, it is generally believed that the tremendous energy of these explosions is of gravitational, rather than nuclear, origin.

More immediate and less spectacular global dangers to humanity arise from use of thermonuclear bombs or even fission bombs in war, principally from the possible magnitude of destruction and radioactive poisoning of the atmosphere (Chap. 13).

Damage. Because the purpose was to produce an effective weapon, it was necessary to compare the atomic bomb with ordinary bombs, in terms of destructive effects. Damage could be classified under the psychological effects of using such a weapon; the physiological effects of the neutrons, radioactive material, and radiation produced; and the mechanical destruction produced by the explosion shock wave. Estimation of the first was outside the jurisdiction of the project. It was estimated that lethal effects might be expected within a 3000-ft radius of the bomb. The radioactivity remaining might render the explosion site uninhabitable for a long time, but this effect would depend on the percentage of activity left behind, an unknown quantity. The principal damage would be caused by the mechanical effects of the explosion, which were difficult to estimate. There were some rough data on the effects of large explosive disasters, and there was more reliable information on the effects of small high-explosive bombs, but no one was certain how to scale these effects upward for high-energy atomic bombs. Serber's report estimated a destruction radius of about 2 miles for a 100 000-ton bomb. Members of the British mission who later came to the project were able to add to the understanding of this topic from their national experience and their recent research.

PROGRAM DEVELOPMENT

Clearly, the greatest problems would be development and engineering. There was still much work to be done in nuclear physics, but enough was known to eliminate great uncertainties. However, the research stage was now past its prime, to be dominated, in turn, by problems of application. The normal meanings attached to research, development, and engineering are altered in the context of wartime science; that was particularly true of the atomic bomb project. Two features determined its general character, the domination of research schedules by production schedules and the nature of the weapon itself. Time schedules for production of ^{235}U and ^{239}Pu were such that the Laboratory had about 2 years' time before explosive amounts of these materials would be available. After that, every months' delay would have to be counted as a loss to the war. Much information had to be obtained at the earliest possible date with whatever difficulty, although later it could be obtained more easily and reliably. Plutonium micrometallurgy was investigated at Chicago, for example, because it was vital to know the density of the new material as soon as possible. The first measurement was made, with great difficulty, from a sample of only a few micrograms. The value of such information depended upon its capacity to influence decisions that could not be postponed. This meant a heavy dependence upon theory and measurements of the type needed to answer theoretical questions. Reliance was placed upon theoretical anticipations partly because of the all-or-none character of the weapon. A purely experimental nuclear explosion would involve dissipation of at least one critical mass of material that might have been used against the enemy. If tests were to be made at all, only one or two would be possible. For so few tests to be meaningful, they would have a high probability of success. The bomb's component parts and phases of operation would have to be designated and tested separately, relying upon theory to supply a picture of its integral operation.

At first the laboratory appeared to be a purely research organization. This appearance was partly a matter of history. Nuclear research was the most advanced part of the program, and its personnel and equipment were most easily available. However, the research character of the organization was partly a matter of considered policy. Normally, the engineer is the "practical man" who translates ideas into practice. Here, not only the ideas but also the standards of practice were new. To keep the center of policy in the research group was not to minimize the importance of the engineering work, but to emphasize its difficulty. Secondary problems undeniably arose from this policy, which displaced the engineer from his normal position, and only through trial and error created a new place for him in the division of project labor.

Theoretical Program

As it emerged from the conferences, the theoretical program's main goal was to analyze the explosion, develop associated calculation techniques, and give increasingly reliable and accurate nuclear specifications for the bomb as new physical data became available. Calculations had to be made for ^{235}U, ^{239}Pu, and a new compound, a uranium hydride, which seemed to have certain advantages over metallic uranium as a bomb material. Calculations also had to be made for various active mass shapes and different combinations of bomb and tamper material. For critical mass calculations, the theory of neutron diffusion in bomb and tamper had to be refined and the energy distribution of fission neutrons and the dependence of nuclear cross sections upon those energies had to be accounted for. For efficiency calculation, the explosion hydrodynamics had to be studied further, taking into account the effects of the large amounts of radiation liberated in the process. The problems of time of assembly, detonation, and predetonation needed further investigation.

The theoretical program also included various analyses and calculations for the experimental program, ranging from ordinary service calculations to design of a slow chain-reacting unit with ^{235}U-enriched uranium. Finally, it included further investigation of bomb damage, of possible autocatalytic methods of assembly, and of amplifying the effect of fission bombs by using them to initiate thermonuclear reactions.

Experimental Physics Program

The experimental physics program formulated during and immediately after the conferences included detailed and integral experiments. Detailed, or differential, experiments attempt to observe the effects of isolated nuclear phenomena. Enough data gained in this way would provide an integral picture of the bomb's operation within a framework of theory. Integral experiments (at least in their early conception) were attempts to duplicate some of the overall properties of the bomb. The two kinds of experiments were intended to supplement each other: to sharpen interpretation of integral experiments and to show possible omissions of elements from the detailed picture. In practice, it proved extremely difficult to devise integral experiments that in any way duplicated the conditions in the bomb. They had, rather, the effect of checking theory in situations in some way similar to the bomb.

A brief outline of the program, as first developed, indicates the experimental knowledge carried over from the previous period.

Neutron Number. The average number of neutrons per fission had never been measured directly, although the Chicago project had measured the number of neutrons from ^{235}U per thermal neutron absorbed. The number of neutrons per fission could be calculated from this measurement and from the ratio of fissions to captures, which, however, was not known reliably in the thermal energy region. The neutron number of ^{239}Pu was completely unknown, although it was expected to differ only slightly from that of ^{235}U. The first experiments planned were measurements of neutrons from ^{239}Pu.

These latter measurements, intrinsically important, were needed as soon as possible to confirm the wisdom of heavy commitments already made for producing plutonium in quantity.

Fission Spectrum. The energy range of neutrons from fission of ^{235}U had been investigated by the British and at Rice Institute and Stanford University. These measurements suffered from the large dilution of ^{235}U by ^{238}U in normal uranium. Work with enriched material had already begun at the University of Minnesota and was to be continued at Los Alamos.

Fission Cross Sections. Fission cross sections had been measured under N. P. Heydenberg at the Carnegie Institution Department of Terrestrial Magnetism, by McKibben's group at the University of Wisconsin, and by Segrè's group in Berkeley. These ^{235}U measurements covered the neutron energy range above 125 keV and the range below 2 eV. When the curve for the fission cross sections over the high energy was extrapolated downward, the thermal energy figure obtained was much larger than the cross section actually observed. Because the extrapolated region covered the important range of neutron energies in a uranium hydride bomb, measurements were planned to investigate cross sections at these intermediate energies and resolve the apparent anomaly. Fission cross sections of ^{239}Pu were already known at thermal energies and at a few energies. Measurements also were planned to cover the entire range of energies up to about 3 MeV.

19

Delayed Neutron Emission. Experiments at Cornell showed no appreciable delay beyond 10 µs in emission of neutrons from fission. One of the initial Los Alamos experiments was to push this time down to 0.1 µs. Theoretically, it was expected that the number delayed even for 0.1 µs would be small.

Capture and Scattering Cross Sections. Little was known about capture and scattering cross sections. Some measurements on normal uranium had been made at Chicago. The Minnesota group had values for elastic and inelastic scattering in uranium for high energies. The Wisconsin group had measured large-angle elastic scattering in many potential tamper materials. Segrè had measured capture cross sections at Berkeley. The principal Los Alamos work planned was on scattering and absorption cross sections of ^{235}U and ^{239}Pu and on the capture and scattering cross sections of various tamper materials.

A new type of measurement, scattering into different solid angles, was also planned. When averaged to give the effective scattering in a given direction, this is the so-called transport cross section.

Integral Tamper Experiments. Several experiments, to imitate scattering by a tamper in the actual bomb, were planned to measure the scattering in potential tamper materials.

The Water Boiler. At the April conferences, the possibility of constructing a slow-reacting unit by using uranium with enriched ^{235}U content in water solution was discussed. Such a unit would provide a useful neutron source for experiments and give practice in operating a supercritical unit. The decision to build it was not reached until later.

Experimental Techniques. A large subsidiary program was initiated to investigate techniques for producing and counting neutrons of a given energy, for measuring fissions in various materials, and for measuring neutron-induced reactions other than fission. Systematic recording of nuclear properties for the experimental program required both accuracy and standardization of several difficult techniques. Therefore, instrumentation was a major activity.

Chemistry and Metallurgy

The DSM project included a large amount of research on the chemistry and metallurgy of uranium. The microchemistry and micrometallurgy of plutonium were investigated at Chicago as soon as small amounts were available. The chemical investigations were necessary for designing methods for recovering plutonium from the pile material and for decontaminating it; that is, separating it from radioactive fission products.

The exact division of labor between the Los Alamos chemistry laboratory and other laboratories had not been settled. It was not known whether ^{235}U, ^{239}Pu, or both, would be used, or whether the bomb material would be metal or a compound. Uranium-233, producible from thorium by a "breeding" process similar to that by which ^{239}Pu is made from ^{235}U, was also a possibility. Mechanical requirements for the bomb material could not yet be specified. A characteristic difficulty was that the time for research with gram and kilogram amounts of material would have to be as short as possible, to avoid delaying bomb production.

Purity requirements for ^{235}U and ^{239}Pu had to be established. Because of the large alpha radioactivity of ^{239}Pu, light impurities had to be almost eliminated. Most light elements to be present in not more than a few parts per million. For ^{235}U, these tolerances could be relaxed greatly. Although it had not been determined whether final purification would be done at Los Alamos or elsewhere, an analytical program was necessary to develop techniques for measuring small amounts of impurity in small samples of material. A radiochemistry program was needed to prepare materials for nuclear experiments and to develop a neutron initiator for the bomb.

The metallurgy program included research and development on metal reduction of uranium and plutonium, on casting and shaping of these metals and compounds such as uranium hydride, and on various possible tamper materials. Investigation of the physical properties of uranium and plutonium was needed, and a search for alloys with better physical properties than those of unalloyed metals was required. The metallurgy group would prepare materials for physical and ordnance experiments, particularly projectile, target, and tamper materials for the gun program.

As a somewhat autonomous part of the chemistry program, plans were made to construct a deuterium liquefaction plant at Los Alamos to supply liquid deuterium for experiments and for eventual use in the thermonuclear bomb if its development proved feasible and necessary.

Ordnance Program

The most difficult problem was to find ways to assemble several critical masses of material fast enough to produce a successful high-order explosion. Subsidiary, but still very difficult, problems were incorporation of active material, tamper, and assembly mechanisms into a practical airborne bomb. These were the problems of the Ordnance Division, which barely existed at the beginning. As a matter of fact, no pre-existing group could have had much success in this work because a new field of engineering was being explored. Experience showed that those successful in this work came from various technical backgrounds, all of which contribute and none of which dominates: physicists, chemists, and electrical and mechanical engineers.

The ordnance program simultaneously investigated alternative methods. The uncertainties of nuclear specification, and the possibility that any one line of investigation might fail, made such a policy unavoidable. Three methods for producing a fission bomb (autocatalysis, the gun, the implosion) were discussed, and the gun and implosion methods were chosen for early development. Autocatalysis was not eliminated, but it would not be considered until some scheme that would give reasonable efficiency was proposed. The gun seemed more practical because it used a known method of accelerating large masses to high velocities. The problem of "catching" a projectile in a target and starting a chain reaction in the resulting supercritical mass was obviously difficult, but it seemed soluble.

The implosion method was much farther removed from existing practice. The required simultaneous detonation over the surface of a high-explosive sphere presented unknown and possibly insoluble difficulties. The behavior of solid matter under the thermodynamical conditions created by an implosion was far beyond laboratory experience. As even its name implies, the implosion seemed "against nature." It was first investigated as something to fall back on should the gun, unexpectedly, fail. Credit for early support and investigation of the implosion method should go to S. H. Neddermeyer, who almost alone believed that it was superior. At a meeting on ordnance problems late in April, Neddermeyer presented the first serious theoretical analysis of the implosion. His arguments showed that compressing a spherical shell into a solid sphere by detonating a surrounding high-explosive layer was feasible and would give both higher velocity and a shorter assembly path than the gun method. Investigation began almost immediately. Implosion subsequently received two priority increases and at the end of the project was the dominant program throughout the Laboratory.

During the April conferences, the discussion of ordnance served

22

mainly to outline the problems. R. C. Tolman had studied the problem of gun design. One member of the Reviewing Committee was E. L. Rose, a gun design expert. Rose showed that by sacrificing durability, an inessential property, the otherwise prohibitive size and weight of a large gun could be so reduced that it and the target could be included in a practical bomb. Other elements discussed were internal ballistics of the gun, external and terminal ballistics (guiding and seating the projectile and initiating the chain reaction), safety, arming and fuzing devices, release, and bomb trajectory from a plane.

Experimental ordnance work did not start for several months. Requiring immediate action were the problems of obtaining test guns and high explosives, building a proving ground, and employing or training personnel to do the research.

Reviewing Committee Report

General Groves had appointed the Reviewing Committee to report on the organization of the Los Alamos project and on the status and program of its technical work. The main question was the status of the ordnance program. The initial concept was that research in nuclear physics should be virtually completed before large-scale ordnance development was undertaken. In March 1943, however, Oppenheimer had written a memorandum on ordnance urging that experimental work be undertaken as early as possible and be recognized as one of the most urgent project problems. Tolman recognized the importance of the issue and recommended that Rose be appointed to the Reviewing Committee as an ordnance expert.

Because the Reviewing Committee report of May 10, 1943, contained recommendations that had an important bearing on further development of the project, its main features are outlined below.

Nuclear Physics Research. After extensive review, the committee approved all of the nuclear physics program, the most advanced part of the work. It noted the newly discovered possibility of using uranium hydride, pointed out that the hydride's existence had been learned of at Los Alamos somewhat by accident, and recommended a more systematic technical liaison between this and other branches of the larger project. It also recommended that study of ^{233}U as a possible explosive material be continued.

Less Developed Parts of the Program. The committee also reported on the programs for investigating the thermonuclear reaction, chemistry and metallurgy, and engineering and ordnance.

The committee recommended that investigation of the thermonuclear bomb be pursued, but mainly along theoretical lines and on a lower priority than the fission bomb. This confirmed established Laboratory policy.

In both chemistry and engineering, the committee recommended substantial revision of policy. One of the principal organizational questions was about jurisdiction over the chemistry purification program. Purification of active material, particularly ^{239}Pu, presented a major technological problem. The chemistry of plutonium was first investigated by its discoverers, Kennedy, Seaborg, Segrè, and Wahl. The investigation was pursued and a method was first practiced by the Metallurgical Laboratory chemists in separating and decontaminating plutonium produced in the piles at Oak Ridge and Hanford. The committee recommended that the purification be done by the Los Alamos Laboratory instead, because Los Alamos would be responsible for the correct functioning of the ultimate weapon, and repurification work would be a consequence of experimental use of plutonium.

The committee's second major recommendation agreed with Oppenheimer's statement that ordnance development and engineering should begin as soon as possible. The committee felt that it was time for close relation of nuclear and engineering research. Although the nuclear specifications indicated a wide range of possible designs for the final weapon, further design determination must depend also upon engineering specifications. Engineering research on development of safety, arming, firing, and detonating devices, portage of the bomb by plane, and determination of the bomb trajectory were needed also.

Both recommendations entailed major expansion of project personnel and facilities. It was estimated that 30 more chemists and technicians were needed for the purification program, with a corresponding increase of laboratory facilities. Ordnance and engineering work would require a doubling of project personnel, with an extensive increase of offices, drafting rooms, shops, and test areas for ballistic and explosives work.

Administrative Recommendations and General Conclusions. The committee strongly recommended Oppenheimer for Director. They recommended that three administrative positions be created: director of ordnance and engineering, to take charge of the recommended program; associate director, to take charge of some major phase of the scientific

24

work and assist the director and act in his absence; and administrative officer, to take charge of nontechnical administrative matters, particularly to maintain cordial and effective relations with the military administration. The committee reported favorably on the competence and work assignments of scientific personnel.

The committee was dissatisfied with the organization and function of the procurement system. They found the procurement officer, D. P. Mitchell, well qualified for the position by technical training and experience. Their principal criticism was about operation of the University of California Purchasing Office in Los Angeles, which they felt had been responsible for serious and avoidable delays. They recommended establishing a second purchasing office in New York under separate contract.

The security policy that the Director established under the authority granted him met with the committee's approval.

The committee's final administrative recommendation, which could not be entirely specific, concerned morale and maintenance of the "special kind of atmosphere that is conducive to effective scientific work." The committee recognized that this was hindered by the isolation and military character of the post, and it was by improving relations between the military and technical organizations that they saw hope for maintaining morale.

CHAPTER 2

The British Mission

In December 1943, the first representatives of the British atomic bomb project came to Los Alamos. The arrival of O. R. Frisch and E. W. Titterton marked the climax of long negotiations among the British, Canadian, and American governments seeking to integrate the three countries' atomic bomb research.

Although Britain's Directorate of Tube Alloys (T. A. Project) had a very high priority in 1942, so many of her physicists and so much of her industrial capacity were engaged in other urgent war work that it was impossible to undertake as large a program as the United States had launched. The British organization decided to limit itself to particular phases of the problem and established research teams in various university and industrial laboratories.

In the summer of 1942, there had been enough progress toward collaboration so that the British reports on the theory of fission and the fission bomb, and reports of experimental measurements of nuclear constants were accessible to Oppenheimer's group in Berkeley. British analysis of the bomb mechanism was somewhat more advanced than ours, so access to these reports was of substantial value. A memorandum by Oppenheimer to R. E. Peierls in November described the Berkeley work and discussed differences between British and American theoretical work. The incompleteness of collaboration at this time is indicated by the fact that there was no mention of the deuterium bomb in the memorandum.

In the fall of 1943, President Roosevelt and Prime Minister Churchill discussed closer collaboration to hasten production of atomic bombs. A Combined Policy Committee, set up in Washington, decided to place several British scientists in American laboratories. Evidence of the genuine cooperation that resulted from this sacrifice on Great Britain's part is that British scientists were assigned to all parts of the American project, especially at Los Alamos, the most highly classified section of all.

At this time Niels Bohr, the eminent Danish physicist, escaped from Denmark and was appointed adviser on scientific matters to the British Government. His advice also was made available to the United States. Bohr and his son Aage came to Los Alamos in December 1943 shortly after Frisch and Titterton arrived. To ensure his personal safety and as a security precaution, Bohr was known as Nicholas Baker and his son as James Baker. Great care was taken to prevent any reference to their real names, even in classified documents. The Bohrs did not become resident members of the Laboratory but made several extended visits as consultants.

Bohr found many of his former students here, and his coming had a very healthy influence on research. He came at the right moment. The exigencies of production and the innumerable small problems that confronted the physicists had led them away from some of the fundamental problems of the bomb. Study of the fission process, for example, had been neglected, and this neglect obstructed reliable prediction of important phenomena, such as the energy dependence of the branching ratio between fission and neutron capture. Bohr's interest gave rise to new theoretical and experimental activities that answered many questions. Some of the most important experiments on the velocity selector were made at his instigation (Chap. 6), and he strongly influenced research on the nuclear properties of tamper materials.

Bohr's criticism and his concern for new and better methods enlivened the discussion of alternative means of bomb assembly. These discussions showed that the "orthodox" implosion was the best method. Bohr participated very actively in design of the initiator. The idea of using a special design to attain maximum mixing of beryllium and polonium was conceived during discussions with him.

Last but not least was his influence on Laboratory morale. It went further than having the great founder of atomic research in the Laboratory, and further than the stimulus of his fresh suggestions. He saw the administrative troubles better than many of those enmeshed in them. His infuence was to bring about stronger, more consistent cooperation with the Army in pursuit of the common goal. And, most important, he gave everybody in contact with him some of his understanding of the ultimate significance of control of atomic energy.

Sir Geoffrey I. Taylor was another very useful British consultant. Los Alamos nuclear physicists generally lacked experience in hydrodynamical investigations. Their work on the hydrodynamics of the implosion and the nuclear explosion was too formal and mathematical. Apart from the contributions of the American consultant John von Neumann (Chap. 7), most of the simple intuitive considerations that give true physical understanding came from discussions with Taylor. His most important general contribution was his understanding of the "Taylor instability," the generalization that when a light material pushes a heavy one the interface between them is unstable (Chap. 5). This principle was important in the theory of jets, in interpreting high-explosive experiments, in initiator and implosion bomb design, and in predictions about the nuclear explosion. Taylor also stimulated serious theoretical investigation of the "ball of fire" phenomena (Chap. 11).

Sir James Chadwick of the Cavendish Laboratory, scientific adviser to the British members of the Combined Policy Committee in Washington, came to Los Alamos early in 1944 to head the British mission. It was uncertain at first whether the British would work under Chadwick on problems of his choosing or be assigned to Laboratory groups. The latter arrangement was adopted, and eventually British scientists worked in nearly all Laboratory divisions. Seven were experimental nuclear physicists, two were electronics experts, five were theoretical physicists, and five were experts in the properties and effects of explosives. Chadwick stayed in Los Alamos only a few months. His successor was Peierls.

Lord Cherwell, Churchill's personal adviser on scientific matters, visited Los Alamos in October 1944.

In addition to those already mentioned, the British mission staff included E. Bretscher, B. Davison, A. P. French, K. Fuchs, J. Hughes, D. J. Littler, Carson Mark, W. G. Marley, D. G. Marshall, P. B. Moon, W. F. Moon (secretary), W. J. Penney, G. Placzek, M. J. Poole, J. Rotblat, H. Sheard, T. H. R. Skyrme, and J. L. Tuck. Although several members came to Los Alamos by way of the Canadian and United Kingdom Laboratory in Montreal, all were attached to the British staff except Mark, who remained in the employ of the Canadian government.

CHAPTER 3

General Administrative Matters
April 1943 to August 1944

LABORATORY ORGANIZATION

The Los Alamos Laboratory's first problems were those common to organizational beginnings: program definition, division of responsibilities, and liaison. Responsibilities were divided, like the program, into experimental physics, theoretical physics, chemistry and metallurgy, and ordnance. Each became an administrative division, consisting of several operating units or groups. Group leaders were responsible to their respective division leaders, who were responsible to the Director. A Governing Board with responsibility parallel to that of the Director was established. It consisted of the Director, division leaders, general administrative officers, and individuals in important technical liaison positions.

The Board provided a means for relating the work of the various divisions and for relating the Laboratory program to other Manhattan District activities. It heard reports of the latest nuclear calculations and measurements, and from these set basic specifications for ordnance and chemistry. As Ordnance provided experimental and design data, the Board set fabrication requirements for the metallurgists to meet.

The progress of procurement and production was frequently reviewed, particularly that of active materials and separated isotopes needed in the program. The Board supervised liaison with other project laboratories on these and related matters.

For the first eight months, perhaps two-thirds of the Governing Board's time was devoted to lay matters. Frequent topics were housing, construction and construction priorities, transportation, security restrictions, personnel procurement, morale, salary scales, and promotion policies. In most of these discussions, the Governing Board again provided a link between the technical program and general administration.

29

The adversities of the first months are illustrated by a few very minor items chosen at random. In the first meeting, on March 30, 1943, it was mentioned with some triumph that a calculating machine had finally been borrowed from the Berkeley Laboratory. The scarcity of transportation is illustrated by the fact that a request for assignment of one pickup truck was brought for decision as high as the Governing Board. In May, the housing shortage was so serious that the Board, itself, assigned the six remaining apartments.

Members of the Governing Board were Bacher, Bethe, Kennedy, D. L. Hughes (Chap. 3), Mitchell, Parsons (Chap. 7), and Oppenheimer. Later, McMillan, Kistiakowsky (Chap. 7), and Bainbridge (Chap. 7) were added.

Shortly after the beginning of the Laboratory, a Coordinating Council of Group Leaders or above was established. In contrast to the Governing Board, this was not a policy-making body although policy problems sometimes were delegated to it. For example, the Council was asked to establish criteria for deciding which Laboratory members should be classed as staff members with unrestricted access to classified information. Because its members were the heads of operating groups, collectively in contact with all members of the Laboratory, it also served as a vehicle of general opinion on technical, administrative, and, occasionally, community affairs. Divisions and groups held their own regular meeting and seminars. These, together with informal discussions and regularly published reports, were the main vehicles of technical information in the Laboratory.

Finally, there was a weekly colloquium for all staff members. Staff members, who had scientific degrees or equivalent training in the field of their work, were therefore presumed capable of giving or receiving benefit in general discussions of the technical programs. The colloquium was less a means of providing information than an institution that contributed to the viability of the Laboratory and to maintaining the sense of common effort and responsibility.

Among all these channels of communication, the colloquium raised the most serious question of policy. From a narrowly technical point of view, it was the least easy to justify, and to military security it appeared to present the greatest hazard. Regular attendance would give any staff member a generally complete and accurate picture of Laboratory problems and progress, which, of course, was its purpose. Withholding essential scientific information from the colloquium would have defeated this purpose and would have compromised basic policy. In practice, the relatively scientific and academic tone of colloquium discussions made it possible to avoid mentioning many matters of relatively small scientific, and relatively great tactical, value. If this was impossible, the tactical value of information was sometimes lessened by omitting quantitative details. Nevertheless, the communication policy was a departure from the customs normally surrounding protection of military secrets.

LIAISON

Establishment of liaison presented unusual difficulties that reflected the complexity of the Manhattan District organization. As the Reviewing Committee had pointed out, some way had to be established for interchanging pertinent information between this and other branches of the project. The isolation of Los Alamos, even from other branches of the project, was a basic Manhattan District policy. Apparently, it was difficult to separate the virtues of this isolation from its vices, and the needed liaisons were achieved only after the most earnest representations.

The procedure established in June 1943 for liaison with the Metallurgical Laboratory at Chicago is fairly typical. Permission was given for exchanging information by correspondence between specified representatives of the two projects or by visits of the Los Alamos representatives to Chicago. Information was restricted to chemical, metallurgical, and certain nuclear properties of fissionable and other materials. The representatives might discuss schedules of need for and availability of experimental amounts of ^{235}U and ^{239}Pu. No information could be exchanged on design or operation of production piles, weapons design, or comparative schedules of need for, and availability of production amounts of, active materials. Three members of the Los Alamos Laboratory were to be kept informed of the time estimates for producing large amounts of these materials. Also, it was agreed that General Groves' office would grant special permission for other Laboratory members to visit Chicago to discuss specific matters.

As the program developed, information on research at the Metallurgical Laboratory, the University of California, and Iowa State College was needed. This work concerned the chemistry and metallurgy of uranium and plutonium and methods for analyzing their impurities. Information also was needed on nuclear research at the Argonne

31

Laboratories at Chicago. It was important to know when material would become available from the production plants at Oak Ridge (Site X), the form in which it would be received, and the processing it would have undergone. It was also essential to know the analytical procedures that the production plants would use in determining the impurities and active content of this material.

The need for accurate information on production time schedules was the most urgent. The estimates received during the Laboratory's first summer were vague, incomplete, and contradictory, so it was difficult to make sensible schedules of bomb research and development. The Governing Board said that scheduling with the existing information was impossible and that unnecessary delays would certainly result from this kind of blind operation. They strongly urged that Los Alamos maintain a full-time representative at Oak Ridge. In November 1943, it finally was agreed that Oppenheimer could visit the production plant at Oak Ridge. When material began to arrive at Los Alamos in the spring of 1944, the situation improved.

Getting information that the Ordnance Engineering Division required was especially difficult. Most of it had to be sought in agencies outside the Manhattan District from whom knowledge of the purpose and even the existence of Los Alamos had to be concealed. Many devices were used: blind addresses, a Denver telephone number, and NDRC identification cards. The office of Dr. Tolman, Vice-Chairman of NDRC, was instrumental in obtaining reports on such subjects as gun design, armor plating, explosives, detonators, and bomb damage. The liaison with Army and Navy Ordnance and with the Army Air Force is discussed later (Chaps. 7 and 19).

Among the more troublesome and less obvious liaison needs were those with the University of Caifornia and within Los Alamos itself. Although the work was to be done under a more or less standard War Department contract, the University of California was, in matters of policy, virtually unrepresented at the site. Security regulations excluded its officers from discussions of technical and administrative policy, so they were allowed to concern themselves almost exclusively with a narrow range of legal and contractual affairs. At Los Alamos there were two administrative offices, the military and the Laboratory. Although the division of labor was defined in a general way, most of the difficulties of dual organization had to be lived through before effective cooperation was established. Because of security policy, the officers charged with

32

administering the community and post were largely ignorant of the Laboratory's work. Therefore, although the Manhattan District was the basic organization in the DSM project, its local military representatives were excluded from the sphere of Laboratory policy. Added to these difficulties were the troubles of life in an isolated and unpracticed community.

Thus a very great administrative burden fell upon the Director. Although his primary responsibility was success of the scientific program, it was equally his concern that this success not be jeopardized by extraneous difficulties. The Reviewing Committee's administrative recommendations were aimed principally at improving this situation. Apart from specific procurement difficulties, the committee's main concern was to improve relations between the Laboratory and Post Administration and to relieve the Director of as many nontechnical responsibilities as possible. At the beginning, project operation was taken over by a temporary organization of scientific staff members and technicians. The important thing was to avoid delaying research work. A host of small problems stood in the way: transportation, warehousing, procurement, planning of laboratory construction, and housing. The enthusiasm with which these jobs were undertaken was notable, as was the *esprit de corps* developed in the process. Nevertheless, there was antagonism between the Laboratory and the military organization. For those who were there, what stands out is not this initial conflict, which only reflected the diversity of American life, but the fact that through common purpose, and by the measure of actual accomplishment, it was reduced to secondary importance.

The staff members were heartened when Oppenheimer read at a colloquium a letter, dated June 29, 1943, from President Roosevelt. It said, in part: "I wish you would express to the scientists assembled with you my deep appreciation of their willingness to undertake the tasks which lie before them in spite of the dangers and the personal sacrifices. I am sure we can count on their continued wholehearted and unselfish labors. Whatever the enemy may be planning, American science will be equal to the challenge. With this thought in mind, I send this vote of confidence and appreciation."

Apart from the business and procurement offices, the Laboratory had only two administrative officers other than the Director. These were E. U. Condon of Westinghouse Research Laboratories, and W. R. Dennes of the University of California. Neither had planned to remain with the project, so Condon left in May and Dennes in July of 1943. The Reviewing Committee had recommended appointment of an associate director and an administrative officer to coordinate nontechnical administrative functions and act as liaison with the Post Administration. Neither position could be filled at the time, but certain urgent requirements were met by appointing new administrative officers. David Hawkins of the University of California took over liaison with the Post Administration in May 1943. D. L. Hughes, Chairman of the Department of Physics of Washington University at St. Louis, Missouri, was made Personnel Director in June. B. E. Brazier, formerly of the T. H. Buell Company, Denver, took charge of construction and maintenance in May. In January 1944, David Dow of the legal firm of Cadwalader, Wickersham, and Taft of New York, was appointed Assistant to the Director in charge of nontechnical administration.

By July 1944, Laboratory administration was organized as follows.

A-1	Office of Director	D. Dow
A-2	Personnel Office	C. D. Shane (Assistant Director)
A-3	Business Office	J. A. D. Muncy
A-4	Procurement Office	D. P. Mitchell (Assistant Director)
A-5	Library, Document Room, Editor	C. Serber and D. Inglis
A-6	Health Group	Dr. L. H. Hempelmann
	Maintenance	J. H. Williams
	Patent Office	Major R. C. Smith

PERSONNEL ADMINISTRATION

Because of the newness of large-scale organized research, there was no class of professional scientific administrators. A choice had to be made between a large administrative organization staffed with persons unacquainted with the peculiarities of scientific research and a system by which most administrative responsibility fell to the scientists. As with the engineering program, it was partly a matter of expediency and partly of policy that the center of gravity remained in the scientific staff. The policy adopted meant increased unity in the Laboratory, but it undeniably entailed a loss of administrative efficiency.

The Personnel Office, in particular, illustrates these remarks. Hughes, the director, was a physicist with administrative experience as Chairman

34

of the Department of Physics at Washington University. The Laboratory was so organized that the Personnel Office depended almost entirely upon the representations of division and group leaders.

In addition to working with Laboratory divisions and groups on employment and salary matters, the Personnel Office had charge of a Santa Fe office for receiving and employment and of the Housing Office at the site. Under its jurisdiction fell personnel security, draft deferment, and placement of military personnel assigned to the Laboratory. The Laboratory became administratively involved in these affairs when, through their effects on staff morale, they affected the success of the work.

Community Affairs and Housing

A most urgent community problem was construction and organization of a school. The old Los Alamos Ranch School had a small public elementary school for its employees' children. It had been believed that this old building would be adequate for an elementary school and that a high school could be established in another original Los Alamos building. This plan soon proved unfeasible, so the Director and the Commanding Officer, Colonel J. M. Harmon, appointed a school committee. The committee made plans for a school building and supervised the curriculum planning and hiring of teachers. They hired W. W. Cook of the University of Minnesota as consultant. Cook and Brazier designed a building to house both elementary and high schools. Construction began late in the summer of 1943, and because of a high priority was complete in time for the opening of a fall school term. The committee was continued as a school board.

The elementary and high schools were operated as free public schools with salaries and procurement expenses borne by the Government through the contractor. A nursery school, for which a building had been provided in the original construction plan, was partially self-supporting. It permitted part-time, or more rarely full-time, employment of women with young children. Its financial deficit also was carried by the Government.

Another matter in which the Laboratory administration was interested was community representation in the civil affairs of Los Alamos. In June 1943, a Community Council with members elected by popular vote was established to supersede an earlier appointed committee. Its function was intended to be purely advisory. However, it was elected only by Laboratory members and their wives and did not represent the entire community. It was advisory, not to the Commanding Officer, but to the Laboratory administration. In August 1943, the new Commanding Officer, Lieutenant Colonel Whitney Ashbridge, and the Laboratory Director approved a more representative council that met with Laboratory representatives and the Commanding Officer. Some regarded the council as a thorn in the side of community administration, and at times it was. However, it sought to guide its deliberations and recommendations by the single standard of the success of the project. Sometimes its recommendations could not be carried out because of manpower and material limitations. Sometimes the limitations derived from Army administrative customs. However, many recommendations were accepted. Under council guidance, a system of small community play areas was built for the children of the Post. Traffic laws were written with the advice of the council, which also acted as a traffic court to administer a voluntary fine system. Other topics frequently considered were operation of the post exchange, messes, and commissary; the milk supply; maid service; public transportation; and the hospital.

A major problem that dogged the Laboratory from the beginning was housing. Los Alamos, originally conceived as a small, more or less stationary, community of research scientists, developed into a large, complex, industrial laboratory. The housing problem put a constant drag upon Laboratory efforts to get and keep an adequate staff.

Construction at Los Alamos was not easy. The population growth had strained power and water supplies. Construction was expensive of critical manpower and materials, and the large group of construction workers further strained community facilities. These difficulties, plus a constantly shortening period of amortization, necessitated cheapening of construction. To the shortage of housing this added a troublesome inequality.

It was twice necessary, and a third time almost necessary, to use outside housing facilities. At the very beginning of the project, Laboratory members were housed temporarily in nearby guest ranches. By the beginning of summer 1943, the original housing was filled, and new housing was not yet ready. From June 19 to October 17, therefore, the project took over Frijoles Lodge at the Frijoles Canyon headquarters of the Bandelier National Monument, 14 miles from Los Alamos. After its acquisition by the Albuquerque District Engineer, Frijoles came under the jurisdiction of the Personnel Office. To run it, the Laboratory hired S.

36

A. Butler, assistant manager of the La Fonda Hotel in Santa Fe, who later became Assistant Personnel Director. Frijoles Lodge was used again from July 17 to August 5, 1944, when the project faced another critical housing shortage.

The Laboratory also had an administrative interest in the community hospital. This hospital for both military and civilian personnel was under the jurisdiction of the Chief of Medical Section of the Manhattan District. Its existence and excellent record contributed importantly to project morale. Another important function was its cooperation with the Laboratory's health and safety program, discussed in detail later.

Security Administration

It would be difficult to exaggerate the security precautions taken at the beginning of the project, particularly those relative to personnel. Moreover, administration of security policy was important not only to safeguard information, but also because of its effects on morale, and the possibility that serious breaches of security might lead to even more stringent external control.

Formal clearance of personnel for work on the project was arranged through the Intelligence Officer at the site. This procedure was slow and cumbersome, especially at first. In September 1943, a plan was approved to supplement clearance where necessary by an interlocking system of vouching for the loyalty and good faith of Laboratory members.

Administration of security matters pertaining to Laboratory personnel and their families was delegated to Hawkins as Contractor's Security Agent. A security committee, composed of Hawkins, Manley, and Kennedy, met with the Intelligence Officer. Recurring topics of discussion were the pass and badge system, monitoring for classified material left unattended, means of preventing classified discussions in the presence of outsiders, and publication and revision of security regulations.

The most irksome restrictions were those on personal freedom. Travel outside a limited local area and contact with outside acquaintances were forbidden except on Laboratory business or in case of personal emergency. Mostly, these restrictions were accepted as concessions to the general policy of isolation. A small group thought they were not strict enough, and no one was satisfied with the definitions of "personal emergency." Removal of these restrictions in the fall of 1944 was a relief

37

after a year and a half of extreme restriction. Another security feature was mail censorship. This was unusual in itself, and amusing in that it began because of the suspicion of unannounced censorship. Not long after the Laboratory started, this suspicion spread as a rumor, and evidence that letters had been opened was presented. The Director, who could not guarantee that censorship was not occurring, made strong representations to General Groves' office. General Groves instituted an investigation, with negative results. However, many members of the Laboratory then urged that official censorship be instituted, and this was done in December 1943. This censorship deterred inadvertent spreading of information about the project, which might contribute to consistent rumors and continuing public interest in its activites. Censoring was conducted by a standard military censorship office in Santa Fe, under the direction of the Intelligence Officer, Captain P. de Silva.

Salary Policy and Administration

The most pressing personnel administration problem in the early months was salaries. Scientists' salaries were determined by the OSRD scale, based upon scientific degrees held and number of years since their conferral, or by the "no loss/no gain" principle that those from academic positions, whose salaries are normally based on a 10-month year, be paid at twelve-tenths their previous rate. One difficulty was that industrial workers had been paid more than those from academic positions. Another was that technicians, lacking academic degrees but often highly skilled, had to be hired at prevailing rates. Although technicians ranked below the younger professional scientists, they often were paid more. Finally, although a general commitment had been made to a policy of length of service and merit increases, there was no administrative mechanism to implement it.

When he arrived in June 1943, Hughes' first major responsibility was to prepare salary policy recommendations based upon a survey of this and other comparable laboratories. Within the regulations of the National War Labor Board, he proposed a salary scale for the various classes of Laboratory employees and a plan for increases. There was no provision for increasing salaries above $400 per month, where they were virtually frozen.

The salary and increase proposal was presented to the Manhattan District and the University of California in June, but approval was postponed because certain formal changes had to be made owing to changes in national policy. A conference was held on the subject at Los Alamos in January, and approval was finally granted on February 2, 1944, after a year of operation. During this period no system of promotion was possible, although the proposed policy was followed in determining the salaries of newly hired individuals.

The chief difficulty concerned the younger scientists hired under the OSRD scale. This scale provided for an annual salary increase, but the Contracting Officer, Lieutenant Colonel S. L. Stewart, would not approve it in the absence of an approved Laboratory salary policy. Inequities, as measured by degree of responsibility and usefulness, were numerous within this OSRD group and between it and those hired on a "no loss/no gain" basis.

Final agreement about salary policy was not reached until the end of the war, but improvement resulted from a reorganization in July 1944, when a new working agreement was reached.

Draft Deferment

Because of the great scarcity of trained scientists and technicians during the war, every effort was made to prevent induction of those whose services were essential and satisfactory. For security, turnover of such personnel had to be kept absolutely minimal. However, the secrecy requirements made it impossible to give Selective Service any real information about the nature and importance of the Laboratory's program, or the work of an individual. Moreover, most of the scientists and technicians were under thirty years of age and thereby vulnerable to the draft.

Because of the importance of the project, paradoxically, deferment of Laboratory employees was a complex matter. Dennes came to the Laboratory empowered to represent the University War Council in deferment matters. He worked with Selective Service to eliminate the confusion that existed during the Laboratory's first days.

The most essential Selective Service liaisons were with the New Mexico State Director of Selective Service and the Selective Service Agencies of the Manhattan District. Each gave the Laboratory the utmost cooperation. As the war progressed and Army and Navy needs increased, deferment requirements became more stringent. The Laboratory depended increasingly upon official Manhattan District certification of its needs. In February 1944, the War Department forbade deferment of men under 22 in the employ of the Department or its contractors. The Laboratory had a small group of men under 22 who were highly trained and essential. Even so, they could not be deferred, so when they were inducted, there was no choice but to reassign them to the Laboratory as members of the Special Engineer Detachment (SED).

Personnel Procurement

Despite staffing difficulties, the working population doubled about every nine months. Although fewer new employees were scientific staff, their absolute number increased monthly almost until the end of the war. At the same time the difficulty of finding competent scientists, especially upper technicians and junior scientists, increased. Senior scientists, needed only in small numbers, were usually well known to Laboratory members. These men often were eager to join the Laboratory, and releases from less critical work or other Manhattan Projects could be obtained through the Washington Liaison Office. Junior men were needed in great numbers, but recruiting trips to universities were forbidden by security regulations. In November 1943, the assistance of Dean Samuel T. Arnold of Brown University was obtained in these matters. It also was arranged that M. H. Trytten of the National Roster of Scientific and Technical Personnel could visit universities and hire young scientists. Trytten assisted the Laboratory for several months. Arnold remained as personnel liaison in Washington through the project.

Military Personnel

A few Army and Navy officers with scientific training worked at the Laboratory at various times, but most military personnel were from the Women's Army Corps (WAC) and the SED. The latter originally was established as a small detachment (about 300 for all Manhattan projects) in which essential personnel could be placed when deferments were no longer possible. In November 1943, when junior scientific personnel were extremely hard to find, the Laboratory was informed that some new graduates of the Army Specialized Training Program could be assigned to the Laboratory in the SED at the beginning of the year.

Although basic Laboratory policy was to hire civilians, they were increasingly hard to find. They were being inducted rapidly into the

Army, where their assignments often were less appropriate to their training than if they were transferred to the SED. There was no choice but to welcome all technically trained enlisted personnel for whom civilian counterparts could not be found. In May 1945, 29% of the SED personnel held college degrees, including several Doctor and Master degrees, mostly in engineering, chemistry, physics, and mathematics. Several competent scientists also came from the WAC detachment, as did many technical and office workers.

Policy regarding enlisted personnel was essentially identical with that for civilians. After their arrival at the site and assignment to the Laboratory, all further matters of placement, job classification, transfer, and promotion within the Laboratory were under the jurisdiction of the Laboratory Personnel Office.

The SED's rapid growth brought many administrative problems connected with morale, accommodations, and working conditions. Most serious was the shortage of multiple-unit housing, which made it impossible to provide quarters for married enlisted men. Further, Major de Silva objected to hiring enlisted men's wives (except as nurses), although they could have been quartered in the dormitories for women workers on the project. Also, security regulations prevented the men from bringing their wives to Santa Fe or other nearby communities. The restrictions against travel and association with persons away from the project, therefore, were harder on enlisted personnel than on civilians, whose wives and children lived at Los Alamos.

Another problem was that military promotion, the responsibility of the SED Commanding Officer, was the only material means available for recognizing responsibility and excellence in technical work. The SED Table of Organization permitted promotion of one-third of the men to each of the grades Technician Third Class, Technician Fourth Class, or Technician Fifth Class, and about one-tenth of those in Technician Third Class could be promoted to Technical or Master Sergeant. As most of the men arrived with a rank no greater than Technician Fifth Class there was ample opportunity for promotion at first. However, the SED Commanding Officer and the Laboratory had to agree on the promotion rate, and for several months no such policy was established.

A third difficulty was conflict of military and technical duties. Although the official working hours in the Laboratory were 8 hours a day for 6 days a week, many Laboratory research groups worked irregular and usually much longer hours. This practice conflicted with barracks duties and formations.

The presence of engineer and military police detachments required some consistency of treatment of military personnel, in accordance with the usual military organization. However, the SED's position was improved. In June 1944, general supervision of SED military administration was given to Major de Silva. As Intelligence Officer, his work brought him into close contact with the Laboratory administration. In August, the regulation forbidding travel and outside social contact was relaxed for military personnel in the Laboratory so that they might visit their wives and families.

In August, Major T. O. Palmer was appointed Commanding Officer of the SED. Much of the credit for maintaining SED morale must be given to him. A system of promotion recommendations by Laboratory groups and divisions was soon worked out. The problem of conflicting duties was not, and perhaps never could be solved. The overtime work done by many groups and individuals required essentially civilian conditions.

Salary Problems

The Laboratory Personnel Department faced an unusual range of difficulties. The greatest single difficulty was salary policy. Both the general salary policy and its detailed administration were under the supervision, and subject to the direct veto, of the Contracting Officer Lieutenant Colonel Stewart. However, he found his responsibility impossible to discharge to the Laboratory's satisfaction. He was stationed in Los Angeles, where his services were urgently needed for procurement matters, and he had only a general and overall acquaintance with Laboratory problems. Either of two conditions would have remedied the situation: (1) that the Laboratory have a strong, well-organized personnel office that could represent its need with enough consistency, detailed justification, and vigor to compensate for the Contracting Officer's remote position; or (2) that the Contracting Officer be stationed at Los Alamos, where he could understand the Laboratory's detailed needs. Finally, partial satisfaction of both conditions tended to solve the salary problem.

By June 1944 it was apparent that administrative reorganization was necessary. Hughes' previous experience and Los Alamos' function had been primarily building a competent scientific staff. The Laboratory's rapid expansion and its ramification in directions other than "research" created new personnel problems. After a year spent in building up the scientific staff and seeking to formulate and work out its personnel policies under increasingly difficult conditions, Hughes returned to his previous position at Washington University. His position was taken by C. D. Shane of the Radiation Laboratory, Berkeley, who brought as his general assistant Roy E. Clausen of the University of California. Armand Kelly, formerly at the Metallurgical Laboratory of the University of Chicago, was brought in as an expert on salary and salary control. Hawkins was made responsible, under Shane, for draft deferment and military personnel matters.

The most serious personnel problem still was salaries. After reviewing the situation, Shane accepted the position as Personnel Director with the understanding that he and his office would have a reasonable degree of autonomy over salaries, not subject to veto by the Contracting Officer except in terms of Federal salary policy and regulations. Agreement with the Contracting Officer to this effect was reached in July 1944.

OTHER ADMINISTRATIVE FUNCTIONS

The Business Office

In February 1943, shortly before the Laboratory administration moved to Santa Fe, the University of California appointed J. A. D. Muncy as Business Manager. Security restrictions put quirks into the operations and added several unusual functions. For security reasons, it had already been decided to locate the Purchasing Office in Los Angeles. It was also considered desirable to have the Accounting Office physically connected with the Purchasing Office. Therefore, a general business office in Los Angeles took over most operations at the point where money was disbursed. Complete records of all transactions were kept in that office, and government and university audits were made there. However, a Santa Fe bank account for emergency purchases, travel advances, and cashing personal checks for the contractors' employees reached considerable proportions. It was, in fact, the second largest account in the bank, and because it was in Muncy's name, he frequently received circulars from charitable organizations suggesting large contributions.

43

The "normal" functions of the Business Office were payroll control, issuance of travel advances and preparation of travel expense bills, procurement of materials from the emergency purchases fund, and maintenance of records for Workman's Compensation and for the California State Employees' Retirement System, to which University employees were obliged to contribute after six months of employment. To avoid giving the bank a list of Laboratory personnel, monthly salaried employees were not permitted to have accounts in the local bank. The Business Office at the site made up the monthly payroll and forwarded it to the Los Angeles office, where checks were written and mailed to banks designated by the employees. However, in 1943, the Laboratory hired a large group of laborers and construction workers who were paid on an hourly basis, and beginning in January 1945, the salaries of machinists and other shop workers were computed on an hourly basis. These payrolls were made up and checks were written by the Business Office at Los Alamos. Approximate monthly payrolls of $50 000, $160 000, and $175 000 in June 1943, 1944, and 1945, respectively, indicate the project's tremendous growth. The payroll for hourly workers in June 1943 was roughly $23 000; in June 1945 it was approximately $130 000.

In keeping payroll records at the site, there was considerable difficulty in getting accurate records of attendance. The Manhattan District considered inadequate the University procedure of having a supervisor certify monthly that all his charges were present with the exceptions noted. The Laboratory considered group leader attendance certifications by days and half-days impractical. Personnel were too scattered, particularly in groups doing field work, and scientists often worked at night, although not on regular shifts. Scientists often worked more than 48 hours a week, and because the contract did not allow for overtime payments it was felt that no deductions could be made for absences. The only procedure used until September 1945 for the scientific and administrative staff was a negative report made monthly by each employee, without any group leader certification. Although this system was never considered satisfactory, more rigorous control probably would have imposed an almost prohibitive administrative burden and would have lowered the morale of scientists who were actually giving more than 48 hours a week to project work.

Reimbursement for travel on project business was handled like the payroll. Although advances were issued from the local account, travel expense bills were forwarded to the Los Angeles office, which mailed checks to the payees' banks.

The emergency purchases fund was used when it was impractical to route a materials request through the Los Angeles Purchasing Office, because of its urgency or because of the character of the materials. Most material purchased on this fund in 1943 fell into the former category, being construction supplies needed immediately. Disbursements from this fund in June 1943 were approximately $23 000; in June 1944 they had dropped to $4000. In 1944 the materials purchased were principally batteries, dry ice, and cylinders of gas, all unsuitable for shipment from Los Angeles. In June 1945, during preparations for the Trinity shot, about $38 000 was spent for miscellaneous items ranging from radio tubes to canvas water bags, plus an increased volume of the normal batteries, gases, etc. Among the unusual purchases made with this fund were 88 cows that apparently suffered radioactive burns during the Trinity test.

One extraordinary duty of the Business Office was to handle the financial end of the temporary housing mentioned in Chap. 1. The cost of opening and operating the ranches used made the expenses to employees considerably greater than they would have been at the site. It was felt that the project should assume this extra cost because housing was not ready at the site as had been promised. Therefore, the Laboratory operated the ranches and billed each individual or head of family for the amount of his living expenses at the site (rent plus $25 per month per person for food).

Five ranches were operated from the end of March to the end of May 1943 at a cost of about $7000. The Business Office also settled claims for damage to the temporary housing occupied by the Laboratory's personnel, with the assistance of the Contracting Officer.

Other unusual Business Office functions arose from the attempt to keep a list of personnel from accumulating outside the project. Personnel were asked not to cash personal checks in Santa Fe, and check-cashing facilities were provided at Los Alamos. By 1945 the daily average of checks cashed was $3000-$4000. All personal long-distance calls and telegrams were charged to a Business Office account, and the daily telephone bill increased from $57 in June 1943 to $745 in June 1944. When New Mexico income tax returns were due, the Business Office assigned a number to each employee and reported to the income tax bureau the amount of income paid to that number in New Mexico during

45

the year. The employee then used his number instead of his name on his return.

The decision to keep the main accounting office in Los Angeles was wise, both because of its proximity to the Purchasing Office and because its staff, which grew to 70 people, would have required its own housing project at Los Alamos.

Because security regulations prevented personnel from taking out new life insurance and because of the extra-hazardous work at the project, the problem of providing insurance was extremely complex and never solved adequately, although the Director's Office and Business Office tried. When the project was organized, technical area employees were covered by an OSRD policy for injury, illness, or death, placed with the Fidelity and Casualty Company of New York. In September 1943, this was replaced by Manhattan District Master Policy 1 with the Sun Indemnity Company, which offered additional benefits including extra-hazardous work insurance. In July 1944, this was replaced by Master Policy 2, with premiums to be paid by the individual or the Laboratory instead of the Government. Master Policy 3 provided for accidents not arising from employment. At about this time, there was considerable discussion about the fact that the extra-hazardous insurance policy was inadequate for people working on radioactive substances because it made no provision for injuries that might not appear for 10 to 15 years after they were received. Eventually, this problem was partly solved by a special arrangement with the University of California. The Government deposited $1 000 000 with the University to be used for payment, with Government consent, of up to $10 000 for injuries resulting from extra hazards listed in a secret letter to the University. The Laboratory provided statutory Workmen's Compensation of the State of New Mexico for all persons permanently assigned to work in New Mexico. Claims paid under Workmen's Compensation through December 1945 totaled only $18 000, of which $12 000 was death benefits for two laborers killed in a moter vehicle accident in 1943. Accident policies essentially the same as the expired Master Policies 2 and 3 were made available to individual employees in September 1944. For some time, there was no coverage for travel on noncommercial, nonexperimental aircraft used by project employees, but eventually this was covered by a personal accident policy with Aero Insurance Underwriters for civilian employees.

Procurement

The community's isolation created acute and serious problems for the Procurement Office. Supplying a large research laboratory from the ground up is difficult; doing it secretly, in wartime, 1200 miles from the nearest large market and 100 miles from the nearest rail and air terminals would seem impossible. Yet the Procurement Office overcame the obstacles of time, space, and security to satisfy the Laboratory's exacting and apparently insatiable demands. The fact that the Laboratory was able to meet its tight time schedule is a tribute to the competence and efficiency of its Procurement Office, guided from the beginning by D. P. Mitchell of Columbia University.

In February 1943, Mitchell, Oppenheimer, and several other scientists discussed purchasing policies with representatives of the Army and the University of California. At University insistence, it was agreed that all purchasing and payment would be administered directly by members of the University staff, and within their entire discretion as to appointees but subject to general supervision by the Contracting Officer. In effect, this meant that although Mitchell was in charge of ordering materials for the Laboratory, the actual purchasing would be done by University appointees. This organizational complication brought with it another security complication. The University's purchasing office must be located in Los Angeles, and its employees would not be permitted to come to Los Alamos to deal directly with those placing orders or to know anything about the Laboratory work.

Mitchell as Procurement Officer was to be guided primarily by the necessity for speed and was not to be held responsible for the kind or quality of items purchased. He was authorized to place orders through requisitions signed by qualified staff members and showing quantity, description of item, date required, urgency, and suggested source. From the orderer's statement of urgency, Mitchell would judge the degree of priority required and the means of communication and transportation to be used. Primarily, he was to supply the needs of the technical staff as promptly and with as little red tape as possible. On the whole, this policy was maintained successfully.

Until the Los Angeles Purchasing Office was established, purchases were made through the Purchasing Office of the Radiation Laboratory at Berkeley.

The Los Angeles office, organized by D. L. Wilt, began operation March 16, 1943. After September 1943, A. E. Dyhre was in charge.

Branch offices in New York and Chicago were set up in April 1943. Except in emergencies, the Laboratory's Procurement Office dealt directly with the Los Angeles Office by mail or teletype. Requisitions for items not available in the Los Angeles area were forwarded to the New York and Chicago branches from Los Angeles. The three offices employed a total of about 300 people, including 33 buyers and 22 expediters, at their peak. The average monthly dollar volume was about $400 000, for about 6000 items. However, in May 1945, over $1 million was spent for more than 10 000 items.

Some local purchasing was permitted. At first, "local" meant a 500-mile radius including Denver, but as security restrictions tightened, "local" became a 100-mile radius, including Albuquerque. Many bulk items, such as fuel and building supplies, could be purchased advantageously from local suppliers. For security reasons, purchases were made in Muncy's name. The Post Supply Section, directed by Major Edward A. White, often was asked to supply items for the Technical Area. A system whereby the Technical Area could requisition on Major White's office was a great help to the technical work.

The first groups of scientists had brought a cyclotron, two Van de Graaff generators, a Cockcroft-Walton accelerator, and some electronic equipment. There was nothing else at Site Y to constitute a laboratory. Within a few months, the Procurement and Purchasing Offices had to equip physics, chemistry, and electronics laboratories and machine shops, and to prepare supply stockrooms for them. The range of materials required was incredibly great—everything from workers' work clothes to 10-ton trucks. The Procurement Office bought such things as rats, meteorological balloons, sewing machines, restaurant equipment, jeweler's tools, and washing machines, in addition to what might be considered standard shop and laboratory equipment. Although assigned AA-1 priority by the War Production Board (WPB), Project Y often had to ask the District's help in securing higher priority or a WPB directive for particular items. Thus, despite wartime difficulties, stockrooms were ready in short order—one for chemical supplies (K stock), one for general laboratory supplies (S stock), one for special electronic supplies, and one for each of the shops. Responsibility for the electronic and shop stockrooms soon was turned over to the various operating groups.

Procurement could never become routine because of the Laboratory's continuous expansion and the constant, necessary changes in the technical program. Time was always the most critical factor, and the

Procurement Office's shipping instructions on requisitions sent to the Purchasing Office often were extremely important. The Procurement Office sometimes had to reorder by air express an item that had been shipped by freight contrary to instructions. Waiting for freight delivery would have meant holding up vital experiments that would cost much more in time and money than the cost of duplicating an order.

Security also caused innumerable difficulties for the Procurement and Purchasing Offices. The employees of these offices could have no direct contact with the using groups at the Laboratory, could know nothing about their work, and therefore could not understand its significance or appreciate the urgency and responsibility of their own work. For security reasons, using groups were almost never able to deal directly with manufacturers and dealers. Questions about design or fabrication had to be transmitted through the New York or Chicago Purchasing Offices to the Los Angeles Purchasing Office, from there to the Los Alamos Procurement Office, and finally to the using groups. The answer had to be returned along this same path to the supplier. No direct shipment could be made to Site Y. All suppliers were instructed to ship goods to Chicago and Los Angeles warehouses, from which they were trans-shipped to Site Y with their original labels removed to prevent unauthorized persons from learning what kinds of things were being received. Originally, the Los Angeles and Chicago warehouses did nothing but transship orders, and the Site-Y warehouse checked shipments and approved invoices. Because Government regulations insisted upon prompt payment of bills to avoid loss of discount, the procedure was changed so that invoices were checked against shipments at Los Angeles and Chicago and approved there. This procedure led to difficulties: goods were received that were neither usable nor returnable, items were missing from shipment but checked on the invoices, and packages of photographic film were opened for inspection.

Laboratory technical groups periodically criticized the Los Angeles Purchasing Office to the Director, who periodically transmitted the criticisms to the appropriate Army and University officials. The University and Purchasing Office justifiably maintained that much of their difficulty was directly traceable to the strict security regulations under which they operated. Statistics that the Contracting Officer, Lieutenant Colonel Stewart, compiled from time to time showed the Purchasing Office to be fairly efficient. During the war many manufacturers stopped publishing catalogs, so those available in no way represented existing conditions. Men had been in the habit of designing apparatus, starting to build it, and then ordering parts. This habit nearly proved disastrous on several occasions. For example, one group designed a special camera for the Trinity test, started construction, and then ordered the parts. They were notified that the lenses they had ordered were not on the market, would have to be ground to order, and might not be ready in time to be useful. Also, the particular plate backs incorporated into the camera design were no longer available and were not being manufactured. Purchasing Offices scoured the country and succeeded in finding about one-third of the required number of plate backs. To secure the rest, it was necessary to go through the Washington Liaison Office to get a WPB directive ordering the former manufacturer of these items to stop his current production and make the necessary plate backs for the Laboratory. The cameras were ready in time for Trinity, but only after a tremendous expenditure of effort by all concerned. The Procurement Office had tried to teach the using groups the importance of determining the availability of materials before completing designs and starting work, but it was difficult to change old habits.

Knowing nothing about the work, the Purchasing Office could know nothing about the uses for which particular items were needed, and so could not understand which specifications were critical. The buyer would see no reason to ask the Los Angeles office to check with the Site-Y office to check with the user—and would place the order. He could not be expected to know that for the scientist's purpose, chemical composition of a material was the one all-important criterion. The only possible solution was to have the staff members make their specifications as complete and explicit as possible without revealing the nature of their work. To eliminate red tape and supply the Laboratory as quickly and efficiently as possible, Mitchell organized his department into two main sections. The Procurement Section was supervised by E. E. Olsen; the Service and Supplies Section, by H. S. Allen. The Procurement Section consisted at first of two groups, Buying and Records. Later, Property Inventory was added. The Buying Group was responsible for checking specifications on purchase requests, suggesting a possible manufacturer or vendor to the Los Angeles office, justifying high urgencies, and answering questions from the Los Angeles Purchasing Office. Essentially, the local buyers existed to give the Los Angeles buyers the information they required to purchase the things needed at the Laboratory. The Records Group maintained correspondence files, purchase

50

requests, and Kardex files of expendable and nonexpendable goods. When one understands that every purchase request required about 60 pieces of paper, including printed forms and teletypes, the importance of the Records Group becomes evident. The Service and Supplies Section consisted of the stockrooms, receiving, shipping, and records.

Certain special procurement channels bypassed the Los Angeles Purchasing Office. These concerned parts for the completed bomb mechanism, materials including uranium and plutonium coming from other branches of the project, and materials obtained directly from the Army or Navy, such as electronic components and completed devices of an electronic nature, guns, propellants, and high explosives.

Library and Document Room

No research laboratory can exist without a library well stocked with standard technical reference works, files of technical journals, and reports of work in progress, especially when it is isolated from all other universities and libraries. The Los Alamos Library, organized and directed by Charlotte Serber, served its purpose well. It was one of the few administrative groups in the Laboratory about which there were no substantial complaints from the scientific staff.

The Library had to provide quickly a comprehensive collection of books and journals on physics, chemistry, engineering, and metallurgy. Normally, it take years to accumulate such a collection. Much of the initial problem was solved by loans, chiefly from the University of California library. The staff members planning the Laboratory requested approximately 1200 books and complete files of 50 journals from about 1920 on. Many were impossible to secure on the market, but the University library was able to supply nearly all of the rare out-of-print titles. For new publications, because of security restrictions, orders were sent to a forwarding address in Los Angeles, and from there to the University library in Berkeley, which placed orders with book publishers and dealers to be sent to the Los Angeles receiving warehouse. From Los Angeles, the books were forwarded to Los Alamos. By July 1945, the Library included approximately 3000 books, 160 journals per month, and 1500 microfilm reproductions of specific articles and portions of books.

Most of the Library's work, however, was reproducing and distributing reports of work in progress. For this purpose, it had two small subgroups known as the workshop and the document room. The workshop typed, reproduced, and assembled technical reports and manuals in very close collaboration with the editorial section and the photography and photostating shop. The document room distributed the completed reports in accordance with security regulations because nearly all of the project work was classified. The Laboratory felt that an employee should have all the information he needed to make his work effective and to avoid breaking security regulations when dealing with outside workers. To this end, the Library's document room kept a list of personnel entitled to have access to all or certain categories of classified documents. This list was kept up to date by advice from group and division leaders. Most documents were kept in the document room to be read there or borrowed temporarily by qualified persons. In addition to a file of Los Alamos reports, the document room kept a file of documents from other Manhattan District projects. By January 1945, there were 6090 reports on file, exclusive of extra copies of the same report and of the approximately 10% of the total circulated each week.

A minor Library duty was to train the secretarial staff in preparing reports for reproduction and handling classified documents. In January 1945, the document room assumed from the Patent Office the duty of issuing patent notebooks, keeping a record of notebooks issued, and collecting them from individuals when they left the project.

Editorial Section

From the beginning, the Laboratory produced many reports. Whole fields of research were amplified by work done here. Papers in these fields were made obsolete, and reports written here became standard reference works for this and other Manhattan District projects. Therefore, experimental results had to be reported speedily and accurately in a form readily accessible to other employees who required the information for their own work. It soon become apparent that the responsibility for editing and reproducing all reports should be centralized to ensure accuracy as well as speed. Early in 1943, D. R. Inglis of Johns Hopkins University was appointed Project Editor. All reports of completed work or work in progress went to Inglis' office for technical and editorial review. An appropriate form of reproduction was then selected. Through Inglis' efforts the Laboratory was assured of technically accurate, and editorially consistent, reports.

Health and Safety

A Health Group, reporting to the Director, and supervised by Dr. L. H. Hempelmann was part of the Laboratory administration from the beginning.

Health problems were classified as standard industrial health and safety problems, definition of health standards for special hazards, establishment of safe operating procedures, and routine monitoring and record keeping. All were part of the Health Group's responsibility, and Dr. Hempelmann acted as chairman of the Laboratory's Safety Committee. By April 1944, this committee felt that it had become too unwieldy to handle the growing safety problems. Mitchell, Procurement Office Leader, became head of a new Safety Committee to supervise all safety installations, inspections, and activities in the Technical Area and outlying sites. This included fire, general safety, and maintenance, as well as technical safety. Dr. Hempelmann represented the Health Group on the Safety Committee. Execution of safety policies was taken over later by the Safety Group under a full-time safety engineer (Chap. 9). Establishing safe operating procedures, routine monitoring, and record keeping remained under the Health Group's general jurisdiction, but wherever possible, these tasks were delegated to the operating groups or appropriate subcommittees of the Safety Committee.

The Health Group was responsible for establishing and disseminating health standards, specifically for safe levels of exposure to radiation and to radioactive and chemical poisons. Its primary concern was to protect the health of Laboratory employees. Secondarily, it sought to protect the legal interests of employees and of the Contractor by keeping records of the hazards to which individuals were exposed, the extent of exposure, accidents, and tests for overexposure. It also made recorded pre-employment medical examinations of all technical personnel and made complete examinations, including necessary tests, of all terminating employees. Ordinary industrial accident records, however, such as those of shop injuries, were kept by the Post Hospital.

Originally, it was assumed that biological and physical research on health problems would be the responsibility of other Manhattan District laboratories. However, reliance on the work of others did not always provide necessary information when it was needed. Research sections were set up as needed within the Health Group or, by its request, in other groups. Development of monitoring apparatus began in the spring of 1944, with many of the instruments being built in the Electronics Group.

53

In August 1944, the Health Group started to investigate biological methods of testing for overexposure to radioactive poisons. During its first year, the Health Group's work was relatively uncomplicated. A semiresearch problem, which appeared almost immediately, was to discover the extent of variation in normal blood counts. Variations at first thought symptomatic of overexposure to radiation were found to be common in normal blood.

Early work was largely on the hazards of external radiation from accelerating equipment and radioactive sources. Heavy-metal poisoning from uranium had to be guarded against, as did other chemical hazards, but these problems were not serious. The Health Group's really serious problems appeared in the early spring of 1944 when the first plutonium arrived in Los Alamos.

Plutonium metabolism is like that of radium in that it is deposited in the bone, where its alpha radiation may cause bone sarcoma. The next heaviest deposition is in the kidneys, where its radiation may destroy kidney function. This, however, happens only at dosages higher than those needed to cause bone injury. Also, plutonium is eliminated from the body much more slowly than is uranium. However, plutonium has much lower alpha activity than radium and is less easily absorbed from the digestive tract. In general, the problems of handling plutonium were comparable with those of handling radium, with allowances for the vastly larger quantity of plutonium processed and the fact that no empirical information was available on the toxic effects of small amounts over a 10- or 20-year period.

The general similarity with the radium hazard had just been discovered. Therefore, Hempelmann and representatives from Chicago and Oak Ridge visited a luminous paint company in Boston to learn how they handled the radium hazard. On his return, three committees were established in the Chemistry and Metallurgy Division to develop methods for controlling the plutonium hazard. An instrumentation committee was to design counters to measure radioactive contamination of laboratories and personnel. A second was to design apparatus and equipment for handling plutonium. Apparatus was designed in consultation with the chemists concerned and was built or procured by the Chemistry and Metallurgy Service Group. A third committee drew up rules and recommendations for handling radioactive materials safely. The procedures were put into effect in March 1944 with the understanding that willful noncooperation would result in immediate dismissal. A

54

section of the Service Group was established under W. H. Popham to enforce these procedures. It provided proper protective equipment, laundered this equipment, monitored the laboratories and decontaminated them when necessary, and kept complete records. It worked very closely with the Health Group.

The Health Group also conducted an extensive educational campaign among the groups working with plutonium. They gave lectures on plutonium toxicology and held many conferences with operating groups about how to apply the general recommendations. The Health Safety Handbook was given to new members of the Division.

Despite these precautions, members of the Health Group and of the Chemistry and Metallurgy Division were not satisfied with the progress of biological studies on plutonium made by the other projects. This dissatisfaction was crystallized when in August 1944, a minor chemical explosion threw several milligrams of plutonium into the face of a chemist. A research program to develop tests for detecting overdosage of plutonium was undertaken (Chap. 9).

Another difficulty was lack of adequate monitoring equipment. Alpha-ray counters lacked either sensitivity or portability and were not received in adequate numbers. This made it necessary to wipe surfaces suspected of contamination with oiled filter paper and to measure the collected activity with stationary counters. Contamination of hands and nostrils was measured in the same way. Because instruments received from Chicago did not meet the local monitoring requirements, the Electronics Group of the Physics Division began development of such equipment in May 1944 (Chap. 9).

Another Health Group activity was to control the danger of poisoning in the high-explosive casting work. Standard protective measures were put into effect, and no serious trouble was encountered. The medical group examined all exposed personnel monthly and gave periodic lectures about the dangers of toxic effects from high-explosive work. Worker education was possible because all the plant supervisors were experienced in this type of work. The number of cases of TNT dermatitis was in keeping with the number of men exposed. This is an allergic reaction that cannot be prevented entirely in any plant operation.

Shops

The principal shop facilities were machine shops, drafting rooms, a glass shop, and photographic shops. Although these were mostly standard service groups, they had many unusual administrative and technical problems.

Originally, a drafting room and machine shop (known as V Shop) was to design and fabricate laboratory tools and instruments primarily for the Experimental Physics Division and the Chemistry and Metallurgy Division. The glass shop was an adjunct of the chemistry groups. Two photographic shops were added in 1943, one mainly for routine recording and duplication, the other as an adjunct of the ordnance research program, responsible for technical photography and optical research (Chap. 15).

After the ordnance program began, additional plans were made for an ordnance drafting room and a large ordnance machine shop (later called C Shop). Several small student shops or special shops were built at various times, of which the largest was the Graphite Shop of the Miscellaneous Metallurgy Group (Chap. 8).

Responsibility for organizing these shops was first assumed by Mack (Chap. 1). The Laboratory was fortunate in obtaining Gus H. Schultz from the University of Wisconsin Shops as foreman of the Laboratory shop (V Shop). He was not only thoroughly familiar with the requirements of a laboratory shop, he also had substantial industrial experience.

V Shop originally covered 8000 square feet, planned for 30 toolmakers and machinists, to meet an expected load of about 1500 man-hours per week. This load was reached in October or November of 1943, by which time the goal had been set much higher.

In July 1943, Mack resigned as Shop supervisor and set up the Optics Shop and research group in the Ordnance Division. His place was taken by E. A. Long, head of cryogenic research in the Chemistry and Metallurgy Division.

By March and April of 1944, about half the V-Shop load came from the Chemistry and Metallurgy Division, whose requirements were rapidly increasing. This problem was met by shifting some of the Metallurgy work to C Shop, and by adding about 500 sq. ft. in area to V Shop in May.

Outstanding fabrication problems solved in V Shop included fabrication of beryllium oxide bricks for the Water Boiler (Chap. 13). The dies were developed in V Shop, as was the technique for facing the bricks. Another was development of the apparatus and technique for welding the thin stainless steel envelopes of the Water Boiler. Yet another was machining and grinding of tungsten carbide. In all cases the primary

56

responsibility was borne by the operating group, but the actual development work was done by shop personnel.

Construction of C Shop began in July 1943 and was completed in October. Its area was 8800 sq. ft., planned for about 40 machinists and toolmakers, to meet a load of about 2000 man-hours per week. Rex Peters was its foreman and C. Cline its supervisor.

The experimental shop ran relatively smoothly and harmoniously, whereas the ordnance shop was full of crises. The experimental shop was organized after a familiar pattern and was staffed and supervised by men with adequate training and experience. C Shop, by contrast, was designed for a type of work not completely anticipated. Equipment and personnel proved inadequate to the demands. By the time the difficulties were appreciated fully, the Laboratory had grown so large that it was impossible to overcome the existing lag.

In May 1944, Cline was transferred to the Engineering Group and W. M. Brower took his place. Whereas Cline had little experience in shop supervision, Brower had handled difficult shop situations in Berkeley and Oak Ridge. Brower was able to represent shop needs and problems in the Ordnance Division councils. He got support for rapid procurement of needed equipment and reorganized some shop procedures. Despite these efforts, the problems of C Shop deepened, and Brower left the Laboratory in mid-August 1944.

The nature of the C-Shop difficulties may be illustrated in three ways. First, very little of its work was routine production. Most items were produced singly or in small lots. Every item had to be given detailed specifications in the engineering drafting room, which created an enormous workload and involved close cooperation between detailers and the scientists who prepared rough drawings. This is a common problem in laboratory shops, where, as a result, machinists become very skilled at working from rough drawings supplemented by informal consultation with users. This was impossible in C Shop because of its size and because few of the machinists had the necessary training. Constant complaints about delay in the drafting room and inadequate checking resulted.

A second sympton of inadequacy was that even a rigid priority system could not prevent delays of urgent work. Constant small irritations were associated with this priority system, such as deciding between two such unrelated programs as the gun and the implosion.

Lack of experience with peculiar fabrication problems added to the

57

difficulties. Hemisphere machining was one example. The implosion program called for many hemispheres of various materials and sizes. A 60-in. lathe acquired for turning large hemispheres proved useless for the purpose. Peters finally solved the problem with a specially rigged boring machine. Eventually, the lathe was needed for other jobs, but the point is that none of Peters' immediate superiors knew how this nonstandard work should be done.

As the above examples suggest, the C-Shop problems were caused by the difficulty of developing an adequate Ordnance Engineering Group. Although the shop had a competent foreman, he was not in a position to overcome the general lack of foresight in obtaining men and equipment. This lack, moreover, was not solely the responsiblity of Peters' superiors in the Engineering Group. These men were inadequately prepared to understand the emerging needs of the ordnance research and development groups, who, in turn, were not yet geared to their roles as weapon designers (Chap. 1).

It is not true, however, that the shop and engineering difficulties were inseparable. They were connected primarily because of organizational arrangements. The original plan, which placed C Shop under the Ordnance Division Engineering Group, was plausible in terms of the contemplated narrow range of the ordnance program. As that program broadened to include the gun program and the rapidly expanding implosion program, such arrangements became less plausible. C Shop then became a service organization working for several semi-independent organizations. Throughout the Laboratory, emphasis began to shift toward development. The line of division between the two big shops became less well defined. Finally, it was clear that the remedy for shop troubles was to place both C and V Shops under unified management. This would allow greater flexibility in division of labor between shops and would give C Shop the strong leadership needed to overcome its constant difficulties and to prepare it for the even more difficult days ahead. Such a step would also simplify Engineering Group problems, being a step away from administration and toward concentration on the increasingly difficult problems of design and development. During the August 1944 reorganization, C Shop was moved from the Ordnance Division to the V-Shop administration of Long and Schultz.

Construction and Maintenance

The procedure used for constructing the original Laboratory buildings was standard for Army installations. Oppenheimer, McMillan, and Manley gave specifications for the original buildings to the Manhattan District Engineer's Office in New York. Plans were drawn by the Stone and Webster Corporation of Boston, originally expected to do the construction. The drawings were transmitted to the Albuquerque District Office of the U.S. Army Corps of Engineers, which let a contract to the M. M. Sundt Company. When the buildings were completed, the Sundt Company transferred them to the Albuquerque District, which then transferred them to the Santa Fe Area Office of the Manhattan District Engineers, in theory the "using service." The actual using service, the technical staff, had no official position in this process, and because during the critical period of actual construction they were still scattered about the country, liaison was totally inadequate. The Albuquerque District remained in formal charge of construction until early 1944, when the Manhattan District assumed complete responsibility.

By May 1943, the original buildings were occupied and were being expanded. The Sundt Company had undertaken two relatively large structures, a new warehouse and an addition to the cyclotron laboratory, but they would not be able to complete the necessary work in time. Ordinarily, the Army was responsible for providing additional construction workers, but in this early period it could not do so, and the University of California hired several carpenters, plumbers, electricians, and laborers. Under contract regulations, these workers could not be employed for any permanent construction, only for maintenance work and to build shacks and "lean-to's." These men first worked for members of the scientific staff, but later were under Brazier, the Technical Area supervisor of construction and maintenance. Brazier was responsible for the preliminary design of the major expansion that began in June and included a new office building, offices and laboratories for the ordnance program, and a heavy machine shop. When Brazier left in January 1944, his staff had grown from about 12 to 264 men.

General Groves wanted all construction to be handled by the Army Engineers, and his final decision was hastened by complaints from the War Manpower Commission, the United States Employment Service, and the American Federation of Labor about irregularities in the project's hiring of construction workers. In January 1944, Brazier's entire staff was turned over to the Army payroll except for three foremen who remained on the University of California payroll. The scientific staff believed that security and efficiency would improve with a separate

59

construction and maintenance group for the Technical Area. Although it was impossible to keep the entire group on the University of California payroll, Charlie Stallings, Melvin Foley, and Dan Pfaff, in charge, respectively, of carpenters, plumbers, and electricians, were kept, and their assistants were assigned permanently to the Technical Area. Under the direction of John Williams, the group was responsible for maintaining, repairing, and installing all scientific equipment or machine tools under the jurisdiction of technical personnel, and also for building and remodeling scientific apparatus and equipment.

The construction and maintenance group under the Army Engineers was responsible for alterations, additions, and repairs to buildings and utilities. The Army also had to establish separate organizations for maintenance and construction of the Technical Area and outlying sites and of the post housing and administrative areas. Separate priority lists were kept, with urgency ratings assigned by those requesting service.

Nearly all major new construction was handled by contractors under the supervision of the Post Operations Division. The original contractor, M. M. Sundt Company, remained in charge until the end of 1943. They were succeeded by the J. E. Morgan Company, which built a section of the housing area during the first three months of 1944. They, in turn, were succeeded by R. E. McKee, who remained in charge of construction with an average force of 700-1000 men. The architect, W. C. Kruger, whose contract was originally issued by the Albuquerque District, was retained by the Manhattan District throughout the life of the project.

Requests for all but the most minor construction were made by group leaders or their superiors, and urgency ratings were assigned like those for orders on the Procurement Division. Such requests were submitted to the office of David Dow, who acted as liaison between the using groups and the Post Operations Division. One of the most frequent problems was the users' inability to foresee their needs, because construction often depended upon the results of experiments in progress.

Patent Office

In accordance with OSRD procedure for protecting Government interests in scientific research, the Contractor was required to "report the progress of all studies and investigations undertaken, disclose to the Government all inventions made in carrying out the work of the contract, and furnish a complete final report of findings and conclusions." Here again, security was an important factor in determining administrative organization. Because few Contractor's representatives were permitted to visit the Laboratory or to know much about its work, they could not make the necessary patent reports. Consequently, the University turned over much of its reponsibility for protecting Government interests to Major Ralph Carlisle Smith, the Patent Officer, who arrived in July 1943 to establish the Patent Office.

The Patent Office was influenced in many ways by security considerations. Because the Patent Divison had to report on all scientific work and was the only office where all this information would be compiled in language understandable to an individual with a general scientific background, the Director and the Security Officer felt that only a few absolutely trustworthy people could be permitted to work there. For some time, Major Smith had no assistants at all, but eventually he obtained a minimal staff of enlisted men and women. Only after a year and a half could he secure two legally trained, scientific assistants.

Security considerations also forced the Patent Office to assume responsibility for all patent matters affecting subcontractors or employees who came here from other projects. The Patent Office assumed responsibility for the early subprojects, such as those at Purdue University and Stanford University (Chap. 1), as well as for the later subcontractors, such as F. Flader and the California Institute of Technology (Chap. 9). Furthermore, employees transferred here from other Government projects were not permitted to communicate directly with their previous colleagues in the patent field, so unfinished patent matters were transferred here for completion.

The Patent Office established methods and procedures for recording completed work and secured the cooperation of the technical staff in keeping necessary records. Numbered notebooks were issued, originally by the Patent Office and later by the Library document room, in which staff members recorded details of experiments and exact dates of the various development stages of invention and discoveries. Completed notebooks, and those turned in by people leaving the project, were filed in the Library document room. Through the Business Manager's Office, patent agreements were secured from every employee, subcontractor, and consultant of the University of California. The Patent Office obtained special patent agreements from military personnel and civilian employees of the War and Navy Departments, and special patent contracts from individuals on loan to the Manhattan District from other

61

employers. Foreign personnel were not required to sign agreements but did prepare records of inventions and executed U.S. applications to the benefit of the U.S. Government. The Patent Division prepared monthly reports on the activities of foreign personnel. These and similar reports of visits by consultants and foreign personnel were sent to General Groves' office. All terminating personnel had to appear before the Patent Officer and assert that they had made no unrecorded inventions and had turned in all original records to the document room.

The most important Patent Office work was preparing patent applications to protect the Government and prevent outside interests from later dominating the pertinent fields of research and development. Major Smith and his staff had to be expert in various technical subjects, as well as experienced patent attorneys. Experiments covered much more than nuclear physics; they included chemistry, metallurgy, ordnance, and explosives and electronics. Patent cases were classified into production, chemistry, and metallurgy of fissionable materials; isotope separation; power reactors; electronic equipment; and the bomb itself with its various developments and improvements. Altogether, about 500 patent cases covering work at this Laboratory were reported to Washington OSRD Headquarters and about 300 in work done on other projects were handled. Of these, several were filed in the U.S. Patent Office.

Because of the pressure of time and the limited staff, it was impossible to write cases in the usual manner. Ordinarily, an inventory or research scientist prepares invention reports of things he considers new and useful and submits them to a patent attorney for aproval and preparation of a formal application. Here, the members of the Patent Office read the daily records and other reports of research workers, inspected the laboratories and test sites, and attended scientific seminars and conferences. In all of these sources, they found ideas and practices that were new and useful, prepared the applications so as to give maximum scope to the invention in their relation to the entire project and associated fields, and submitted them to the inventors for final approval. Because the technical staff were pressed for time, this unorthodox procedure was extremely helpful. Another complication was that because of the nature of the work, many developments had to be covered before there was any physical embodiment proving that the inventions were workable; that is, before any "actual reduction to practice," in legal jargon. The test shot at Trinity was the first reduction to practice for many inventions whose success was long before anticipated by completing and filing a series of patent applications. Completed cases were transmitted by Army courier to the OSRD Washington Patent Headquarters, headed by Captain R. A. Lavender, USN, and filed with the U. S. Patent Office.

CHAPTER 4

Technical Review
April 1943 to August 1944

INTRODUCTION

The role of theory in forming basic decisions in the DSM project is shown by the fact that even after Los Alamos was established there was still no absolute experimental confirmation that the bomb, in terms of its basic nuclear processes, was feasible. In April 1943, it was still possible, although extremely unlikely, that an efficient nuclear explosion might be ruled out on any of three counts. First, the neutron number had not been measured for fission induced by fast neutrons, but only for "slow" fission. Second, the time between fissions in a fast chain might be longer than assumed. Finally, plutonium fissioning had been studied by observing fission fragments, but this did not prove that the neutron number was the same as for ^{235}U.

The first physical experiment completed at Los Alamos, in July 1943, was observation of neutrons from ^{239}Pu fissioning. The neutron number was measured from an almost invisible speck of plutonium and found to be greater than that for ^{235}U. As mentioned earlier, this result justified the decision to construct the plutonium production pile at Hanford.

The other early confirming experiment (measurement of delayed neutrons) also gave the expected favorable results. It showed that delays in neutron emission were negligible. The third possibility, that the neutron number might be radically smaller for fast neutron fissions than for slow, was not investigated until the following year. Assurances on this score were considerable, however.

NUCLEAR SPECIFICATIONS OF THE BOMB

With completion of the Laboratory and the preliminary experiments described above, the nuclear physics program entered its main course,

64

through experiment and calculation, to provide nuclear specifications for the bomb. This work proceeded by a series of successive refinements. At each stage, the information gained served to determine the Laboratory's work more concretely.

At the beginning, it was important to estimate reliably the amount of active material per bomb that would be needed. Otherwise, there was no way to determine what size of uranium separation and plutonium production plants to develop.

In the next stage, more accurate information was needed as a basis for concurrent ordnance work. The gun-design program required estimates of the size and shape of the projectile and the muzzle velocity it must be given. Also, the effective mass of active material would limit the overall size of the bomb, which would, in turn, determine the type of airplane to be used in its delivery. In the last stage, after certain basic engineering specifications were determined, it was necessary to find, for instance, the exact mass and shape of active and tamper material and to help make the project an actuality.

Such separation into stages gives a false chronology. Even the preliminary stage did not end with the feasibility experiments described above. As understanding of requirements became more reliable, the Laboratory continued to influence the ^{235}U and ^{239}Pu production plans of the Manhattan District. Also, the choice of a tamper, the uranium hydride possibility, and the mechanism of assembly—gun or implosion—remained undetermined for some time.

Despite these overlappings so characteristic of war development and of the entire Manhattan Project, one can trace the gradual shift from nuclear physics, through the difficult problems of the bomb assembly mechanism, to final development.

One of the first problems was more precise determination of critical masses. As explained in Chap. 1, in a qualitative way, the critical mass depends on the diffusion rate of neutrons out of the active mass as compared with the rate at which they are generated in it. Therefore, an essential tool was the statistical theory of neutron diffusion. Ordinary diffusion theory is valid in the range where the mean free path of diffusion particles is small compared to the dimensions of interest. This was not true of the bomb. The number of neutrons in a given small region depends not only on that in adjacent regions, but on the entire distribution throughout the mass. An integral diffusion theory had to be used and methods were found to apply it in practical calculation. This problem was one focus of development. Another was refinement of certain rough assumptions that were necessary in making earlier calculations. One such assumption was that neutrons were scattered isotropically. The correction was to take account of angular dependence. Another assumption was that core and tamper gave the same mean free path, which they generally do not. Still other assumptions subject to correction were that the neutrons had the same velocity, that the various cross sections were independent of velocity, and that there was no energy loss through inelastic collisions.

Most of these refinements took account of new dependencies; also, the errors from theory and experiment had to be limited to comparable magnitude. Therefore, the early experiments, other than those already described, were centered around measurement of cross sections, their energy dependence, and the number and energy spectrum of fission neutrons. The most time-consuming measurements were those of the fission cross sections of ^{239}Pu and ^{235}U as a function of energy from the thermal to the high-energy end of the fission spectrum. Relative and absolute fission cross-section experiments permitted plotting fission cross-section curves as a function of energy for both ^{235}U and ^{239}Pu from thermal energies to several million electron volts. These results were used in more accurate critical mass and efficiency calculations and were partly responsible for abandonment of the uranium hydride program; partly because they showed that the energy dependence that would make the hydride an efficient weapon did not occur, and partly because the evidence they provided for considerable radiative capture at thermal energies made the critical mass and efficiency estimates of metal uranium bombs more optimistic. Investigation, suggested by the behavior of fission cross sections at low energies, led to the discovery that radiative capture in ^{235}U was indeed significant and was even greater in ^{239}Pu. Because the neutron number had been measured at thermal energies for total absorption (capture plus fission), not fission alone, and because capture would become less important at the high energies of neutrons operative in the bomb, the effective neutron number in both materials was higher than had been assumed. Therefore, the hydride program continued after the spring of 1944 only at low priority.

Although the hydride program was unsuccessful, learning about its limitations contributed in several ways to the whole program. For example, the assumption that the fission cross section was inversely proportional to neutron velocity made clear the importance of inelastic

66

scattering in the tamper. In the first approximation, it was assumed that only neutrons scattered back elastically would contribute importantly to the reaction. But if decreasing neutron energy was compensated for by increasing fission cross sections, this assumption could not be made safely. In a lengthy series of backscattering and transmission experiments with a list of potential temper materials, the scattering cross sections were measured for neutrons of various energies and scattering angles, and the energy degradation of scattered neutrons also was measured.

THE GUN METHOD

For the first six months, the gun method of assembly was the focus of ordnance activity. The implosion program was considered a standby, and its facilities were an adjunct to those of gun development. Until August 1944, major activity involved the plutonium gun, which was further from standard practice than the ^{235}U gun. The gun had several unusual features. The assembly velocities required to prevent predetonation of a plutonium bomb were near the upper limit of standard gun design, 3000 ft/s. The gun also had to be light, and it was expendable. The tube had to be as short as possible for inclusion inside a practicable airborne weapon. This meant operation with the highest possible peak pressure. Finally, the gun had to have the highest attainable reliability.

The first guns were designed and being produced by the Naval Gun Factory by September 1943, and they were received at Los Alamos in March 1944. Proof firing to test propellant behavior and investigate problems of projectile and target design began in September with a 3-in. Naval antiaircraft gun. Proof firing was also undertaken at 20 mm to investigate the "blind" target assembly (so-called because of the projectile seated in it) and to prove the polonium-beryllium gun initiator.

In August 1944, when the plutonium gun assembly was abandoned, the high-velocity gun had been thoroughly proved and the techniques of proof well developed. Development of the ^{235}U low-velocity gun could proceed, therefore, without new basic difficulties while the main effort was directed to the mounting difficulties of the implosion program.

THE IMPLOSION

The proposal for the implosion assembly was to use the plastic flow

67

tamper and active material under high-explosive (HE) impact. A hollow subcritical sphere of these materials would be compressed into a solid supercritical sphere. The first acknowledged advantage of the implosion over the gun was its much shorter assembly time. This was especially important for plutonium assembly because of its expected high-neutron background, which, given slow assembly, would make predetonation a serious danger. From this early conception there were several steps of evolution to the implosion mechanism finally used in the Trinity explosion and Nagasaki bomb. These steps arose from the results of hydrodynamical calculations, from the discovery of a still higher neutron background in plutonium than had been anticipated, and from the difficulty of achieving symmetry in the imploding shock wave. In the first experiments with cylindrical and spherical implosions, the ratio of HE charge to hollow liner was such that the liner was just assembled into solid form. This was necessary in the first instance to permit terminal observation. It was recognized that larger charges would give faster assembly and might give some advantageous compression if the implosion were symmetrical enough. But observation of results was impossible. Therefore, and because the decisive virtures of the fast implosion were not realized, it was deferred.

One difficulty with the fast implosion as early conceived was the uncertainty of a successful chain reaction at the time of initiation. There was some discussion about a modulated initiator that would be "turned on" at the time of complete assembly, but this represented a serious added complication in bomb design.

The first decisive change in the concept of the implosion came from a rough quantitative analysis of the assembly velocities attainable with very large charges of HE, which suggested that because of the focusing effect of the converging material, a strong steady source of neutrons could be introduced into the bomb (by deliberately leaving the material impure) and yet the experimentalist could still beat the chain reaction and attain complete assembly. It was only one step from this to the realization that much of the kinetic energy of the imploding material would be transferred into potential energy of compression at the center of the converging mass. This remarkable phenomenon, compression of "incompressible" solid matter by the extreme pressures produced by the implosion, was too far from ordinary experience to be grasped immediately or easily.

These two steps made it clear that implosion has qualitative advantages over gun assembly, and that the many difficulties in its development would be worth overcoming. At an October 28, 1943, meeting, the Governing Board reviewed the implosion program and decided to strengthen and push it.

It is convenient to treat the theoretical and experimental aspects of implosion separately here, for their convergence lay much farther ahead than was anticipated.

Assembly time and neutron background are complementary; to increase the latter, the former must be decreased proportionately. At the very beginning, however, they had hardly been given equal weight. Raising the chemical purity standards posed a difficult problem, but the chemists could accomplish difficult things. To increase velocities by the gun method, however, required a gun weight increased as the square of the velocity.

Quantitative investigation of the implosion hydrodynamics proved very difficult. An approximate method adaptable to hand calculation gave uninterpretable results. In the spring of 1944, the problem was set up for IBM machine calculation. These machines, recently procured to calculate odd-shaped critical masses, were well adapted to solving the partial differential equations of the implosion hydrodynamics.

Naturally, at the beginning of this new line of investigation, thought was given to imploding uranium hydride. It is about half as dense as uranium, and the space occupied by the hydrogen would be recoverable under sufficient pressure. Samples of hydride prepared at Los Alamos were investigated at P. W. Bridgman's high-pressure laboratory at Harvard. Pressure density data up to 10 kilobars (kbar), still very low pressure from the implosion viewpoint, indicated that the hydride was not easily compressible.

While theoretical investigation was familiarizing the Laboratory with implosion's enormous potentialities, its empirical study was beginning. Some data were obtained from terminal observation, some from flash photography of imploding HE cylinders, and some from flash x-ray photography of small imploding spheres.

Whereas implosion theory assumed a symmetrical converging detonation wave, the only feasible method of detonating the HE was to initiate one or several diverging waves. The workers hoped that with several detonation points symmetrically spaced around a sphere the difference would not be essential. Terminal observations gave some indications of asymmetry of collapse, but it was difficult to ascertain their cause. The

69

first successful HE flash photographs of imploding cylinders showed very serious asymmetries in the form of jets that traveled ahead of the main mass. Interpretations of these jets included the possibility that they were optical illusions.

While the radical difficulty of achieving a smooth implosion was becoming clear, other problems also were growing. The HE casting plant, under construction since the end of 1943, was still not in operation, and men to staff it were hard to find. The number of experimental charges that could be produced was extremely limited. Also, development of instrumentation to investigate medium- and large-scale implosions was a mountain of uncompleted effort.

To understand the gloom that prevailed in the late spring and summer of 1944, one must recall that by this time plutonium delivery schedules were well established, and weapon amounts would become available by the summer of 1945. Therefore, there was about a year in which to complete the program. The virtues of implosion had been demonstrated theoretically, yet no success in development had been achieved. Furthermore, at this same time, events in another part of the Laboratory program threw full responsibility for the use of plutonium onto the implosion program.

METALLURGY

A positive feature of the hydride was the interest taken in its preparation and fabrication. Among the first studies the metallurgists undertook was the art of preparing high-density hydride compacts. Thus, although after a year or so it was known that the hydride would not yield an efficient weapon, it could be fabricated easily and was used in making experimental reactors.

The main goal of metallurgy in this period was to develop techniques for preparing the large amounts of active and tamper materials necessary for the bomb. Apart from early work with the hydride, effort concentrated on uranium. This subject was already fairly well developed in other branches of the project, but the Los Alamos requirements were more exacting. There was more emphasis on high chemical purity and yield. A bomb-reduction technique ultimately satisfied these requirements.

One reason for the early work on uranium was its hoped-for resemblance to plutonium, as yet nonexistent in workable amounts.

70

When the first such amounts of plutonium appeared in March 1944, techniques for its reduction were already being developed. By the end of the first period, satisfactory bomb-reduction methods had been perfected. Scientific study of plutonium properties was limited by the time available and the pressure to develop usable methods. Usability standards, moreover, were much harder to meet than those for uranium. According to the original requirements, all operations would have to be performed so as to avoid contamination with even a few parts per million of light elements. This necessitated development of heavy-element refractories. The substantial relaxation of purity requirements that came when the plutonium gun program was abandoned guaranteed success; by this time the original high-purity goals had nearly been reached, and some simplification of techniques became possible. A discovery important in final bomb design was the existence of allotropic forms of plutonium. In July 1944, experimental proof of the alpha (room-temperature) and beta phases was obtained.

In addition to metallurgy of active materials (uranium hydride, uranium, and plutonium), several powder metallurgy techniques were developed for fabricating materials with important nuclear properties (notably boron and beryllia) to attain the highest possible densities. The main pressure for boron production came, again, from the hydride gun program, for which it would be difficult to dispose enough critical masses of hydride into gun and target.

The Laboratory tried to procure large amounts of boron enriched in ^{10}B, which constitutes about 20% of normal boron. Urey refined his method for separating ^{10}B, at the request of the Los Alamos Laboratory. A pilot plant was built in the fall of 1943 to develop the method and provide experimental amounts of the separated isotope. Estimates in February 1944 set the needed production rate at a figure comparable to production of separated uranium. Standard Oil of Indiana started to build a plant, but construction difficulties and a decreasing probability that boron would be used in large amounts decreased its scheduled capacity by 25%.

Even after there was reasonable assurance that a hydride bomb would not be used, especially not a hydride gun, production of the ^{10}B isotope was maintained because of its potential usefulness in an autocatalytic bomb if one could be developed. This isotope was, indeed, very useful in small quantities in counters and as a neutron absorber.

71

The metallurgists developed high-density beryllia compacts for the Water Boiler tamper. Actual production was by the Fansteel Metallurgical Corporation under subcontract.

CHEMISTRY

The chemists' principal work was on uranium and plutonium purification, analysis, and recovery. The first purification work began when there was very little plutonium and only microchemical investigation had been undertaken. Until the first plutonium arrived at Los Alamos in October 1943, work was limited to study of various stand-ins, including uranium. After that and until arrival of the first Clinton plutonium, stand-in work, microchemical work, and the results of similar work at other branches of the project provided the only information available. Nevertheless, by August 1944, there was strong assurance that purity specifications could be met on a production basis. The "wet" purification stage was in essentially final form, as were the recovery processes for recycling plutonium in chemical and metallurgical residues. The final "dry" purification stage had been outlined, but exact procedures were not finalized. One of the most serious difficulties was preventing contamination from dust, etc. that would undo the purification. This made construction of an air-conditioned laboratory building necessary. Discovery of the serious danger from plutonium poisoning (Chap. 3) necessitated development of enclosed apparatus wherever possible.

The extraordinary plutonium purity requirements necessitated supersensitive analytical techniques. In some cases, only a few parts per million of impurities had to be measured. To add to the difficulties, samples assayed had to be small, especially when analytical techniques were first being developed. Spectrochemical methods were developed by or were in close liaison with the Metallurgical Laboratory. Gasometric methods for oxygen and carbon analysis were developed at Los Alamos.

By August 1944, the difficult plutonium purification and analysis procedure was virtually complete. Corresponding processes for ^{235}U were relatively simple. Relaxation of purity requirements made it unnecessary to pursue these researches further. Transition to production methods was difficult because of increase in knowledge of the serious dangers of plutonium poisoning.

72

The radiochemists cooperated with the experimental physics and ordnance groups. A main contribution was preparation of thin foils of various materials and specifications. They developed several new techniques for preparing these foils. They also developed very sensitive neutron counters. Early in 1944, a radon plant was constructed to help look for neutrons associated with alpha radioactivity (Chap. 6) and as a source of material for a possible radon-beryllium initiator. Another possible choice for the initiator was polonium. Research polonium was prepared by irradiating bismuth in the Clinton pile and purifying it at a special plant. Except for polonium research, the radiochemists' main activity in the summer of 1944 was to design and construct a "mechanical chemist," a remote control plant for extracting and handling the highly radioactive radiolanthanum to be used at the Bayo Canyon RaLa site.

THE DISCOVERY OF ^{240}Pu

Perhaps there is no better illustration of the interrelation of research and development at Los Alamos than the series of developments that led to discovery of ^{240}Pu in the Clinton product. Plutonium's value as a bomb material was doubted until its neutron number was first measured in the summer of 1943. Even then, there were still serious difficulties: 1 gram of plutonium emits 2×10^9 alpha particles per second. To keep the neutron background from (α,n) reactions low enough to make fast-gun assembly feasible required high purity; for three light elements, less than 1 part per million.

Spontaneous fission measurements were first undertaken at Berkeley to ascertain the neutron background from this source of ^{235}U. At Los Alamos, these measurements were refined and extended to ^{239}Pu and other materials. In the summer of 1943, a report from France stated that Joliot had found a neutron emission associated with the alpha radiation of polonium, but not from the action of this radiation of light-element impurities. Although this report was believed incorrect, it was recorded in the Minutes of the Governing Board with the general intention of looking into all questions concerning spontaneous neutron emission.

As a result of the Joliot report, work began on highly sensitive neutron counters, and a radon plant was obtained because radon was the alpha emitter that could be most highly purified. Heavy neutron emission from alpha emitters might make a modulated initiator impossible. It might also mean a prohibitively high neutron background in plutonium itself.

Increasingly reliable spontaneous fission measurements showed that spontaneous fission of plutonium was slow enough so that the neutron background was not serious. Meanwhile, however, another piece of research entered the story. Fission cross-section measurements at low energies, whose justification was to obtain data for use in calculating the uranium hydride critical mass, showed resonances in the ^{235}U fission absorption spectrum. This led to expectation of sizable radiative neutron capture. For ^{239}Pu this meant production of a new isotope, ^{240}Pu. Because this isotope would be produced by absorption of two neutrons in ^{238}U, its concentration in the pile plutonium would increase with heavier irradiation.

In the summer of 1944, when the Clinton plutonium made by chain reactor arrived (much more heavily irradiated than that made by cyclotron bombardment), the existence of ^{240}Pu was verified, as was the fear that it might be a strong spontaneous fissioner. Neutron background in the plutonium that would be produced at full power was increased so that, to prevent predetonation, assembly velocities would have to be much greater than those possible with the plutonium gun. The only alternative to abandoning the gun method for plutonium was to find ways to separate out the offending isotope. This would mean another major investment in a separation plant and could hardly be accomplished in time. Implosion was the only real hope, and from current evidence, not a very good one. Nevertheless, the Laboratory had strong reserves of techniques, trained manpower, and morale. It decided to attack the implosion problem with every means available, "to throw the book at it." The program was taken out of the Ordnance Division and divided between two new divisions, one to be concerned primarily with implosion dynamics investigations, the other primarily with development of adequate HE components.

THE WATER BOILER

The Laboratory enjoyed its first major success when the Water Boiler began to produce divergent chain reactions on May 9, 1944, and many experiments were run during the summer to determine nuclear quantities of interest. The Water Boiler was an integral experiment that provided a

74

general check of theory. In fact, the theorists had predicted its critical mass almost perfectly. Although they point out that this exactness was fortuitous in view of blind assumptions they were forced to make, their prestige in the Laboratory got a well-deserved boost.

CHAPTER 5

The Theoretical Division
April 1943 to August 1944

ORGANIZATION

The Theoretical Division was formed (Chap. 1) to develop nuclear and hydrodynamical criteria for design of the atomic bomb and to predict its detailed performance. At first, most of its effort was devoted to calculating the critical mass and nuclear efficiency.

With implosion's rise to prominence, the organization, under H. A. Bethe as Division Leader, was formalized into the following groups in March 1944.

T-1	Hydrodynamics of Implosion and Super	E. Teller
T-2	Diffusion Theory, IBM Calculations, and Experiments	R. Serber
T-3	Experiments, Efficiency Calculations, and Radiation Hydrodynamics	V. F. Weisskopf
T-4	Diffusion Problems	R. P. Feynman
T-5	Computations	D. A. Flanders

In June 1944, R. Peierls took charge of the Implosion Group, replacing E. Teller, who formed an independent group outside the Theoretical Division. The new group acquired full responsibility for implosion IBM calculations. In July, Group O-5 (Chap. 7) joined the Theoretical Division on a part-time basis because its work in the Ordnance Division was almost completed (Chap. 14).

DIFFUSION PROBLEMS

An early task in the theoretical program was to develop a means for accurately predicting the critical mass of active materials. The essential and most difficult factor was the theory of neutron diffusion. Others were principally matters of evaluating data from scattering and fission experiments to obtain the appropriate cross sections and the number of neutrons emitted per fission. The critical mass is the amount of material from which neutrons disappear by leakage and nuclear capture at the same rate at which they are born from fissions that occur in the mass. To calculate this requires knowledge of the average way in which neutrons distribute themselves in the mass. This is the problem of neutron diffusion theory.

It was possible at the outset to write the integral equation whose solution would give the exact neutron distribution, taking account of the variation in neutron velocity, the dependence of scattering and fission cross sections on velocity, and the anisotropic nature of scattering. This equation, written simply on the basis of conservation considerations, was formulated by Boltzmann and bears his name, but as it stands, it has no known exact solution.

However, two kinds of approximate solutions were possible. One was ordinary differential diffusion theory, in which the neutron diffusion was treated by analogy with heat diffusion. These calculations were relatively simple, but the results were known to be inaccurate. In fact, the neutron diffusion problem does not meet the requirements of differential diffusion theory, among others, of a small change of neutron density per mean free path. This condition is satisfied approximately in a large pile; but in the bomb, the critical size is of the order of magnitude of the mean free path. The other attack was based on exact solution of the integral equation of diffusion for one special case. This solution, found at Berkeley by S. P. Frankel and E. C. Nelson, was completed in the first months at Los Alamos. Much of the Theoretical Division's subsequent work was its effort to find valid solutions under less restrictive conditions. The conditions are that

(a) neutrons have a single velocity,

(b) the core and tamper nuclei are treated as stationary, and all neutron collisions with them are treated as elastic,

(c) neutrons are scattered isotropically, and

(c) the mean free path of neutrons is the same in core and tamper.

The "extrapolated end-point" method of solution was worked out for untamped spheres and later extended to equal mean free paths in core and tamper. The method was developed independently, but it was later found to be an extension of Milne's procedure for solving certain astrophysical problems.

The extrapolated end-point method permitted calculating the critical mass of a solid ^{235}U sphere with an effectively infinite tamper. Three problems were to allow for the finite tamper thickness, to calculate other than spherical shapes, and to find ways to calculate the critical mass when the mean free paths in core and tamper were not identical. The extrapolated end-point method could not be applied except as an approximation of uncertain accuracy. In essence, it applies differential diffusion theory to a fictitious scattering material whose boundary or end point extends a calculable distance beyond the actual boundary of the material. It is strictly valid only under the conditions enumerated.

The first problem above was solved by temporary methods, such as replacing the finite tamper with an infinite tamper and a fictitious neutron absorber. The second problem was more difficult.

It was met by resorting to the inexact methods of diffusion theory (using the extrapolated end-point method) by which odd shapes could be calculated. The ratio of the critical masses of an odd-shaped body to a spherical body thus obtained could be applied as a correction to the accurately known critical mass for a sphere.

The third problem, unequal mean free paths in core and tamper, led to a much longer series of developments. The first, an effort to use variational principles to solve the original integral diffusion equation, was applied successfully to spherical and cylindrical shapes, slabs, etc. This served to check the accuracy of the extrapolated end-point method. The agreement was very close. However, extending the variational techniques to unequal mean free paths in core and tamper proved extremely laborious. In June 1944, a technique developed at the Montreal Project was introduced. It was based upon an expansion of the neutron density in spherical harmonics and was easier to apply than the earlier variational method, and it gave very accurate results.

By mid-1944, the neutron diffusion problem had been solved under restrictions (a), (b), and (c) above, but with (d) eliminated. Solutions for particular cases were still rather expensive.

Meanwhile, assumption (c) was being studied. Very accurate results were obtained by assuming isotropy and substituting for the scattering cross section the so-called transport cross section, a kind of weighted average that gives the effective scattering in the initial direction of motion of the scattered neutron. Assumption (b) was entirely reasonable unless

78

inelastic scattering in tamper materials had to be considered seriously. With this exception, the main remaining limitation was assumption (a). Most of the work described so far was done by Group T-2, but every group participated at some point.

Group T-4 was investigating methods for reducing the many-velocity problem to a series of one-velocity problems. The problem came about naturally during the investigation of the uranium hydride bomb, in which energy degradation of neutrons from elastic collisions with hydrogen was one of the essential characteristics of the chain reaction. Very early, methods were found for treating the hydride problem, with a continuum of velocities, under unrealistic assumptions, such as an infinite medium of core material in which there was a sinusoidal distribution of neutrons. If core and tamper were of different materials, they could not be treated at first. However, by July 1944, a method applicable to a spherical core and tamper allowed treatment of a continuum of velocities if there was no inelastic scattering in the tamper medium. Unfortunately, this inelastic scattering was not negligible in the tampers being considered. This difficulty was soon overcome to the extent of allowing for three or four neutron velocity groups instead of the continuum.

In hydrogenous material, one could not assume that neutrons were scattered isotropically [assumption (c)]. However, it was found semiempirically that this fact was adequately accounted for by using the transport cross section as the all-metal diffusion medium.

Other means of accounting for the continuum of velocities were adopted in special problems, such as calculating the distribution of thermal neutrons in the Water Boiler.

THE WATER BOILER

One of the first practical critical mass calculations was to estimate the critical mass of the Water Boiler. As for the hydride calculations, the slowing down was an essential factor. In fact, the Boiler would be of small critical dimensions only because it slowed neutrons to thermal velocities, taking advantage of the larger thermal fission cross section of ^{235}U. The standard method, Fermi's "age theory," for calculating the thermal neutron distribution in piles was inaccurate when applied to a small enriched reactor because it required a very gradual slowing of the neutrons. This condition was satisfied for a carbon moderator, with mass

79

12 times that of the neutrons. It was not satisfied for a hydrogenous moderator such as water, because the neutrons and hydrogen nuclei are of the same mass, and energy loss can occur rapidly. A method developed at Los Alamos used the differential diffusion theory but assumed that neutrons were of three velocities (fission energies, intermediate, and thermal). The final predicted value was almost exactly correct, although rather unplanned. Water Boiler cooling and shielding, temperature changes versus degree of criticality, effect of sudden changes in control rod position, etc., were examined theoretically, as were fluctuations in its operation during experiments. Most groups participated in these calculations, but the main work was that of R. F. Christy in Group T-1.

THE GUN

Critical mass calculations for the gun assembly were complicated primarily by its odd shape. The problem was to estimate not only the number of critical masses in the completed assembly, but also the amount of active material that could be disposed safely in the two parts before assembly. It was also necessary to know how the system went from its initial subcritical to its final supercritical position to be able to calculate the probability of predetonation. The early rough specifications had been based on critical mass estimates from differential theory. By February 1944, the Ordnance Division wanted more reliable specifications and Group T-2 specified the actual bore for the ^{235}U gun. The ^{239}Pu gun assembly was specified shortly later. Group T-2 gave essentially complete specifications for both gun assemblies by the summer of 1944, fortunately after cross-section measurements by the Detector Group showed slightly lower average ^{235}U values than those used in earlier calculations.

THE IMPLOSION

The idea of something like an implosion, as an alternative to gun assembly, had occurred to several people before the beginning of Los Alamos. Its first history at Los Alamos belongs mainly to the Ordnance

Division, which made initial calculations of attainable assembly velocities.

The Theoretical Division entered the picture in the fall of 1943 when von Neumann proposed the fast implosion, and its potential as a qualitatively better weapon than the gun was appreciated. This development is covered in Chaps. 7, 15, and 16. Here emphasis is upon the theoretical problems. Implosion studies were the responsibility of Group T-1, with the assistance of other groups, particularly T-2.

The first problem studied was the time of assembly when (as von Neumann proposed) large amounts of explosive were used. The energy required for plastic deformation was small compared to the total explosive energy, so to a first approximation, the kinetic energy of the mass moving inward could be assumed to be conserved.

Hydrodynamical calculations were required to predict quantitatively the motion of an imploding compressible liner. The partial differential equation describing the implosion was too difficult for hand calculation when a realistic equation of state was used, so the first effort was to find simpler approximate equations of state. The first method was based on a multiphase model in which the state of the imploding material was assumed to change discontinuously. Unfortunately, the results proved very difficult to interpret.

IBM machines had been ordered for calculating critical masses of odd-shaped bodies. They arrived in April 1944, and preparations had been made to use them for numerical integration of the hydrodynamical equation. Preliminary calculations had to be made to determine the initial conditions at which to start the IBM calculations. It was necessary to derive the equation of state of uranium at high pressures, a calculation based on the Thomas-Fermi model of the atom. Results at low pressures were obtained from P. W. Bridgman's data, and the intermediate region was determined by interpolation. The first IBM results were extremely satisfactory, so the unrealistic multiphase implosion model was dropped.

Then a new problem brought the division's work into closer relation with experimental implosion studies. Calculations of implosion dynamics had required an inward-moving spherical shock wave, but such a wave had proved impossible to create. The erratic results of multipoint detonations, especially the jets observed, directed theoretical attention to interference of detonation waves. It was found that a diverging spherical wave accelerates materials less rapidly than a plane wave, and still less rapidly than a converging wave. In an implosion with many detonation points, the explosive waves are divergent to start with, but it had been assumed that their interaction would make them convergent. Theoretical examination showed that this smoothing was by no means assured, and that the fact to be concerned about was the high pressure developed at the point where detonation waves collided. The most obvious way to avoid these difficulties was to use explosives arranged so that they would produce converging waves initially. J. L. Tuck had suggested use of such lens configurations, and this was an argument for its adoption. However, it was a completely untried, undeveloped method that no one wished to use unless absolutely necessary.

G. I. Taylor visited Los Alamos in May 1944 and presented arguments to show that an interface between light and heavy material is stable if the heavy material is accelerated against the light material, and unstable otherwise. There was a possibility of serious instability in the implosion where light high explosive would be pushing against heavier tamper material, or where a light tamper might be pushing against the heavy core. A similar difficulty, leading to mixing, was also foreseen in the nuclear explosion as the core became less dense when expanded against the compressor tamper.

These two developments started a trend of thought that radically altered the whole implosion program. The behavior of the symmetric implosion was soon understood from the IBM results, but it was doubtful whether a symmetric implosion could be achieved. Thus, for the rest of the year, explosive charge design moved toward the lens program, whereas design of the inner components was more conservative.

Calculations of asymmetry development permitted a preliminary statement of tolerable asymmetry. A 5% velocity variation was considered the maximum allowable.

CALCULATIONS

Efficiency calculation was perhaps the Theoretical Division's most complex problem. The theory of efficiency had to follow the neutron chain reaction, and neutron distribution in the bomb in a medium of fissionable and tamper material being rapidly transformed by the reaction in both its nuclear and dynamic properties. Every factor involved in the critical mass calculations was involved here, but in a dynamical context that made dubious some of simplifying assumptions underlying those calculations.

The first efficiency calculations were made at Berkeley (before Los Alamos) for small excesses over the critical mass. They were preceded by investigation of the hydrodynamical behavior of the core and tamper during the chain reaction, and this led to the theory of the shock wave that travels into the tamper and the rarefaction wave that travels into the core from the core-tamper interface. The effects of these phenomena on efficiency were calculated. Neutron diffusion was treated by differential theory, which allowed simple estimates of the efficiency dependence on tamper properties such as mean free path, absorption, and density.

The next step, applicable to bombs having a far greater than critical mass, was based on the decreased multiplication rate for small expansions of the exploding bomb.

Once the efficiency in these two cases had been estimated, a semiempirical formula was developed that fitted the Los Alamos calculations for large excess masses and reduced the limit of small excesses to the earlier formula developed at Berkeley. This formula, by Bethe and Feynman, made efficiency estimates easy when the critical mass (the radius to which the bomb core must expand before neutron multiplication is stopped) and initial multiplication rate were known.

Use of the Bethe-Feynman formula for intermediate excess masses was justified as follows. Given small excess masses, the effect of the ingoing rarefaction and outgoing shock waves on the mean density approximately canceled. The same was true for large excess masses, because the waves were reflected back and forth many times before the multiplication was stopped, and the multiplication could be seen as a function of the average pressure. A plausible argument was that because this hydrodynamical independence held at both extremes, it also held in the intermediate cases.

Restrictions and unproved assumptions involved in all these calculations include the following.

(a) Radiation effects can be neglected.
(b) Neutron multiplication can be calculated by an adiabatic approximation.
(c) Tamper and core have the same neutron scattering per unit mass.
(d) Material density in core and tamper is the same.
(e) Absorption in the tamper is equivalent to that in an infinite nonabsorbing tamper.

(f) Depletion effects in the material are unimportant.

Of these assumptions, (f) was easiest to allow for. Depletion effects were negligible for small efficiencies and could be calculated for larger ones. Rough methods were found for estimating the effects of relaxing (c), (d), and (e). Assumption (b), the error involved in the adiabatic approximation, was investigated in detail. In this approximation, the total number of neutrons in the expanding bomb is assumed to increase at a rate proportional to itself, the rate at any instant being calculated from the excess over critical at that instant, assuming that the core and tamper nuclei are at rest. This is the same assumption as (b), discussed earlier under diffusion problems. The nuclei are relatively at rest and the assumption is good. But during the explosion, the bomb material acquires a very high mean mass motion and the assumption is questionable. A correction factor was found by considering the nonadiabatic theory of small expansions of a slightly supercritical bomb. Except for assumption (a), by the end of 1943, it was possible to predict the efficiencies of proposed weapon designs.

After this, efficiency studies shifted to more specific problems, such as developing criteria for choosing tamper material and investigating the efficiencies of implosion bombs. A third study was how radiation affected the course of the explosion and the attainable efficiencies.

Group T-3 investigated the factors affecting tamper choice. Apart from radiation, the virtues of a tamper were (1) its neutron-reflecting properties, and (2) its effect on the hydrodynamics of the explosion. Point (1) could be understood perfectly by knowing the number of neutrons the tamper scattered back into the core, the time delays in this backscattering, and the energy of the returned neutrons. Calculating these effects depended on knowledge of elastic scattering cross sections as a function of the angle of scatter, and of inelastic and absorption cross sections. Point (2) involved calculating the extent to which various tampers tended to hold the active material together by their inertia during the explosion, and the shock wave behavior in the tamper.

Group T-3 and the D-D Group of the Experimental Physics Division made extensive calculations to interpret the latter's scattering data on various tamper substances. Their calculations were limited because they had to bridge the gap between a detailed theory for which the differential constants were unknown and a semi-integral type of experiment in which only certain average effects were measured.

The effects of radiation on the nuclear explosion were the most

84

problematic factors to be considered. Knowledge of radiation effects was important in predicting weapon design efficiency and in choosing a tamper. Different tampers have different degrees of transparency to radiation, a property that affects the course of the explosion and its efficiency. During the initial bomb expansion, the active material is heated exponentially by release of fission energy. The tamper is heated, but far less rapidly. In the time available, the only effective mechanism for transferring heat from core to tamper is the outgoing shock wave.

Early in 1944 Group T-3 began to rework earlier efficiency calculations. They set aside the assumption that the multiplication rate depended only on the average pressure over the core and tamper and examined in detail its dependence on the shock and rarefaction waves. They first assumed that these were plane waves but later used an "informed guess" as to the effects of convergence and divergence. Only much later were these effects actually calculated so that it was possible to set aside assumptions (c) and (d) and to consider an arbitrary combination of core and tamper materials. Another refinement was replacement of differential diffusion theory by more exact methods.

In May 1944, Taylor's stability considerations created a new worry about efficiency. According to Taylor's principle, when the hot core material pushed against the cold tamper, the interface would be unstable and then core and tamper would mix, possibly, lessening tamper effectiveness. Investigation showed that this effect probably would not be large because the loss of active material that leaked into the tamper would be partly compensated by the remaining tamper fragments. Also, by the time instabilities became serious, radiation would have moved the interface between light and dense material out into the tamper, and the mixing would be mainly tamper with tamper.

Another aspect of implosion bomb efficiency is predetonation. Initial pressure and density distributions in the implosion are nonuniform, whereas in the gun assembly they are uniform. This difference, however, was shown to be unimportant. The great difference was in the larger neutron background of ^{239}Pu and in the dependence of the predetonation probability on the course of the implosion. For a long time it was hoped that an efficient weapon that used only a steady neutron source would be possible. In such models, the efficiency had to be regarded as a random variable with a rather large dispersion, depending upon the particular moment when a neutron managed to start a divergent chain reaction. This involved developing a statistical theory of chain reactions in which not only the average number of neutrons per fission but also the random variation of this number from fission to fission was important.

THE SUPER

Work on the deuterium bomb or Super project was secondary, but continued throughout the course of the project. From its conception, before Los Alamos, this work was directed by Teller, later joined by Fermi. The first idea of such a bomb, at least in relation to the Los Alamos program, evolved in a lunchtime discussion between Fermi and Teller early in 1942.

Fundamental understanding of the fast thermonuclear reaction had been reached by the time Los Alamos began. In the first rough calculations, Teller had ignored radiation, which drains off energy at a rate that increases rapidly with temperature. These calculations indicated that the reaction would take place if ignited by the explosion of a fission bomb as "detonator." In fact, they indicated that the reaction would go too well, and that the light elements in the earth's crust would be ignited.

Energy transfer was well enough understood in the summer of 1942 to indicate that a Super could be made. At the Berkeley summer conference, Teller argued that such a bomb was feasible.

Konopinski's suggestion, which turned out to be very important, was to lower the deuterium ignition temperature by admixture of artificially produced tritium (^{3}H). Tritium's apparently greater reactivity led to this proposal. This idea was not followed up immediately because of the obvious difficulty of manufacturing tritium and the hope of igniting pure deuterium. Eventually, new ignition difficulties made introduction of artificial tritium seem necessary.

Another topic of the Berkeley conference was the effect of secondary nuclear reactions. Products of the deuterium-deuterium (D-D) reaction were, with about equal probabilities, a ^{3}He nucleus and a neutron, or a tritium nucleus and a proton. Bethe noted that the reaction of tritium with deuterium (T-D), although secondary, was very important. The T-D reaction releases nearly five times as much energy as the D-D reaction, and the reaction cross section was likely to be larger.

Because of the Berkeley Super discussions, its investigation continued, D-D and T-D cross sections were measured, and when the Los Alamos Laboratory was being planned, a Super research program was included.

86

Measurement of the D-D cross section was undertaken by Manley's group at Chicago; that of the T-D cross section, by Holloway's group at Purdue.

At Los Alamos there was no systematic theoretical work on the Super until the fall of 1943. E. A. Long's group started a Cryogenic Laboratory with the object of building a deuterium liquefaction plant. Professor H. L. Johnston at Ohio State University worked under subcontract on liquid deuterium properties (Chap. 8).

In September, Teller proposed more intensive investigation of the Super. Experimental cross sections had been revised upward, so the bomb would be feasible at lower temperatures. Also, there was some evidence that the German interest in deuterium might be directed toward producing a similar bomb. Work was resumed but not with high intensity because Teller and his group were occupied with more urgent problems.

The Governing Board re-evaluated the Super program during a February 1944 meeting. Theoretical difficulties indicated that it might be difficult to ignite deuterium because of energy dissipation. If igniting deuterium was too difficult, Konopinski's proposal to lower the ignition temperature by admixture of tritium could be used. A small percentage of tritium would bring the ignition temperature down from about 20 keV to around 5 keV.

The practicability of using T-D mixtures was limited because tritium was so hard to obtain. It could be produced from the reaction of neutrons with ^6Li, yielding tritium and ^4He. The very small tritium sample used in cross-section measurements at Purdue had been produced by cyclotron bombardment. Larger scale production would be possible in a pile such as Hanford's, but only the small percentage of excess neutrons not needed to keep the pile productive could be used.

Because of the remaining theoretical problems and because the Super might have to be made with tritium, development time would be much longer than originally anticipated. However, it was decided that work on so portentous a weapon should be continued in every way possible without interfering with the main program. Tolman, at this meeting as General Groves' adviser, affirmed that although the Super might not be needed as a weapon for the war, the Laboratory had a long-range obligation to continue this investigation.

Although no final decision was made at the meeting, subsequent policy was defined. Teller's group confirmed the difficulty of igniting pure deuterium. In May 1944, Oppenheimer discussed tritium production with

General Groves and C. H. Greenewalt of the DuPont Company. They decided to undertake experimental tritium production using surplus neutrons in the Clinton pile.

DAMAGE

In the summer and fall of 1943, the shock wave produced in air by the explosion, the optimum explosion height, and effects on the diffraction by obstacles such as buildings and of refraction caused by temperature variation were investigated. There was some calculation of the energy that might be lost through evaporation of fog particles in the air. The size of the "ball of fire" after the explosion and the time of its ascent into the stratosphere were estimated. The theory of shallow and deep underwater explosions was investigated and led to model experiments.

An important question resolved at this time was the dependence of damage upon the characteristics of a shock wave in air. Damage from small explosions is roughly proportional to the impulse, which is pressure integrated over the pulse duration; that is, the average pressure of the pulse times its duration. Investigation showed that existing blockbusters were near the size limit at which further increase of the pulse duration increases the damage. Damage from large explosions such as those contemplated depended only on the peak pressure. This was important because the peak pressure depended on the cube root of the energy, whereas the impulse depended on its two-thirds power. For this reason, large bombs are relatively less effective than small ones in creating physical damage. Calculations showed that the peak pressure of bombs of about 10 000 tons of TNT would fall below the level of "C" damage at a 3.5-km radius.

Another important point clarified at this time was the optimum detonation height. Reflection of shock waves by solid obstacles were known to increase shock wave pressure. However, this effect was much greater for oblique incidence than elementary considerations had indicted. In fact, oblique incidence up to an angle of 60 or 70° from the vertical gives a greater pressure increase than normal incidence. Therefore, it was concluded that the damage radius could be extended by detonation at 1- or 2-km altitude.

EXPERIMENTS

Work ranged from that in which the theorists played a large, semi-independent role to ordinary service calculations, particularly analysis of experimental data. For this and to consult on experiment design, every experimental group had theorists assigned to it. Extensive calculations were necessary in all experiments whose results depended upon nuclear constants in a complex statistical way. Then it became necessary to relate the measured quantities with these constants by theory, after using this theory to decide whether a given experimental design would yield enough accuracy to justify its execution and to interpret the data obtained. The theorists played this part in most of the experimental determinations of nuclear quantities described in Chaps. 6 and 7.

An example of theoretical influence on experimental design was the "Feynman experiment," which was never performed but whose principle was embodied in several experiments. This was the proposal to assemble near-critical or even supercritical amounts of material safely by putting a strong neutron absorber (the ^{10}B boron isotope) uniformly into the core and tamper. An absorber whose absorption cross section was inversely proportional to the velocity of the neutrons absorbed would decrease the multiplication rate in the system by an amount directly proportional to the absorber concentration. Thus a supercritical material could be made subcritical by adding boron. By measuring the rate at which the neutron died out in this system, one could calculate the rate at which it would increase if the boron were absent.

The Theoretical Division made many analyses of Ordnance Division experiments. In preparing for the RaLa experiments, for example, Group T-3 analyzed gamma-ray attenuation in a homogeneous metal sphere surrounding the source and calculated how it would be increased with compression during implosion of the sphere. The theory of the magnetic method of implosion also was investigated in the Theoretical Divison.

Group T-1, and later Group F-1, made safety calculations for the Oak Ridge Y-12 and K-25 plants. The Group Leader, E. Teller, was appointed consultant for the whole Manhattan District on the dangers of possible supercritical amounts of material being collected together in the plants producing separated ^{235}U. The Computations Group, T-5, ran many calculations for other groups and for related investigations of the mathematical theory of computation.

89

CHAPTER 6

The Experimental Physics Division

March 1943 to August 1944

ORGANIZATION

The Experimental Physics Division was one of the first organized. These were the initial groups.

P-1	Cyclotron Group	R. R. Wilson
P-2	Electrostatic Generator Group	J. H. Williams
P-3	D-D Source Group	J. H. Manley
P-4	Electronics Group	D. K. Froman
P-5	Radioactivity Group	E. Segrè

Two new groups under H. Staub and B. Rossi were created in July and August 1943. The first was to develop improved counters; the second, improved electronic techniques. In September they were combined as the Detector Group, P-6, under Rossi. Group P-7, the Water Boiler Group, under D. W. Kerst, was created in August. R. F. Bacher was Division Leader from the time of his arrival in July 1943.

EQUIPMENT

When the first members of the experimental physics groups arrived in March 1943, the buildings to house the accelerating equipment were not completed.

The bottom piece of the Harvard cyclotron was laid on April 14, and in the first week of June there were initial indications of a beam. The early work, with an internal beam on a beryllium target probe, gave an intense neutron source. Early in 1944, an external beam was developed. The two electrostatic (Van de Graaff) generators were moved onto their foundations in Building W in April. The "long tank," which gave 1 microampere (μA) at 4 MeV at Wisconsin, produced a beam on May 15. The "short tank," which had operated at 2 MeV, with higher current, gave a beam on June 10. Both were used to produce neutrons from the Li(p,n) reaction, covering (by proper exploitation) the energy range from 20 keV to MeV. After providing some useful information, the short tank generator was rebuilt to give higher energy, and thereafter ran satisfactorily at 2.5 MeV. It was again ready for use in December 1943.

Because Building Z, which housed the Cockcroft-Walton accelerator, was completed later than Buildings X and W, installation was not begun until the end of April. The first beam was obtained June 7. This accelerator, used to produce neutrons from the D(d,n) reaction, was usually called the "D-D source," and P-3 was called D-D Group. This source provided neutrons up to 3 MeV.

That all the accelerating equipment was installed and put into operation in such a short time affirms that group members worked long, hard hours. While the accelerating equipment was being set up, plans and instrumentation for experiments were going ahead. At the cyclotron, a 5- by 5- by 10-ft graphite block was set up to give a flux of thermal neutrons. It was later rebuilt and enlarged. By use of modulation, the cyclotron ranged from thermal energies up to the kilovolt range, where it overlapped the low-energy range of the electrostatic generators. Building G was built as an adjunct to Building Z to house a graphite block for standardization of slow-neutron measurements. Less spectacular, but equally necessary, was setting up of equipment for the electronics laboratory, photoneutron source work, spontaneous fission investigation, research on counter equipment, and the Water Boiler.

This division's program was almost entirely in neutron and fission physics. Except for spontaneous fission, the reactions to be studied were induced by neutrons of various energies. Together with photoneutron sources, the cyclotron, the Van de Graaffs, and the Cockcroft-Walton gave neutrons of reasonably well-defined energies from the thermal region up to 3 MeV. The greatest uncertainties were in the 1- to 20-keV range. All the experimental arrangements involved use of fission, neutron, and radiation detectors, together with the necessary electronic equipment for registering data.

PRELIMINARY EXPERIMENTS

Preliminary experiments were required to prove conclusively that the atomic bomb was feasible. One was to measure the time delay in neutron emission after fission, the other was to confirm the theoretically plausible belief that the number of neutrons per fission was essentially independent of the energy of incident neutrons.

The average time required for a neutron generated in a fissionable mass to produce its successors in the chain reaction was of primary importance in determining the final bomb efficiency. This time consists of the time of flight and the emission time. The first, approximately 10^{-8} s, is the time between emission of a neutron after fission and a new fission caused by absorption of this neutron. The emission time is the lifetime of the compound nucleus plus the time between splitting apart of the fission fragments and emission of neutrons from them. From theoretical arguments, both these times should be negligible (about 10^{-15} s), but it was imperative to confirm the theory experimentally because it was critically important that this time be small.

Before leaving Princeton, the Cyclotron Group had begun instrumenting a "Baker experiment" to determine the emission time after fission. This experiment, their first at Los Alamos, took advantage of the very high fission fragment speed to measure very short emission times. After 10^{-8} s, the fragments are about 10 cm from the point of fission if there is no material in their path. In this experiment, ^{235}U foil was wrapped around a neutron counter, and two cases were compared, one in which the fragments were permitted to travel out from the counter, and another in which they were stopped near it. For geometrical reasons, the chance that a neutron will be counted decreases rapidly with the distance at which it is emitted. These two cases gave the same neutron count within the limits of experimental error, so it was possible to conclude that most neutrons were emitted in less than 10^{-9} s, and that the percentage emitted in more than 5×10^{-9} s was negligible. This result was confirmed later by a different method using apparatus made primarily to measure the neutron number. Later, the same group demonstrated that the fission time was also less than 10^{-9} s.

The second unverified assumption, that the neutron number was the same for fissions from slow and fast neutrons, was not tested accurately until the fall of 1944 (Chap. 12). The theoretical assurance was strong. A more urgent confirming experiment was measurement of the ^{239}Pu neutron number. At the beginning of the project, it was not experimen-

92

tally certain that ^{239}Pu fission would produce neutrons. To some extent, therefore, the entire plutonium production program was still a gamble.

The first nuclear experiment completed at Los Alamos was comparison of the ^{235}U and ^{239}Pu neutron numbers using a barely visible speck of plutonium, all that then existed. In July 1943, the Electrostatic Generator Group compared neutrons emitted from known masses of uranium and plutonium by counting the number of protons recoiling from fast neutrons in a thick paraffin layer surrounding the fissionable material. The fissions are produced by less energetic neutrons. Ionization pulses from the proton recoils were observed with samples of normal uranium containing ^{235}U, with ^{239}Pu, and without any fissionable material. The numbers were compared by simultaneously recording the fission rates in a monitor chamber. To determine the relative number of neutrons per fission from the relative number per microgram, it was necessary to measure their relative fission cross sections for the particular energy spectrum of the neutrons used. This was done by comparing the two materials in a double fission chamber.

The same group ran another experiment simultaneously to check the first experiment. This used a thorium fission detector, and the primary neutrons used to cause fission had energies well below the thorium fission threshold. Despite the small amount of plutonium available, these experiments showed that the plutonium neutron number was greater than that for uranium, and gave a value for the ratio of these numbers, which was not materially improved later. The Cyclotron Group checked this ratio at about the same time.

THE NEUTRON NUMBER

The assumption that the neutron number is independent of the energy of the neutrons that initiate fission needed experimental confirmation. In the spring of 1944, the Cyclotron and Electrostatic Generator Groups compared ^{235}U fissions from thermal neutrons with those from 300-keV neutrons and found the ratio of neutron numbers to be unity within wide limits of experimental error. This ratio was remeasured later with smaller experimental errors, and the value of 1.0 was confirmed for both ^{235}U and ^{239}Pu.

93

The neutron-number measurements described above are relative; that is, they involve comparing one neutron number with another. The only absolute measurement was Fermi's at Chicago. Because this value was questionable (Chap. 1), it was checked early. The graphite block at the cyclotron gave a strong flux of thermal neutrons to produce fissions in the sample. The number of fast neutrons (from fission) was determined by measuring the resonance activity acquired by indium foils and calibrating this measurement by comparison with the activity induced by a radon-beryllium source of known output. The number of fissions was counted simultaneously, and the number of neutrons per fission was obtained for both ^{235}U and ^{239}Pu. The radon-beryllium source used was calibrated by the Standards Subgroup of the D-D Group. Even without standardization, the ratios of ^{235}U and ^{239}Pu neutron numbers gave a check of previous relative measurements. In Chicago, meanwhile, Fermi, also checking the absolute neutron number of ^{235}U by two methods, counted 2.14 and 2.18 neutrons per fission, which agreed with the Los Alamos result.

SPONTANEOUS FISSION MEASUREMENT

Before coming to Los Alamos, the Radioactivity Group had been making spontaneous fission measurements in Berkeley. Their practical importance arose from the need to minimize the neutron background in the bomb material. In particular, the neutrons from spontaneous fissions would set a limit to this background below which it would be useless to reduce the background from (α,n) reactions in light-element impurities. The sample size was limited by the need to avoid spurious counts in the ionization chamber from coincidence of several alpha pulses, simulating the large pulse of a single fission. Because spontaneous fission decay is a very slow process, the data were taken over long periods, with great care taken in design and operation of equipment.

At Los Alamos, the Radioactivity Group built new ionization chambers and designed new amplifiers to use the larger samples of material becoming available. The Pajarito Canyon Field Station was set up several miles from Los Alamos to get away from the high-radiation background at the Laboratory, which would have masked completely the low counting rate from spontaneous fission.

94

In the fall of 1943, the Laboratory received a report that Joliot had found neutrons associated with the alpha radioactivity of polonium, presumably characteristic of alpha emission. Because of the difficulties of purifying polonium, it was believed that Joliot had overestimated the purity of his material and that his neutrons were really from the (α,n) reaction of light-element impurities. Such a "Joliot effect," if real, might materially affect the Laboratory program. Plutonium, as an alpha emitter, might have a neutron background that could not be brought down to tolerance by chemical purification. Thus, a polonium-beryllium initiator might be unusable because of neutrons associated with the polonium alpha radiation.

Radon was the alpha emitter that could be purified most easily, so the Radiochemistry Group built a radon plant and procured 2 g of radium for "milking." They found no spontaneous neutrons. This work was dropped after polonium purification and direct measurement of spontaneous neutrons from plutonium were achieved.

By December 1943, there were indications that some of the ^{235}U fissions probably were not spontaneous but were caused by cosmic ray neutrons. Although ^{238}U fission rates determined at Berkeley and Los Alamos agreed fairly well, the Los Alamos ^{235}U rate was higher. Because many cosmic ray neutrons are too slow to cause fissions in ^{238}U, the results would be explained by the higher cosmic ray intensity at the Los Alamos elevation of 7300 feet than at sea level in Berkeley.

The early estimate of 200-ft/s minimum assembly velocity for the ^{235}U gun method was based upon the Berkeley spontaneous fission measurements, which indicated about 2 neutrons/g/s from the source. After the discrepancies were observed, it was found at the Pajarito station that a boron-paraffin screen reduced the number of "spontaneous" fissions observed in both ^{235}U and ^{239}Pu. To estimate the spontaneous fission rate of ^{239}Pu in a reasonable time, a new system, built in the spring of 1944, permitted measurement from 5 mg of plutonium. In July 1944, a significant difference was found between the spontaneous fission rates of plutonium from cyclotron irradiation and the much heavier irradiation of the Clinton pile. Fermi suggested that the higher rate in the latter material might be caused by ^{240}Pu produced by the (n,γ) reaction in the pile. A reirradiated sample gave still higher counts. This was the first direct observation of the new isotope.

For economic operation, the Hanford plutonium would be heavily irradiated, so its neutron background was predictably too high for the

95

gun assembly. This fact forced abandonment of the plutonium gun assembly program, and necessitated success of the implosion.

ENERGY SPECTRUM OF FISSION NEUTRONS

Pre-Los Alamos work on the energy distribution of neutrons emitted by the fission process indicated that the mean energy was about 2 MeV, but that appreciable numbers of neutrons had energies less than 1 MeV and so could not cause fission in ^{238}U. The cloud chamber data from Rice Institute (Chap. 1) involved large corrections. Stanford data on ion chamber pulse size distribution (Chap. 1) looked reasonable theoretically. These results, showing neutrons tailing off from 1 MeV, agreed with older experiments on range and effective energy, as obtained from slowing-down and from hydrogen cross-section measurements. The photographic emulsion technique used at Liverpool (Chap. 1) showed a much sharper maximum at about 2 MeV. All the measurements suffered from being made with large masses of dilute material that could cause neutrons to lose energy from inelastic scattering before being measured.

Another early problem, therefore, was to determine the fission spectrum more accurately. The photographic emulsion technique seemed most promising for covering a wide range of neutron energies in one run and for minimizing the scattering material. It was straightforward and involved no appreciable corrections if carefully executed. The Electrostatic Generator Group had exposed plates at the University of Minnesota. One of their early tasks at Los Alamos was to set up equipment and train personnel to read the plates. Early results showed that, in shape, the high-energy end of the spectrum agreed with the British and Stanford results, but the low-energy side disagreed with both. The plates were calibrated with the D-D source and electrostatic generator Li(p,n) source. Measurements agreed generally with the Rice Institute cloud chamber measurements. Meanwhile, the Stanford method was carefully reviewed because its maximum and mean energies were lower than those from cloud chamber and photographic plate data. This could be explained by inelastic scattering and consequent distortion toward lower energies. The final Stanford report was written in the summer of 1943, and then personnel of the group came to Los Alamos.

At Los Alamos, a detailed comparative study of the advantages, difficulties, and limitations of the various schemes for neutron spec-

96

troscopy was made. Several experiments were done by the Electrostatic Generator and Detector Groups, and they eventually gave converging results.

As a corollary to obtaining quantitative knowledge of the fission spectrum, much effort was spent in designing mock-fission sources; that is, neutron sources with a neutron spectrum comparable to the fission spectrum. Such sources were later used in semi-integral experiments to measure average cross sections under conditions simulating an actual fission bomb. A satisfactory mock-fission source was achieved in May 1944 by allowing the alpha particles from a strong polonium source to fall onto a mixture of neutron-producing substances. The mixture was proportioned to give a reasonable reproduction of the fission spectrum. Photographic plate determinations indicated that $NaBF_4$ gave an excellent mock spectrum.

FISSION CROSS SECTIONS

The critical mass and efficiency of the bomb depend upon the cross sections for fission, capture, and elastic and inelastic scattering to all energies for which there are appreciable fission neutrons. Work was required to determine the absolute cross sections at various energies and to measure their variation as a function of the incident neutron energy. Fission cross-section measurements were emphasized early because of interest in the uranium hydride bomb. The practicability of this bomb depends on the hypothesis that the delaying effect of neutron slowing by hydrogen was compensated for by a corresponding increase in the fission cross section with decreasing neutron energy. If this hypothesis were true, the explosion rate would remain the same as in a metal bomb although the critical mass would be decreased considerably. Evidence for the inverse dependence of cross sections on neutron velocity was the earlier work at Wisconsin (Chap. 1), which showed approximately $1/v$ dependence from 0.4 to 100 MeV. The same dependence law was also verified between thermal velocities and 2 eV. However, when the latter dependence was extrapolated to higher energies, and the high-energy curve to low energies, the two failed to cross. In fact, between 2 eV and 100 keV a 12-fold increase in the coefficient of $1/v$ had to be accounted for. Because the practicability of the hydride bomb depended upon the actual shape of the curve in this region, it was important to know approximately where the break occurred.

97

Boron absorption measurements by the Electrostatic Generator Group in August 1943 showed that the break was between 25 and 40 eV. This was the first indication that fission cross sections do not follow a simple law in the epithermal region. Because the break occurred at this low energy, the possibility of a hydride bomb was not excluded.

Next, the relative fission cross sections of ^{239}Pu and ^{235}U were measured as functions of neutron energy. It was important to compare the ^{239}Pu fission cross section with that of ^{235}U, which was known to be good bomb material. Such experiments were relatively easy, requiring only simultaneous counting of reactions that occurred when two or more foils of known masses were immersed in the same neutron flux. In this way, the cross sections for all available fissionable materials were measured relative to the ^{235}U cross section. However, the absolute value and change with energy of ^{235}U were questionable. Such experiments were relatively easy, requiring only simultaneous counting of reactions that occurred when two or more foils of known masses were immersed in the same neutron flux. In this way, the cross sections for all available fissionable materials were measured relative to the ^{239}U cross section. However, the absolute value and change with energy of ^{235}U were questionable. Such relative measurements, for ^{238}U, thorium, ionium, ^{239}Pu, protactinium, ^{239}Np, and the (n,α) reaction in boron and lithium, were made over an energy range from about 100 keV to 2 MeV.

In addition to the importance of knowing the absolute cross sections for the primary materials, the other cross sections were useful tools for neutron detection. Elements such as ^{238}U, protactinium, and ^{232}Th, which had fission thresholds at high energies, were useful when a particular fraction of the neutron spectrum was to be examined. Boron and lithium proved to have approximately 1/v absorption cross sections and were useful for measurements in the 1- to 20-keV gap left by Los Alamos accelerating equipment. Further fast-neutron reaction cross sections were measured for gold, phosphorus, sulfur, indium, etc. Even then it was realized that reactions leading to radioactive isotopes would provide useful experimental information if the energy responses were known, because there would be many experimental arrangements in which bulky detection chambers could not be used.

Two difficult problems were associated with cross-section measurements, and one was not yet solved by the end of the war. These were (1) absolute measurement of neutron flux over a wide energy range so that some easily detectable reaction products, such as those from ^{235}U

98

fissions, could be established in terms of an absolute and accurate flux standard, and (2) production of monoergic neutrons of energies from 1 to 50 keV with high enough yields for performing necessary experiments with good energy resolution.

The Detector Group finally solved the first problem for the 400-keV to 3-MeV range by using Li(p,n) and D(d,n) neutron sources and an electron-collection parallel-plate ionization chamber, with which the number of recoil protons from a thin tristearin film could be counted accurately. This experiment depended partly on earlier accurate determination of the (n,p) scattering cross section, at Minnesota. It also depended on theoretical interpretation of the differential bias curves obtained in electron collection.

High counting rates (large alpha background in chambers with ^{239}Pu and other types of background) had led the Detector and Electronics Groups to develop new counting techniques involving electron collection and new fast amplifiers. This caused a minor revolution in the counting techniques and electronic equipment that the Physics Division used.

The second problem was solved partially early in 1944 when the short electrostatic generator was rebuilt. High currents and energy regulation to within 1.5 keV from this machine permitted using the back-angle neutrons from the Li(p,n) reaction down to less than 5 keV. Development of the so-called long counters increased the possibility of bringing the absolute fission cross-section measurements down to a few kiloelectron volts, where they were still extremely uncertain. Checks by independent methods gave lower cross-section values in the 30-keV region. Further investigation of counters and construction of an antimony-beryllium source of 25-keV neutrons confirmed the lower value. This was another blow to the hydride gun program.

Absolute fission cross-section measurements at several energies from 250 keV to 2.5 MeV were undertaken. They were based on comparison of fission cross sections with hydrogen scattering cross sections. The results provided a reliable standard for other measurements in which the relative values were more reliable than the absolute values.

The Cyclotron Group measured the fission cross section of ^{235}U in the region below 1 keV early in 1944. Monoergic neutrons were obtained using the "velocity selector." In this method, the neutrons are separated into velocity groups depending on the time of flight between source and detector over a path several meters long. The velocity selector equipment, from Cornell, was extensively rebuilt before cross-section measurements were obtained. The Radioactivity Group used photoneutrons for other fission cross-section measurements at isolated energies during this period.

CAPTURE CROSS SECTIONS, THE "BRANCHING RATIO"

Earliest measurements of capture cross sections were primarily by the Radioactivity Group. The principal method was to measure radioactivity induced by neutron capture. The capture cross sections of fissionable materials, possible tamper materials, and other materials that might be present in the bomb assembly were of interest.

Capture cross sections were measured in a wide range of potential tamper materials, some very rare by ordinary standards, but cheap compared with active material. Platinum, iridium, gold, and the very rare element rhenium were investigated. The April 1944 Experimental Physics progress report summarized the nearly two dozen elements and isotopes measured.

The Electrostatic Generator Group surrounded a photoneutron source with spheres of potential tamper material and measured the neutron attenuation. This method is better than measuring induced radioactivity because it requires no absolute flux determinations or interpretation of induced activity, and the resultant nucleus need not be radioactive. However, the spheres must be large, allowing degradation of the neutron energy through inelastic scattering, and knowledge of the transport mean free path is required. The long counter was used because its sensitivity was nearly independent of neutron energy. By August 1944, many substances had been examined, and preparations were being made to check the data, a job not completed until the spring of 1945 (Chap. 12).

The capture cross sections for active materials were investigated intensively when two independent sources showed that radiative capture might be competitive with fission. One source was the Cyclotron Group's low-energy fission cross-section measurements. They discovered sharp resonances; that is, relatively narrow energy bands in which the fission cross section increased because of resonance. This result implied that the relatively well-defined resonant energy would be associated with a complementary uncertainty in the duration of the state. This duration might be long enough to permit radiative energy loss as a significant alternative to fission. The second source was neutron-induced radioactivity measurement by the Radioactivity Group. They measured a

100

number of activation cross sections relative to the ^{235}U fission cross section. This method gave consistently higher results than others gave when the known absorption (capture-plus-fission) cross section of ^{235}U was used. It was determined that these difficulties would be removed if the existence of a competing process, such as the (n,γ) reaction, with a large probability compared to that of fission could be shown. The ratio of these probabilities is called the "branching ratio" of radiative capture to fission.

The Electrostatic Generator Group began measuring the ^{235}U branching ratio early in 1944. They measured the ratio of the boron and lithium absorption cross sections to the fission cross section. Because the capture in boron and lithium causes only nonradiative disintegrations, no radiative capture cross section is involved. The ratio of the boron absorption cross section to the ^{235}U *absorption* cross section had been measured already in Chicago, so the ratio of the Chicago and Los Alamos figures would give the ratio of absorption to fission, whose difference from one would be the desired branching ratio. A value of 0.16 obtained for thermal neutrons in ^{235}U indicated radiative capture. This apparently unfavorable result was advantageous if one made the theoretically plausible assumption that the branching ratio decreased at the high energies predominant in an exploding bomb. The advantage arose from the fact that no appreciable energy dependence of the neutron number had been detected, and as all the Los Alamos neutron number measurements were relative to the Chicago measurement of neutrons emitted per neutrons *absorbed*, a finite radiative capture implied a higher ratio of neutrons emitted per *fission*. Therefore, the high-energy effective neutron number had been underestimated. To test the expected behavior of the branching ratio with energy, the experiment was immediately extended to the fast neutron region, and the boron to fission cross-section ratio was measured over a wide energy region. No definite high-energy branching ratio could be obtained, however, until better fission and absorption cross sections were available in the relevant neutron energy region. Extending the experiment resulted principally in determining the boron, and lithium, to fission cross-section ratios. The Radioactivity Group also measured the branching ratio, by comparing the ratio of the ^{235}U fission cross section to the capture cross section of gold (and manganese), and the ratio of the ^{235}U absorption cross section to the absorption cross section of gold (and manganese). The branching ratio

101

was about the same as that obtained by the Electrostatic Generator Group.

A third branching ratio measurement was incidental to Cyclotron Group neutron number measurements in early summer of 1944 (Chap. 6). The ^{235}U value agreed reasonably with those described above. A ^{239}Pu branching ratio value indicated that it was much larger than that for ^{235}U, in fact about 0.5. This indicated a large gain in the effective neutron number for high energies if the branching ratio fell off with high energies as expected.

The branching ratio as a function of energy was only measured indirectly much later (Chap. 12). However, the Radioactivity Group measured the branching ratio in ^{238}U at high energies to determine the neutron absorption by ^{238}U remaining in the separated ^{235}U.

SCATTERING EXPERIMENTS FOR TAMPER SELECTION

At first very little was known about the scattering properties of potential tamper materials, but a tamper had to be chosen as soon as possible. The idea existed that inelastic scattering (in which the neutrons, although not captured by the tamper nuclei, lost part of their energy to them by excitation) would be unimportant because it would probably reduce neutrons to a very low energy so they would not contribute materially to the explosive chain reaction. Also, very little was known about the variation of scattering with neutron energy. The most important part of the fission spectrum was thought to be at high energies near 2 MeV. Experimenters believed that tamper usefulness would be determined by the number of neutrons reflected back to the core. So, it was decided that pertinent information could be collected most rapidly by comparing the backscattering of trial tamper materials for D (d,n) neutrons from the Cockcroft-Walton. This could be done using a nondirectional detector with a paraffin "shadow cone" to reduce direct beam, or using a directional detector. The shadow cone method greatly reduced the range of the measurable scattering angle. It was thought that a directional detector could give an average over the angle from 120 to 180° in the geometry possible.

The D-D Group undertook these measurements by both methods in August 1943. The first directional detector was a spherical ionization chamber with a large directionality factor. The first scatters measured

102

were 1-in.-thick by 10-in.-diam disks of lead, iron, gold, and platinum. The latter two, vulgar wonders in an atomic bomb laboratory, brought a great stream of visitors from other groups. Lead showed up best per unit weight, but because of its relatively low density was not much better than gold or platinum.

Although the geometry used for backscattering covered the 120 to 180° angles, it was discovered (as theory caught up with experiment) that a very small range of angle near 130° was weighted predominantly. Therefore, if backscattering were not uniform, the data obtained could be misleading. Several councils of war about this state of affairs led to extension of the measurements.

The incident was important to the Laboratory's growth. An essential part of the design of such an experiment is enough preliminary analysis and calculation to show that a given experimental arrangement actually can yield the data sought. Most of these elaborate calculations were delegated to the Theoretical Division because of their special skill. It was recognized that closer liaison between the Theoretical Division and the D-D Group was needed for this work.

By the end of October 1943, backscattering measurements on a large list of substances were completed, and many instrumental improvements had been made. After the first survey, the list of possible tamper materials was restricted to tungsten, carbon, uranium, beryllia, and lead. At about this time measurements of the fission spectrum indicated that the important energy range was nearer 1 than 2 MeV. Furthermore, the first experiments indicated that inelastically scattered neutrons *could* play an appreciable role in tamper functioning. Recognition of their possible importance was aided by interest in the uranium hydride bomb. The increase in cross section with decreasing energy that made this bomb seem feasible also suggested that neutrons slowed by inelastic scattering might still contribute to an explosive chain reaction.

Therefore, preparations were made to study scattering as a function of energy and scattering angle, taking account of inelastically scattered neutrons. The D-D and Electrostatic Generator Groups cooperated on this, beginning in November 1943. Backscattering data were obtained at 1.5 and 0.6 MeV, as well as 3 MeV. An experiment also was run to get specific information on the degraded neutrons as a function of primary neutron energy for the elements still being considered as scatterers.

Materials studied were carbon, lead, uranium, beryllia, and tungsten, in order of promise. During May-July 1944, uranium nitride, lead

103

dioxide, cobalt, manganese, nickel, and tantalum at several energies also were studied.

An integral experiment on the hydride bomb was begun in this period. The D-D source was to be surrounded by a modifying sphere mocking the hydride core as nearly as possible. Integral tamper properties around this core, as well as neutron distribution in tamper and core, would be investigated. A new fission detector, a spiral ionization chamber with a spiral of depleted ^{238}U to give a large surface area in a small volume, was developed for this work. It was up to 85% efficient.

THE WATER BOILER

The first chain-reacting unit built at Los Alamos was the Water Boiler, a low-power pile fueled by uranium enriched in ^{235}U. It was the first pile built with enriched material, the so-called alpha-stage material containing about 14% ^{235}U. The necessary slowing or moderation of fission neutrons was provided by the hydrogen in ordinary water. The active mixture was a solution of uranyl sulfate in water. The tamper was beryllia.

The Water Boiler was to provide a strong neutron source of experiments and to serve as a trial run in the art of designing, building, and operating such units. It was an integral experiment to test a theory similar, in some respects, to that involved in designing a bomb. It was the first of a series of steps from the slow reaction first produced in the Chicago pile to the fast reaction in a sphere of active metal. It laid the foundation for instrumental and manipulatory techniques required in later, more exacting steps. Unfortunately, Los Alamos workers did not have the full benefit of experience gained by those at Chicago, so there was unnecessary delay before the first chain reaction was started.

Water Boiler calculations absorbed much of the Theoretical Division's time. Calculation of the critical mass depended upon applying diffusion theory to a complex system of active solution, container, and tamper. To conserve material, it was important to find the optimum solution concentration. The number of hydrogen nuclei had to be large enough to slow the neutrons to thermal energies and small enough not to capture too many of them.

The Water Boiler was isolated for safety reasons. It was first planned for 10-kW operation. The radioactivity of fission fragments from

104

intermittent operation was estimated at 3000 curies (Ci). The minimum safe distance from unprotected people was calculated on the assumption that a mild explosion could disperse this activity into the atmosphere. Isolation was also desirable because of possible high instantaneous radiation in case of an uncontrolled chain reaction.

In September 1943, while design of the Boiler and the building to house it were still preliminary, Fermi and Allison came from Chicago to discuss the problems of such a unit. They pointed out many difficulties in operating the Boiler as a high-power neutron source. Some had been anticipated but their acuteness had not been appreciated fully. One problem was gas evolution that would cause unsteady operation. Decomposition of the uranium salt and consequent precipitation would result from the large amount of radiation to which it would be subjected. Heavier shielding than had been planned would be necessary.

These discussions led to omitting all features necessary for high-power operation and going ahead with the design of a low-power Boiler. Provisions were made, however, for later installation of equipment for high-power operation. The main omission was equipment for chemical decontamination, unnecessary for operation at trivial power outputs. The Boiler could no longer be used as an intense neutron source, but it could be used to investigate a chain-reacting system with a much higher ^{235}U enrichment than previous piles.

The building to house the Boiler, associated laboratories, and later critical assemblies in Los Alamos Canyon at Omega Site (Fig. 3) was completed in February 1944. Design problems included a heavy concrete wall to separate the Boiler from remote-control equipment; a thermostated enclosure to maintain constant Boiler temperature; recording and monitoring equipment, including ionization chambers and amplifiers; control rods and their associated mechanisms; a support for the tamper and container, the container itself, together with means for putting in and removing solution; and design of beryllia bricks for ease in fabricating and stacking the tamper. Specifications for the tamper and active solution were worked out with the Radiochemistry and Powder Metallurgy Groups (Chap. 8), after the original choices of material and size were made by the theorists.

Tests of the fluid-handling and counting equipment late in April 1944 ended with use of normal uranyl sulfate solution. Enough enriched material had arrived to permit determining the critical mass, after the chemists purified and prepared the solutions. Successful operation of the

Water Boiler as a divergent chain reactor was a small but important step toward controlled use of nuclear energy from separated ^{235}U or plutonium.

The Water Boiler, like other controlled reactors, depended upon the very small percentage of delayed neutrons. These allowed keeping the system below critical for prompt neutrons and near critical for all, including the delayed neutrons. Although the delayed neutrons are only about 1% of the total, in the region near critical the system's time dependence (rate of rise or fall) is only about the duration of the delay period. Prompt chains die out constantly, to be reinstated only because of the delayed neutrons.

A Water Boiler experiment proposed by Rossi and bearing his name was designed to determine the prompt period. This period depends on the time it takes for the neutrons to be emitted after fission, on the fission spectrum, and on the scattering and absorption characteristics of core and tamper. It was essential to measure the prompt period in a metal assembly as accurately as possible. Its measurement in a hydrogenous assembly would not give direct information relevant to efficiency calculation, but would provide experience and instrumental development and also would be a check on theoretical predictions.

The Rossi experiment counts neutron coincidences. The presence of a prompt chain in the reactor is presumed whenever a neutron is counted. A time-analyzing system then records the number of neutrons counted in short intervals immediately after the first count. This gives a direct measure of the prompt period.

Another method that gave less interpretable results was to change the degree of criticality rapidly by a motor-driven cadmium control vane. A third experiment was to measure the spatial distribution of neutrons in the solution and tamper by placing small counters in various positions in the Boiler and tamper. This experiment served as a check of calculation from neutron diffusion theory.

A fourth, Rossi-related experiment was to measure fluctuations in the Boiler neutron level. These measurements were of interest relative to the variation of the neutron number from fission to fission, which was, in turn, related to the statistical aspects of the chain reaction in the bomb, particularly the predetonation probability. The first measurement gave the count fluctuation in a counter relative to the average number of counts. This gave information about the neutron number fluctuation as soon as the effective number of delayed neutrons was measured.

106

About mid-1944, the Water Boiler Group planned to make critical assemblies with uranium hydride and to rebuild the Water Boiler for higher power operation.

MISCELLANEOUS EXPERIMENTS

The Radioactivity Group made several special investigations of the fission process. The range of fission fragments in heavy and light materials was measured, as were the energies and number of long-range alpha particles discovered in the fission process. Gamma rays emitted in fission were investigated because knowledge of their number and quantum energies was important to understanding the fission process and because these rays might be used in experiments designed to test the bomb. The gamma-ray measurements were made at the Trinity test (Chap. 18).

A D-D subgroup responsible for calibrating the neutron emission from various natural sources built a graphite column, measured the diffusion length of the graphite, standardized the pile for indium resonance neutrons, and began standardizing various natural sources. They also measured gold and indium capture cross sections for column neutrons and the indium half-life. They conducted various safety experiments on handling and transport of active material and worked to improve source standardization, because the accuracy of such experiments as the neutron number measurements was important to the Cyclotron Group.

Another Experimental Physics Division program was to develop provisions for isotopic analysis. In late 1943, there was uncertainty about the amount of ^{235}U in the enriched samples received at Los Alamos. Enriched samples were made up in normal form and were also diluted with various amounts of normal uranium. One set of samples was then sent to Berkeley to be assayed by the neutron assay method, and another was sent to New York for mass spectrography. The results disagreed by almost 10%. Segrè examined the Berkeley method carefully and found no explanation for the discrepancy. Bainbridge examined the mass spectrographic method and he, too, found no explanation. Therefore, both methods were set up at Los Alamos in the Radioactivity Group. Then the Chicago and New York Laboratories in three independent isotopic determinations on a sample known as E-10 got close agreement. This sample was thereafter used as a secondary standard at Los Alamos with the neutron assay method, which proved valuable in assaying the active parts of the gun assembly and later the ^{239}Pu assemblies. By May 1944, a study of uranium isotopes in the mass spectrometer had given satisfactory resolution. Analyses of normal material and of sample E-10 agreed well.

In late summer of 1944, the ^{239}Pu mass spectrometer was set up and preparations were made to test the ^{239}Pu sample reirradiated at Oak Ridge to assay its ^{240}Pu content. This work was actually done in the Research Division after it was created in August 1944. The sample showed a peak at the 240 position, and the relative abundance of ^{240}Pu to ^{239}Pu agreed well with the figure that could be calculated from the value of the branching ratio and the rather uncertain irradiation. The discovery of ^{240}Pu, following from spontaneous fission measurements, was confirmed.

INSTRUMENTATION

Every successful (or unsuccessful) experiment implies instrumental development and construction not easily appreciated by an outsider. Experimental nuclear physics deals with the realm of the small and the fast. Both the time scale and the amplitude of the phenomena studied must be transformed to make them susceptible to direct control and observation.

Although much of the modulation and control equipment for the accelerators was built at Los Alamos, these developments were less novel and extensive than those of observation and measurement. Except for photographic plate and cloud chamber techniques, all experiments involve counting ionization chamber pulses. A typical experiment uses four distinct steps: the counter or detector, the amplifier, the discriminator and scaler, and finally, the mechanical recorder. The ions produced by the particles being studied are moved to collecting electrodes by an electrical potential applied across them. This registers as a minute electrical pulse in the microvolt range. This pulse is then amplified to volts and fed into the counting system. The discriminator selects pulses of interest. Because their frequency usually is too high to be recorded directly by mechanical means, a further electronic step, the scaler, is used. The scaler "demultiplies" the frequency of incoming pulses so that, for example, it gives one pulse for every 64 incoming pulses. The pulses coming out of the scaler then are recorded mechanically.

Certain developments in the counter or detector have been mentioned; for example, fission detectors, in which ions are produced in an ionization chamber by fission fragments from a sample of fissionable material. These include threshold detectors, which use the materials that fission only for neutrons above a certain energy. Another important development was the Electrostatic Generator Group's long counter, which had a flat response of neutrons of different energies. Still another important instrument that used the well-established hydrogen cross section was the proportional hydrogen recoil counter. Finally, the boron trifluoride proportional neutron counter was an ionization chamber filled with boron trifluoride. Boron-10, about 18% of normal boron, undergoes an (n,α) reaction that produces ^7Li. The number of alpha particles counted corresponds directly to the number of neutrons absorbed. Such counters are surrounded with paraffin to slow the neutrons to energies with high-reaction cross sections. High efficiency was obtained by using high-pressure counters, purifying the trifluoride, and using the separated ^{10}B isotope when it became available.

The most extensive development in counting technique came with fast detectors and amplifiers. Proper counter design and certain gas mixtures reduced electron collection times to as low as 0.2 µs. This and other counter development was done mostly by the Detector Group. Using electron collection required very fast amplifiers developed by the Detector and Electronics Groups. The amplifiers had rise times between 0.05 and 0.5 µs.

Much was done to develop discriminators. For example, differential discriminators (multichannel ones that could classify and simultaneously record pulses of different heights) were developed. A new scaling circuit that increased the scaling reliability became standard Laboratory equipment.

Several electronic techniques were developed for special purposes. Many experiments involved measurement or control of phenomena occurring at specified time intervals; for example, the Rossi experiment, measurement of the length of stay of neutrons in the tamper, or the velocity selector for selecting monoergic neutrons from the cyclotron. Another type of timing circuit developed for the implosion study permitted velocity measurements in explosives. Use of oscillographs for recording implosion data led to extensive development of amplifiers,

109

circuits for printing timing marks on film, sweep circuits, and circuits to delay starting a sweep for a specified time.

Another important Electronics Group job was production of portable counters and other health instruments.

CHAPTER 7

The Ordnance Division
June 1943 to August 1944

ORGANIZATION AND LIAISON

Before its formal organization in June 1943, the Ordnance Engineering Division occupied two or three small rooms in Building U. It was concerned with procurement, gun design, and instrumentation, but its main activity was discussion and analysis of the work that lay ahead, thereby labeling and organizing the elements of the new field.

In May, Captain W. S. Parsons (USN) made a preliminary visit. His transfer to Los Alamos was then requested by General Groves, recommended by Conant and Bush, and approved by the Governing Board. He returned in June as Ordnance Division Leader.

The original groups of the new division were these:

E-1	Proving Ground	E. M. McMillan
E-2	Instrumentation	K. T. Bainbridge
E-3	Fuse Development	R. B. Brode
E-4	Projectile, Target, and Source	C. L. Critchfield
E-5	Implosion Experimentation	S. H. Neddermyer

Parsons selected a competent chief engineer to head Group E-6, George Chadwick, who for 20 years was Head Engineer of the Navy Bureau of Ordnance. Although Chadwick never lived at Los Alamos, from June to September 1943 he was a prospective head engineer, and he worked with the Bureau of Ordnance and the Navy Gun Factory in designing and fabricating the first experimental guns, consulted at Los Alamos about design of the Anchor Ranch Proving Ground, and in August, was asked to help hire machinists and draftsmen from the Detroit area. At this time, he decided not to take the Los Alamos position. The connection with Chadwick in Detroit remained, however (Chap. 7).

111

After a brief interval in which J. L. Hittell directed the Engineering Group, P. Esterline held the position from December 1943 until his resignation in April 1944. Late in May, L. D. Bonbrake took over from R. Cornog, who served after Esterline. The difficulties in finding the right person for Chief Engineer are discussed later.

In the fall of 1943, Groups E-7 under Ramsey and E-8 under Hirschfelder were added to the division.

Early in 1944, with the rapid expansion of the ordnance and implosion programs, division administration was under two Deputy Division Leaders: E. M. McMillan for the gun program and G. B. Kistiakowsky for the implosion program. McMillan was replaced by Lieutenant Commander A. F. Birch as Group Leader of E-1. Kistiakowsky added a new group (E-9) under K. T. Bainbridge to investigate and design full-scale high-explosive assemblies and prepare a full-scale test with active material. L. G. Parratt took Bainbridge's place as E-2 Group Leader. Each of the division's two branches were advised and assisted by a steering committee. Although formally equivalent, the branches were not organizationally equal. The gun program was proceeding smoothly at a constant level of activity. The implosion program, however, had serious organizational and technical problems springing from its rapid growth in size and importance.

In late June and early July, the implosion project was reorganized further when Neddermeyer became chairman of the Implosion Steering Committee, Kistiakowsky became Acting Group Leader of E-5, and two new groups were formed. Major W. A. Stevens headed the first, E-10, whose functions were maintenance and construction for the implosion project and operation of the S-Site plant. L. W. Alvarez, headed the second, E-11, which would develop the RaLa tests and investigate electric detonators.

When Parsons returned to Washington after his first trip to Los Alamos, he arranged for all his connections with the Navy Department to be handled through Lieutenant Commander Hudson Moore of the Research and Development Section of the Navy Bureau of Ordnance. Moore's most important activities were with the Naval Gun Factory in fabrication of experimental guns. He also handled procurement of miscellaneous ordnance materials from Navy stores and liaison with the Navy Proving Ground (NPG) at Dahlgren, Virginia. At the same time, Parsons arranged, for security reasons, that all Navy equipment be shipped to E. J. Workman, head of Section-T OSRD Project, at the University of New Mexico, Albuquerque.

The other principal Ordnance Division liaison was the "Detroit Office." In August, Chadwick was asked to help procure personnel. He set up an office in Detroit to try out design engineers he hired and to pay machinists hired before the new shop facilities were ready.

Also in August, Bush approved using the Section-T OSRD Project at the University of Michigan, to develop radio proximity fuses for the bomb. Section-T funds were to be replenished by transferring Navy Department funds to OSRD.

In October, it was decided that flight test models would be fabricated under the University of Michigan procurement setup using Section-T funds. H. R. Crane, head of the Michigan Project, would place the orders, but the University of Michigan would do only general supervision and accounting. Chadwick's office in Detroit, rather than Crane, would be responsible for inspection and follow-up work. This arrangement was not wholly satisfactory because Crane had an interest in these models; fuse units that his project designed and fabricated were to be incorporated in them.

Financing of the Detroit office was arranged by contract between Chadwick and the University of California until November 10, later extended to March 1944. Chadwick was appointed OSRD representative to facilitate his work on fabrication contracts let by the University of Michigan. In May 1944, Detroit Office financing was assumed by the University of Michigan. In June, Lieutenant Colonel R. W. Lockridge assumed charge of the Detroit office, and in July he was appointed OSRD representative to unify the Detroit-University of Michigan relationship and bring those activities more closely under Manhattan District control.

GUN DESIGN, PROVING, AND INTERIOR BALLISTICS

The only assembly method first considered sound enough in principle to warrant extensive proving and engineering was the gun method. The proving facilities, gun manufacture, and bomb design were focused on use of a single gun to fire active material into a target. However, there was intensive study of other gun types, notably, use of two or more guns on the same target and the possibilities of jet propulsion. Possible use of high explosives to replace slow-burning propellant in multiple guns was explored. None of these schemes proved attractive enough to be serious competitors of the single-gun method.

In April 1943, so little was known about the physical and nuclear properties of the active materials that only very rough estimates could be made of how much material must be fired from a gun, how fast it must be fired, and, in the case of plutonium, how much acceleration it could stand. It was assumed that the material could be made strong enough through alloying and that the density of plutonium was essentially the same as that of uranium. Then, the sizes of critical assemblies were computed from the existing knowledge of nuclear cross sections, and the required velocity of assembly was determined from the purification standards.

The gun performance required was within standard ballistic experience. In fact, the 3000-ft/s velocity for a plutonium projectile was chosen as a practicable upper limit, because, obviously, much higher velocities would be desirable. Otherwise, the gun design bore little resemblance to standard ordnance problems. There was no concern, in the assembly of critical masses, about the usual questions of in-flight stability of the projectile, absorbing the recoil energy, tube erosion, or muzzle pressure. Instead, the requirements were for as lightweight tubes and as reliable interior ballistics as possible. About all that standard ordnance could contribute was general formulas for gun strengths and its theory of propellants, both of which had to be used far outside the range of accumulated experience. The situation was a little disturbing when engineering of the 3000-ft/s gun began, because the standard piece that came closest to its performance had proved unreliable.

Getting these fantastic guns made and proved called for expansion of Ordnance Division personnel, facilities, and liaison. This expansion was instituted by Captain Parsons. In May 1943, division attention was centered on getting the 3000-ft/s gun made and proved because the proposed design was farthest removed from standard practice. The principal departures were (1) the tube should weigh only 1 ton instead of the 5 tons usually characteristic of the same muzzle energy; (2) consequently, it must be made of highly alloyed steel; (3) the maximum pressure at the breech should be as high as practicable (75 000 psi); that is, the gun should be as short as possible, and (4) it should have three independently operated primers.

114

The Naval Gun Design Section undertook the engineering in July 1943. Pressure-travel curves were obtained from the NDRC through R. C. Tolman. These were computed by the ballistics group at Section 1 of the Geophysical Laboratory under the supervision of J. O. Hirschfelder, who subsequently joined the staff at Site Y and continued to supervise the work of the Interior Ballistics Group. The curves were drawn for maximum breech pressures of 50 000, 75 000, and 100 000 psi and were submitted to the Navy Bureau of Ordnance.

This was a unique problem involving special steel and its radial expansion, design and breech, primers and mushrooms for extra-high pressures, insertion of multiple primers, and many smaller details. The absence of rifling and special recoil mechanisms were the only details in which this gun could be considered simpler than standard guns. Nevertheless, the drawings were completed and approved very soon, and the required forgings were ordered in September. There was some delay in preparing the steel because of difficulty in meeting the physical specifications. Fabrication of the guns at the Naval Gun Factory required about 4 months at high priority. The first two tubes and attachments were received at Site Y on March 10, 1944.

These were of two types. Both had adapter tubes surrounding them so that the recoil could be absorbed in a standard naval single-gun mount. On the type-A gun, this adapter did not strengthen the tube and was fitted to the gun proper only at the breech. On type B, the adapter supported the gun tube so that it was much stronger than the bare tube would be. Type A was to allow tests of the wall strength and deformation in the high-alloy gun tube; type B was for interior ballistic studies.

While these guns were being procured, work went into making installations, acquiring personnel, perfecting techniques for testing the guns, and establishing procurement channels for accessories such as propellants, primers, cartridge cases, and rigging gear. Early plans called for a standard-type proving ground with centralized control of all explosives operations. The Proving Ground Group did the proving; and the operation, loading, and care of the guns was directed by an experienced ordnance man, T. H. Olmstead, from the Naval Proving Ground at Dahlgren. Although this plan became impractical when high explosives work became more elaborate, the gun work at the original Anchor Ranch Proving Ground was adequate. The buildings included the usual gun emplacements, sand butts, bombproof magazines, control room, and shop. Novel features were incorporated in recognition of the special proving problem. The uncertainty about whether high-alloy tubes might fragment when overloaded, plus the program for eventually firing the tubes in free recoil, greatly increased the proving hazards. Therefore, the ground level of the gun emplacements was put above the roof of the bombproofs, which were installed in a ravine. To conceal the guns and targets from public view and permit instrumentation in all weathers, shelters were provided that could be rolled away for actual firing. Construction started in June 1943 and continued at high priority until completion in September. The first shots were fired from emplacement No. 1 on September 17, 1943, at 4:11 and 4:55 p.m. A second emplacement was completed by the following March in anticipation of receiving the special guns.

The proof firing between September and March was done chiefly with a 3-in./40 Naval antiaircraft gun with unrifled tubes. These rounds were primarily to test various propellants, to study elements of projectile and target design on the 3-in. scale, and to smooth out instrumentation, which was directed by K. T. Bainbridge. Most of the standard proving-ground techniques were adapted to this work, and some new ones were developed. The familiar photographic methods, microflash, Fastax, and NPG projectile cameras proved extremely useful in studying the condition of the projectile as it left the bore and in detecting "blow by." Muzzle velocities were determined from the projectile camera records, as well as by magnetic coil and Potter chronograph. A photoelectric system also was developed for this measurement. Copper crusher gauges and piezoelectric gauges were used to determine powder pressure. Electric strain gages were used on the gun barrel and certain targets, and there were occasions for using many other standard ordnance methods such as star gauging, terminal observations (on recovered projectiles), and yaw cards.

The success of these standard methods was usually above standard, particularly in photography. One nonstandard technique developed specifically for the interior ballistic problem was to follow the projectile by continuous microwaves during its acceleration in the tube. By the time the type-A and -B guns arrived, the proving ground routine, instrumentation techniques, and propellant performance were well established, at least for work at 3-in. scale. At Los Alamos' request, the Explosive Research Laboratory at Bruceton, Pennsylvania, was studying burning of propellants at very high pressure, thereby increasing preparation for the special gun.

In February, Commander F. Birch assumed direction of Anchor Ranch, with McMillan as Captain Parsons' deputy for the gun. Testing of the type-B gun's interior ballistic behavior began on March 17, 1944. By this time, however, specifications for a lower velocity gun with a muzzle velocity of only 1000 ft/s to be used with ^{235}U became final. Three of these guns were ordered from the Naval Gun Factory in March. Some were to be radially expanded, and required that a special gun mount be designed. Nevertheless, they presented a much simpler problem to the Navy Bureau of Ordnance.

Because of the well-prepared experimental background, testing went smoothly and rapidly. "WM slotted tube cordite" proved the most satisfactory propellant at the high pressure involved. Other propellants proved inferior. In particular, the 5-in./50 Navy powder behaved erratically, as it had before, and this was traced to wormholing of translucent grains. The Mark IV primers withstood over 80 000 psi. The propellant performed properly at −50°C. The interior ballistic problem was solved, but the tube was eroded so badly that it had to be returned to the Naval Gun Factory in April. Attention was then given to mechanical strength and deformation of the type-A gun. By this time, the proving ground was operating at very high efficiency. Installation of a drum camera greatly facilitated record taking, and many pressure, strain, velocity, and time interval measurements were made on one round. By early July, the design was proved, and only by running the maximum breech pressure up to 90 000 psi was it finally possible to deform the gun permanently.

By early July, however, it was clear that the 3000-ft/s gun would never be used. The necessary presence of ^{240}Pu in the Hanford plutonium (Chap. 4) decreased its minimum assembly time far below what was possible by gun-assembly methods.

PROJECTILE AND TARGET DESIGN

Although development of projectile and target designs for the gun assembly should have moved faster than the gun development, uncertainties surrounding the problem were relatively more serious. Not only were the physical properties of plutonium entirely unknown, but the idea of producing a projectile and target assembly that would start the chain reaction at the right time and in a compact geometry was entirely new.

117

Experience in armor penetration greatly discouraged early suggestions that the projectile be stopped by the target. The gun assembly coi.cept as of April 1943 was one in which the projectile passed through the target freely.

Before work on these developments began, the plan was complicated by further uncertainty about the amount of active material that could be disposed safely in the projectile or target. This was particularly important for the hypothetical uranium hydride gun because the critical mass would be small and, for effectiveness, many would have to be assembled. Although planned primarily for the hydride gun, the critical mass calculations for odd metal shapes were not accurate enough to rule out a possible need for such methods in the metal gun model. Development of these mechanisms remained a difficult, important undertaking until February 1944 when the hydride gun was abandoned.

Development of projectile and target designs was centered in the Projectile and Target Group. Other groups, notably the Proving Ground Group and the Metallurgy Groups concerned with heat-treating steel, contributed very active interest and vital assistance. All efforts concentrated on problems pertinent to a 3000 ft/s assembly velocity.

ARMING AND FUSING

The atomic weapons could not be armed and fused satisfactorily by straightforward application of the established art, partly because of the enormous investment represented by a single bomb. Triggering devices that fail on an average of only 1% of the time were hardly acceptable. However, the great value of the single bomb dwarfed the expense of multiple triggering by very fine equipment that would be forbidding in a more commonplace weapon. Also, the Los Alamos scientists thought the bomb should detonate many hundreds of feet from the ground. However, no fusing equipment had been developed explicitly for this purpose because the requirement was unique to the size of the explosion. Just how high above the ground the bomb should explode for maximum total damage was not known. This height was determined mathematically by extension of the theory of blast damage plus knowledge of the expected size of the explosion. None of this theory was available in April 1943.

Development of arming and fusing devices began in May 1943 in Group E-3 under R. B. Brode. The planned fusing system was the same for both the gun- and implosion-type bombs. It called for a performance guarantee that allowed less than one chance in ten thousand of failure to fire within about 100 feet of the desired altitude. Two general lines of development were started. One centered on possible use of barometric switches to fire the bomb; the other, on how to adapt the newly developed electronic techniques to the fusing problem. The radio proximity fuse, radio altimeters, and tail warning devices performed, in some measure, the desired function of detecting distant objects, and their suitability for use in a bomb had to be determined. Clocks were considered as the third, but unlikely, possibility because they would require careful setting just before the bombing run when the chances for human error would be great.

The barometric switches were simple mechanical devices, whereas the various electronic systems were highly complex. However, it was not certain that a reliable barometric indication could be obtained in a falling bomb. Therefore, the group had not only to design a sensitive, sturdy barometric switch that could be put into production, but also to prove these switches in action. The latter effort was the most extensive. It was necessary to fit model bombs with radio transmitters (informers) whose signals were modulated by the action of the barometer, to drop these from airplanes, and to follow the bomb flight photographically, as well as with the radio receiver. With proper timing cross checks, this procedure led to correlation of the barometer recording with the actual elevation. This work began on a small scale in December in cooperation with the Naval Proving Ground at Dahlgren. In March 1944, full-scale bombs were dropped from full height in tests at Muroc. By then, however, it seemed probable that the pressure distribution on the bomb surface was not so sensitive to absolute elevation as would be desired and that barometric firing should be used only if the electronic devices could not be developed. Therefore, barometric switch development became secondary, but the field experience gained in testing these units was of primary importance to development of the weapon. It was through this early effort and cooperation with the Instrumentation group (E-2) that the problems of instrumentation in the field, liaison with the Air Forces, and operations at distant air bases were solved and reduced to the routine so necessary for successful proof of the completed bombs (completed, that is, except for active material).

Before February 1944, the height above ground at which the bomb should be fired was still uncertain, but 150 ft was estimated. At that height the amplitude-operated radio proximity fuse was feasible. Brode had had a major part in developing these fuses for projectiles and immediately began adapting them to the bombs. Setting up a radio proximity fuse laboratory at Los Alamos would involve a large increase of personnel. Accordingly, design development, manufacture, and tests of radio proximity fuses and "informers" were undertaken in liaison with Section T, OSRD. The work was to be done at the University of Michigan under the supervision of H. R. Crane. Field tests were to be made at Dahlgren. This program, begun in the summer of 1943, was entering a major proof phase in February 1944, when theoretical work predicted that if the bomb were efficient enough, the best detonation height might be 3000 ft. Because the amplitude-operated sets would not function properly at such elevations, a new line of electronic development was immediately necessary. Liaison with the University of Michigan Laboratory was continued, however, to continue radio proximity fuse development in case the eventual burst height would be several hundred feet, to continue production of "informers," and to assist in the new lines of attack.

A new type of electronic device (the radio altimeter) was considered in February. It was decided to follow up both the frequency-modulated "AYD" and the pulse-type "718." AYD modification was assigned to the Norden Laboratories Corporation under an OSRD contract through Section T, with approval of the Navy Bureau of Ordnance. The 718 was just getting into production, but the development group at the Radio Corporation of America was still operative. It was intended to negotiate with RCA to use this group for fuse development, until it was learned that they also were developing a tail warning device for the Air Corps that might be adaptable to the fusing problem. Production of these "APS/13" units was just being started, but through the cooperation of the Signal Corps, a substantial part of the pilot production was made available for Brode's work. The third such unit to be made was delivered to Los Alamos in April. It was tested in May by diving an AT-11 plane, and the results were very encouraging. Two full-scale drop tests in June strengthened the conviction that the APS/13, now nicknamed "Archie," was the answer to the electronic fusing problem. The modified AYD persisted in showing difficulties that discouraged its use, even below 1000 ft where it had worked. More and more extensive field tests through June and July included final work on barometric devices and preliminary study of the electronic sets.

120

Concurrent design research and proof of service units involved establishing vibration and temperature tests for clocks, switches, batteries, and electronic equipment in more or less standard procedures for acceptance of airborne equipment. In April 1944, overall design of the arming and fusing system was undertaken. This system included pull switches, banks of clocks and barometric switches for arming, and four modified APS/13 units operating independently to initiate the firing circuit at the desired altitude. This system was selected by preliminary field tests and elimination of as many uncertain elements as possible. Field proof of the entire system began in August 1944 and was a major part of the work the following year. The scarcity of tail warning units forbade their wholesale use in bomb drops. Accordingly, modification development was assisted by using barrage ballons for testing the units (as assembled in models of the bombs) at Warren Grove, New Jersey.

Attention also was given to underwater detonation 1 minute after impact with the surface. This program was hardly under way, however, when model tests predicted that shallow underwater delivery was ineffective. Full attention was then given to the air blast bomb, for which a propeller-activated arming switch also was developed but was discarded as mechanically unreliable in case of icing or misalignment. The only propeller arming actually used was in the four standard Navy nose-impact bomb fuses A.N. 219 in the forward end of the Fat Man. They were to provide good self-destruction (at least) if the primary fusing system failed.

ENGINEERING

Primary responsibility for integrating the weapon was assigned by setting up an Engineering Group, E-6. Competent design engineers would reduce the performance specifications to accepted fabrication practice. These specifications would include the mechanical construction, ballistic and aerodynamic behavior, electrical wiring and incorporation of the arming and fusing equipment, and special provisions for handling the assembly mechanism and active material. The group also would provide design service for the various experimental ordnance programs, procure special materials and shop services from outside industry, and supervise the Ordnance Machine Shop (C Shop).

121

Design group operation actually started about the time the ordnance building (A building) was completed. Because research was just getting started, there was little idea of what the weapon specifications would be, except for aerodynamic performance. Therefore, the primary Design Group activity was designing bomb models and procuring dummy bombs for test drops.

The secondary responsibilities for C Shop and for outside procurement of materials and machine work increased daily. Demands for shop work were perpetually overloading the Los Alamos facilities despite the procurement services in Detroit. Manufacture of hemispheres for implosion experimentation was one of the great problems, so they were procured from Detroit shops through the Detroit office and, in June 1944, from the Los Angeles area.

Although detailed specifications for the weapon were lacking, certain preliminary designs of mechanical and high-explosives assembly, fuse assembly, and molds for charges were made during the first winter. The gun model details were less tentative. Details of the implosion model were expected to become more definite with time, but as experimental information increased, they grew less definite and more complex, with increasing alternatives and additional elements. The picture was changing so rapidly and design contributions sprang from so many Laboratory divisions, that the original organization for engineering development was rapidly becoming inadequate.

It was evident that the level of coordination needed for making a weapon of the implosion system exceeded that represented by a single operational group. Esterline pointed this out when he resigned as group leader in April 1944. His successor, R. Cornog, made every effort to rectify the lack of coordination. However, other organizational plans eventually led to formation of the Weapons Committee (Chap. 9) in March 1945. Meanwhile, engineering problems on the internal structure of the implosion bomb were taken over by a new group under K. T. Bainbridge. This group, formed in March 1944, combined design, supervised by R. W. Henderson, and experimentation on the explosion system. One of the group's responsibilities was the full-scale test of the bomb, covered in Chap. 18. This reorganization of engineering efforts brought implosion design and implosion research together. Work on the gun and external bomb assembly continued in the Engineering Group with close cooperation by the Explosives Development Group.

The new group detailed developments for boosters, Primacord branching, and detonation systems in the light of current research on these components. However, there was little activity on the active core and tamper designs because research was still in the differential stage and there was no acceptable plan for disturbing the active material. The active design work and coordination with experiments on the nuclear physics of the bomb were done in G Division after the August 1944 reorganization (Chap. 15).

Another development that needed special organization was design and manufacture of lens molds for high explosives. Here again, the need for coordinating theory, experimentation, casting practice, design, machining, and procurement went beyond the scope of any one operating group. Accordingly, a Molds Committee was formed in the summer of 1944 and a Mold Design Section was organized under V-Shop administration.

The engineering difficulties were not all related to the persistent uncertainty of final specifications. One basic problem was the developmental character of almost all Los Alamos engineering. Such engineering requires flexibility in meeting the constant change of specifications incident to new experimentation and settling general design principles as a framework within which more detailed specifications may later be fitted. In production engineering, however, emphasis is all on the details of design and problems of tooling and mass production. It was the misfortune of Los Alamos and its engineers that they were drawn primarily from industry and were accustomed to larger, less complex operations than they found here. With differing degrees of directness in different cases, this difficulty was responsible for the large turnover of the engineering staff. This problem was solved by a type of coordination unfamiliar to production plants.

The combination of design and service functions within the Engineering Group reflected the inappropriateness of production methods to a research and development laboratory. The degree of procedural formality necessary in preparing detailed drawings for mass production is unnecessary and burdensome, if applied in development work. Operation on this basis was a frequent source of difficulty, and, by overloading the group with service problems, tended to impair its principal function.

The security policy often was blamed for misunderstandings about details of machine work procured from outside shops. Liaisons were generally so roundabout that they easily led to difficulties. The isolation of Los Alamos, over 40 miles from a railroad, also contributed to delay, particularly in handling heavy equipment. As the project approached its final phases, handling and working of full-scale targets, bombs, guns, and high-explosive systems became a major part of the work. This required not only the equipment for handling the material but the plants and tools for making and assembling the objects themselves. Because provision for this heavy work was incidental to more obvious achievements, it was easy to overlook the important part played by making these provisions in the allotted time—on top of an isolated mountain.

IMPLOSION RESEARCH

Implosion research grew from being the concern of one small group into the major problem of the Laboratory, which occupied the attention of two full divisions. The program was started during the conferences of April 1943 with Neddermeyer's specific proposals (Chap. 1). Neddermeyer had developed an elementary theory of high-explosives assembly, but there was no established art that could be applied to even part of the mechanical problem. Implosion research differed from the gun research for which many mechanical and engineering features and methods of proof were at least relatively standard. Coupled with this undeveloped state of execution was a backwardness in the art of conception. As a result, no well-dated chronology of the appearance of "ideas" can be made. Development was in the form of a spiral. Rough conceptions that appeared early were reintroduced later with greater concreteness in altered contexts. For example, possible electric detonation and the conceivable benefits of compressing active material were considered in the spring and summer of 1943, but because of the lack of development very little could be done.

Another factor that affected the implosion program was that it began as a dark horse and did not immediately win Laboratory support commensurate with the difficulties that had to be overcome.

After the April conferences, Neddermeyer visited the Explosives Research Laboratory at Bruceton to acquaint himself with the experimental techniques applied to the study of high explosives. Certain types of equipment and installations used at Bruceton were considered desirable for the early implosion work, and plans were made for including them at the Anchor Ranch Proving Ground. While at Bruceton, Neddermeyer had his first implosion test fired and was encouraged by the result.

124

The need for personnel experienced in handling and experimenting with high explosives became urgent and, because of the Laboratory's continuing expansion, remained unmet. There was enough general experience in the Implosion Group, however, to start a firing program as soon as the first explosive arrived. The first Los Alamos implosion tests, made in an arroyo on the mesa just south of the Laboratory on July 4, 1943, used tamped TNT surrounding hollow steel cylinders.

All the early research on implosion, before installation of more elaborate techniques, was on recovery of the imploded object from small imploding charges. In these "critical" implosions, the charge-to-linear mass ratio was chosen to just bring the hollow liner into a solid configuration. Because of the simplicity of the setup, these tests were done mostly with hollow cylinders, although spheres also were used. The principal questions were about symmetry and velocity of assembly. In using shells of active material, the symmetry problem was critically important. Early tests indicated that the problem could be solved and possibly would become easier to solve as the mass ratio was increased. The velocities obtained were admittedly very low, however. These implosions gave a quantitative advantage over the gun method because of higher velocity and shorter assembly path. Use of larger amounts of explosive prevented recovery, and results had to await development of experimental techniques. Therefore, the first part of the program was limited to what is called, by contrast with later developments, the slow implosion.

Interest in implosion remained secondary to that in the gun assembly. Using larger amounts of explosive to increase the velocity also was considered, but the impossibility of recovery and the incomplete instrumentation kept such things in the "idea" stage for several months. The decisive change occurred when J. von Neumann visited in the fall of 1943. He had experience in use of shaped charges for armor penetration. Von Neumann and Parsons first advocated a shaped charge assembly by which active material in the slug following the jet would be converted from a hollow cone to a sphere with a lower critical mass value. Von Neumann was soon persuaded, however, that focusing effects similar to those responsible for the high velocity of Monroe jets would operate within an imploding sphere.

It cannot be said that these radical proposals were unanticipated in the Laboratory. However, their qualitative peculiarities had not been grasped or urged as a decisive advantage. Until this time, implosion had been conceived of as a means of squeezing solid matter into a solid sphere, with quantitative advantages over the gun to be weighed against its much smaller certainty of success. In von Neumann's conception it developed definite qualitative advantages. One was no longer concerned with squeezing a solid, but with the hydrodynamical behavior of a "fluid" of active material under the influence of an intense converging shock wave.

The emphasis in implosion research thus changed rapidly from "critical" assemblies, in which the final object was intact, to the most violent action that could be produced with existing explosives. This placed a tremendous burden on the small Pentolite-casting plant that had just been completed and increased the severity of experimental aims because of the additional interest in compression. There was an immediate transition from the thought that the implosion problem could be solved by modest means to a situation for which there were no adequate experimental techniques. The following year was characterized by a succession of new ideas and applications for the experimental procedure and by a rapid expansion of facilities and personnel.

Another major consequence of the new emphasis was the planning and eventual realization of an adequate high-explosive production plant and research program. To develop these and generally assist the implosion dynamics research, the Laboratory acquired the consulting services of G. B. Kistiakowsky. In February 1944, Kistiakowsky joined the staff as Captain Parsons' deputy for implosion. In April, he assumed full direction of the rapidly increasing administrative problems.

The period preceding February 1944 was spent in vigorous development of experimental techniques. Recovery from weak charges was the most rapidly exploited technique because it required no elaborate instrumentation. The required spheres were obtained from Detroit and Los Angeles. Although the test conditions were admittedly far from those in a fast implosion, the recovery technique proved useful in interpreting the process and in revealing the importance of, and possible difficulties in obtaining, symmetry. By February, there was evidence of possible trouble from the interaction of detonation waves and the spread of detonation times in multipoint detonation. The solution was not considered outside the range of existing experimental techniques.

Other techniques were being perfected to the point where they would be quantitatively useful. The difficulties were principally the need to record events *inside* an explosive and to time them within an uncertainty of approximately microseconds. In November, a program for photo-

126

graphing the interior of imploding cylinders by high-explosive flash light (a method developed at Bruceton) was started. Some qualitative results were soon obtained, but the method was not refined and secondary blast effects were not eliminated until spring. Practically the same is true of the flash x-ray method of studying small spheres. The principal problem, precision of time correlation between the implosion and the x-ray discharge, was solved, in cooperation with the Ordnance Instrumentation Group, by extensively modifying the commercial x-ray machines. The Instrumentation Group also designed and built rotating-prism cameras using ultracentrifuge techniques. These were adapted to proving ground use, taking rapidly repeated photographs of cylindrical implosions in December and January, but they were never used effectively for their intended purpose. Much later, they were used successfully in lens investigations (Chap. 16). Also in December, field preparations were started for taking electronic records of objects imploded in a magnetic field. The first shot of this type, the "magnetic model," fired January 4, 1944, was encouraging. The magnetic method was designed to take advantage of the fact that motion of metal in a magnetic field alters the field. The inward motion of imploding metal would induce a current in a surrounding coil, and proper interpretation of this current would give information on the velocity and other characteristics of the implosion. Perfecting the electronic records delayed final proof of this method until spring.

Quantitative data from the x-ray, high-explosive flash, and rotating-prism camera techniques showed their usefulness for determining velocities and symmetry at small scale, but also indicated a need for controlled quality of high-explosive castings and boosting systems, as well as improved detonation simultaneity. Improving these services involved producing castings of uniform density and composition and instituting quality control, including x-ray examination and density measurement of charges. In view of the impending large-scale production of heavy charges, development began on methods of casting and examining such charges for controlled quality.

Although the Anchor Ranch range had been designed to accommodate both the gun and implosion programs, expansion of the implosion program soon crowded the facilities. Casting and detonating large charges required a large casting plant and several widely separated test sites. The casting plant, begun in the winter of 1943, included an office building, steam plant, casting house, facilities for trimming and shaping

127

high-explosive castings, and magazines for storing high-explosive and finished castings.

Administratively, this S Site (sawmill site) was one of the most difficult Laboratory undertakings. Finding men with experience in high-explosive casting, or even general experience in handling explosives, was almost impossible. Supervisory personnel were equally difficult to obtain. Almost the only available channel was the Army. The Site was staffed almost entirely by men in the SED. A few had appropriate industrial backgrounds, more were young soldiers with some scientific training, usually in chemistry, and the rest were relatively unskilled. Originally scheduled for April 1944, steady operation on a reasonable scale did not actually begin until August. Because of increasing demand and the unavoidable lag in S-Site expansion, it was early 1945 before the small, original Anchor Ranch casting room was fully replaced as a source of experimental charges.

The first new experimental method adapted successfully to work at larger scale was the high-explosive flash technique, used on both cylinders and hemispheres. Extending flash x-ray methods to larger scale and using a grid of small ion chambers instead of photographic recording began in the fall of 1944 (Chap. 15). The radiolanthanum (RaLa) method, that of including a strong source of gamma radiation in the imploding sphere and measuring the transmitted intensity as a function of time, was being discussed, and electronic instrumentation and preparation for handling the highly radioactive radiolanthanum were under way. Possible use of the betatron to produce penetrating radiation for work with large spheres was discussed, but the technique was not used until later (Chap. 15). A full-scale active test in a closed vessel was considered, model tests were started, and procurement was investigated. Use of such a containing vessel would permit recovery of active material, in case of failure. This later grew into the "Jumbo" program, which, with other engineering problems, was centralized under the Ordnance Instrumentation Group and later under a special group (Chap. 16).

Preparations for an implosion test with active material began in March 1944. The main problems were choice of a test site, investigation of methods for recovery of active material if the nuclear explosion failed, and design of instruments to measure blast effects and nuclear effects of a successful explosion (Chap. 18). Areas in New Mexico, Colorado, Arizona, Utah, and California, as well as several islands off the coasts of California and Texas, were considered for testing sites. Nearness to Los

Alamos had to be weighed against possible biological effects, even in sparsely populated areas of the Southwest. Before August 1944, most exploration was done by map, automobile, and plane, and still more was planned. Investigation of the recovery problems was finally begun. Recovery from a large containing vessel strong enough to withstand the shock of the high explosive was the principal means considered. The possibility of setting off the bomb inside a large sand pile that would prevent dispersal and permit recovery of active materal in case of failure also was investigated. Other methods were investigated, but during this period the main problem was to design "Jumbo," the containing vessel.

The main activities of spring and early summer were developing new methods, such as RaLa, the magnetic method, and the counter x-ray method, which should be useful with larger scale implosions; increasing production and quality control of case explosives and detonation trains; increasing investigation of detonation simultaneity, including preliminary investigation of electrical detonation systems; and exploiting techniques for implosion study established during the winter. Implosion studies started giving regular, reliable results during the summer of 1944 and, because they were concerned with end results, received great attention from the rest of the Laboratory, particularly the Theoretical Division. In fact, this early work laid the foundation of a new branch of dynamics, the physics of implosion. It had been firmly established that the earlier results on implosion velocities were essentially correct and lower than theory predicted for normal impact by a detonation wave. The dynamics of confluent materials had also been investigated thoroughly.

In July 1944, the Laboratory faced the fact that the gun method could not be used, but at that time not one experimental result gave good reason to believe that a plutonium bomb could be made. There was, however, a large investment in plants and proving grounds and a wide background of experience in improving explosives and timing, which made it possible to launch an even more ambitious investigation of the implosion. The new development was to be centered on possible use of explosive "lenses" that could be designed to convert a multiple-point detonation into a converging spherical detonation wave, thereby eliminating troublesome interaction lines. Preliminary studies of such systems had been made in England and at Bruceton, and adapting them to the implosion problem became the principal objective of the implosion groups. Experimental lens-mold design, the most difficult initial step,

129

occupied several months. Meanwhile, the effort to eliminate the interaction jets from nonlens implosion continued.

DELIVERY

The decision to study and use explosive lenses required more research on both the explosive elements and the mechanical systems. Furthermore, eventual production of these special explosive systems had to be provided for. To be prepared to use the first quantities of plutonium, all this had to be accomplished within a year. Most of the Laboratory was accordingly reorganized to concentrate manpower and facilities on the implosion problem.

Division X was formed, under Kistiakowsky, to experiment with explosive systems and their fabrication and to set up an adequate production system for all special charges (Chap. 15). The more or less established methods of implosion investigation, such as the small-scale x-ray and the high-explosive flash methods were also put under Division X. Development of new, or as yet unproved, techniques for investigating implosion dynamics and the responsibility for design development of the active core of the implosion were made the objectives of G Division under Bacher (Chap. 15). The urgency of these divisions' directives is appreciated, because there was no approved design for either explosives or mechanical systems at the time.

The Delivery Group worked on everything from the completed bomb to its final use as a practical airborne military weapon, so there was an overlap between their responsibilities and those of the groups engaged in final bomb design, particularly the Fuse Development and Engineering Groups. However, the Delivery Group's special responsibility was to see that cooperating groups functioned smoothly as a team with an eye to their eventual collaboration in combat delivery. They also were responsible for liaison with the Air Forces' activities, including choosing and modifying aircraft and supervising field tests with dummy bombs. These activities later included planning and establishing the advance base where the bombs were assembled, assembling and loading the bombs, and testing and arming them in fight (Chap. 19).

The delivery program began in June 1943 when N. F. Ramsey (still working with the Air Forces) investigated the bomb-carrying capacity of available planes. At that time, the main possible weapon was the

130

plutonium 3000-ft/s gun model with a 17-ft overall length. The only plane to meet this requirement was the B-29, which, even so, could carry the bomb only by joining the bomb bays. Possible wing-carriage by other planes had been considered and rejected. It appeared that there might be difficulty in obtaining a test B-29, so the British Lancaster, also capable of carrying the bomb, was investigated. In terms of maintenance standardization, however, the British Lancaster would have been difficult to operate from American bases. Therefore, it was decided that the B-29 must be used.

Preliminary ballistic tests were made in August 1943 at the NPG at Dahlgren, ostensibly for the Air Corps. The dummies dropped were scale models (14/23) of the "Long Thin Man," the 3000-ft/s plutonium gun. These models, consisting of a 14-in.-long pipe welded into the middle of a split standard 500-lb bomb, showed extremely bad flight characteristics. In subsequent months, further tests were made at Dahlgren and much better flight characteristics were developed. Preliminary models of a proximity fuse developed at the University of Michigan also were tested.

During Ramsey's first visit to Los Alamos in September 1943, implosion was just being urged by von Neumann. Preliminary estimates of a 9000-lb, 59-in.-diameter bomb were made. The Bureau of Standards bomb group was asked, through the Bureau of Ordnance, to make wind tunnel tests to determine the proper fairing and stabilizing fins for such a bomb.

In the fall of 1943, plans for full-scale tests began. For the B-29 modification, two external shapes and weights were selected as representative of current plans at Los Alamos. These were, respectively, 204 and 111 inches long and 23 and 59 inches in diameter, the "Thin Man" (gun) and "Fat Man" (implosion). In November 1943, Ramsey and General Groves met with Colonel R. C. Wilson of the Army Air Force to discuss plans for the modified B-29. In December, the first full-scale models were ordered through the Detroit Office, and Ramsey and Captain Parsons visited the Muroc Air Base to make necessary test station plans.

Tests began at Muroc early in March 1944 to determine the suitability of the fusing equipment, the stability and ballistic characteristics of the bombs, and the functioning of the aircraft and bomb release mechanism. The Thin Man model proved stable, whereas the Fat Man was underdamped, which caused violent yaw and rotation. The Michigan proximity fuses failed almost completely. The release mechanism proved inadequate for the Thin Man. Four models "hung up" with delays of

131

several seconds. The last model tested released itself prematurely while the plane was still climbing for altitude. It dropped onto the bomb bay doors, which had to be opened to release it, and seriously damaged them. This accident ended the test until the plane was repaired and the bomb-release mechanism revised.

Tests at Muroc were not resumed until June. The interval was devoted to analyzing the Muroc data; planning a functional mockup of the plane and bomb suspension for handling, loading, shaking, and cold tests; and constructing V Site for investigation of the possible need for heating equipment in the B-29 bomb bay.

Effort was devoted to design and procurement of 23/59-scale Fat Man models for possible B-29 flight tests. Negotiations began for use of the high-velocity Moffett Wind Tunnel for bomb ballistic experiments. Although stable and statistically reliable Fat Man models were subsequently designed without wind tunnel testing, only ballistic experiments under controlled conditions would have yielded definite information on the safety factors involved. This testing was difficult to arrange, and it finally was deemed unnecessary.

Two new bomb models were designed. One was the case for the 1000-ft/s ^{235}U gun assembly, which by contrast with the much larger Thin Man was code-named "Little Boy." It was shorter so that it could be carried in a single B-29 bomb bay. When the plutonium gun assembly was abandoned in midsummer, it became unnecessary to join the bomb bays. The second model was the "1222" Fat Man, 12 pentagonal sections of Dural bolted together to form a sphere surrounded by an armor steel shell icosahedral structure with a stabilizing shell attached. Mechanical assembly of this device required about 1500 bolts.

The June Muroc tests showed that although the Fat Man model tested still had unsatisfactory flight characteristics, field modifications suggested by Captain David Semple, USAAF, to increase the drag gave a stable model. This modification involved welding angularly disposed trapezoidal drag plates into the box tail of the bomb. No release failures occurred, and the fusing mechanisms proved promising.

When by midsummer it was certain that the plutonium gun assembly would not be used, only Little Boy and Fat Man remained. A new Fat Man, model "1561," was developed to improve flight characteristics and simplify mechanical assembly. It was a spherical shell of two polar caps and five equatorial segments machined from Dural castings. The assembled sphere was enveloped by an ellipsoidal shell of armor attached at the equator. The tail was bolted to the ellipsoid. The electrical detonating and fusing equipment was mounted between the sphere and the outer ellipsoid.

CHAPTER 8

Chemistry and Metallurgy
May 1943 to August 1944

INTRODUCTION

The basic problems of the Chemistry and Metallurgy Division were purification and fabrication of active, tamper, and initiator materials for the bomb and several service activities for the rest of the Laboratory, which were largely determined rather than determining. This was true, not because the work was routine or subordinate, but because it was successful. The record of the chemists and metallurgists is one of wide-ranging exploration of techniques combined with extraordinary cleverness in meeting or avoiding technical problems, sometimes on short notice.

Until April 1944, the Chemistry and Metallurgy Division had only a loose structure, with Purification, Radiochemistry, Analysis, and Metallurgy groups headed, respectively, by C. S. Garner, R. W. Dodson, S. I. Weissman, and C. S. Smith. Then the division was extensively reorganized. J. W. Kennedy, Acting Division Leader from the beginning, became Division Leader, and C. S. Smith became Associate Division Leader in charge of metallurgy. The following groups were created.

CM-1	Health and Safety, Special Services	R. H. Dunlap
CM-2	Heat Treating and Metallography	F. Stroke
CM-3	Gas Tamper and Gas Liquefaction	E. A. Long
CM-4	Radiochemistry	R. W. Dodson
CM-5	Uranium and Plutonium Purification	C. S. Garner
CM-6	High-Vacuum Research	S. I. Weissman
CM-7	Miscellaneous Metallurgy	C. C. Balke
CM-8	Uranium and Plutonium Metallurgy	E. R. Jette
CM-9	Analysis	H. A. Potratz
CM-10	Recovery	R. B. Duffield

Group CM-11, formed in June 1944 under A. U. Seybolt, was concerned with uranium metallurgy. This division's work could not be defined completely until the division of labor between Los Alamos and other Manhattan laboratories was decided. The metallurgy program, however, was clear from the beginning, as was the need to set up analytical methods for refereeing all questions of chemical purity, whether purification occurred here or elsewhere. Special service functions included preparing thin-film targets of various materials, purifying thorium for threshold fission detectors, and making metal parts for apparatus.

The special Reviewing Committee (Chap. 1) had favored locating the purification work at Los Alamos, and in May 1943, the necessary planning was undertaken. Headquarters would be at Los Alamos, and facilities would include a large dust-free laboratory building. When this building was adequately staffed, most purification research, and later all final purification, would be done at Los Alamos. Meanwhile, it would be done at the Metallurgical Laboratory, at the University of California at Berkeley, and at Iowa State College. A coordinator would be needed to establish proper demarcation between the work at this site and that at the others.

At the end of July, C. A. Thomas, Research Director of Monsanto Chemical Company, accepted the position. His job was not to coordinate the research programs of the various projects, but simply to establish communication among otherwise isolated laboratories and adjudicate their conflicting requirements for scarce materials. At about this time, the new building, designed by Brazier with the advice of Thomas and his staff, was erected. Although it was built of the same temporary materials as other Los Alamos buildings, it was both dustproof and air-conditioned. It was largely completed and staff members were moving in by December 1943.

Thomas immediately set up a program for extracting polonium, either from lead dioxide residues or from bismuth that could be irradiated in the piles at Clinton or Hanford. Research on the former problem was undertaken at the Monsanto Laboratories; that on the latter, at Berkeley.

In the investigation of plutonium chemistry as distinguished from purification proper, a Berkeley group provided information on the oxidation and valence states of plutonium, whereas the earliest reports on its density and crystal structure came from the Metallurgical Laboratory. The Metallurgical Laboratory measurements were made before investigations at Los Alamos definitely established that there was more than one allotropic form of the metal. However, in February 1944, it was suggested that the difference in structure in barium- and calcium-reduced plutonium, reported by Chicago workers, might be caused by the existence of at least two such forms.

Both the Metallurgical Laboratory and Los Alamos worked on the bomb method of plutonium reduction and on development of spectrographic analysis methods for many elements, particularly the cupferron-chloroform extraction method with copper spark analysis. Work on the latter continued at Chicago, with the final development being done at Los Alamos.

Thomas arranged that the Metallurgical Laboratory should be primarily responsible for procuring reagents of much higher purity than those commercially obtainable, and refractories for use by the many metallurgical groups. Securing an adequate supply of satisfactory refractories became increasingly important as Los Alamos metallurgy expanded. Difficulties were magnified because the initial procurement arrangements were not satisfactory. Under Thomas' auspices, however, arrangements for developing and producing these refractories were initiated in January 1944, and eventually it was decided that a group under F. H. Norton at the Massachusetts Institute of Technology would undertake the research problems. The University of California would study the use of cerium sulfide. Cerium metal was produced at Iowa State College, the bulk of it going to M. I. T. Subsidiary work was also done at Brown University.

Despite careful liaison, Los Alamos metallurgy was sometimes delayed because of the time lag between changes in requirements for refractories and corresponding changes in output of the fabrication groups at other sites. To overcome this time lag, the local refractory research group was enlarged in April 1944 to produce standard refractories. In June, it was decided to send the Berkeley, Ames, and M.I.T. output to Los Alamos to help meet their sharp rise in demand for refractories. Despite all these efforts, procuring enough of the proper types of refractories continued to be a problem.

With the discovery of ^{240}Pu, there was no further need to coordinate purification work. It had become clear that ^{239}Pu could be chemically purified, although with great difficulty. The division of labor among the various sites, moreover, was well worked out.

The chemistry of ^{235}U and its attendant liaison presented much simpler questions than had plutonium. Los Alamos had two main problems to examine, processing of the tetrafluoride for experiments and for weapon production, and problems concerning the Water Boiler, such as decontamination of solutions. Tennessee Eastman at Oak Ridge undertook purification of ^{235}U to the tolerance limits specified by the Los Alamos Laboratory. Los Alamos chemists wanted to know what processing the material had undergone before shipment and the nature of the analysis done at Oak Ridge. They also specified the chemical form in which the material was to be shipped, such as sulfate, nitrate, or tetrafluoride. Other questions concerned isotopic concentration mixing of lots with different concentrations, and assay methods. Los Alamos and the Clinton Laboratories at Oak Ridge cooperated on production of radiobarium-radiolanthanum for the implosion studies (Chap. 17). During the Water Boiler work, particularly decontamination of Water Boiler solutions, Los Alamos relied heavily on the corrosion experts at the Metallurgical Laboratory and at Clinton, and DuPont assisted in obtaining stainless steel for the apparatus.

Scheduling Los Alamos work, particularly purification, involved increased personnel. From a group of about 20 in June 1943, the Chemistry and Metallurgy Division grew to about 400 in 1945. Personnel procurement was difficult because many of the most suitable men were employed in other branches of the project. In the absence of an overall supervisor whose decision about allocating these men would be binding, the difficulties became almost insurmountable. The inadequacy of the metallurgical staff was particularly serious because metallurgical work for ordnance experimentation could not be done elsewhere.

From the completion of the chemistry building in December 1943 to April 1944, about 20 men came to Los Alamos from Berkeley, Chicago, and Ames, where they had been doing research on purification. Four men came from the California Institute of Technology.

URANIUM PURIFICATION

Because, in terms of the gun assembly method for producing a large-scale explosion, ^{235}U purity requirements were three orders of magnitude less exacting than those for plutonium, the chemists concentrated on the more difficult problem. Furthermore, it seemed entirely possible that a uranium purification procedure might be merely a byproduct of that for plutonium.

It was primarily the role of uranium as a stand-in for plutonium that led to the first work on uranium purification. In December 1943, the very exacting microchemical investigations of plutonium purification were curtailed because gram amounts of plutonium from the Clinton pile were expected within 2 or 3 months. It was believed that microchemical experience with such a stand-in as uranium would be more useful. This work, by the Uranium and Plutonium Purification Group in cooperation with the metallurgists, was aimed at plutonium, rather than uranium, standards of purity.

Several methods of uranium purification were investigated during the stand-in work. All used a series of "wet" and "dry" chemistry steps. The original procedure provided a carbonate precipitation (with ammonium carbonate), a diuranate precipitation (with ammonium hydroxide), and a +6 oxalate precipitation as the "wet" purification steps. Igniting the resulting uranyl oxalate to the oxide U_3O_8, reduction to UO_2 with hydrogen, and conversion to the tetrafluoride by heating in the presence of hydrogen fluoride constituted the "dry" part of the process.

The Purification Group (CM-5) studied this basic procedure until about August 1944. However, early that year, the Radiochemistry Group (CM-4) also studied uranium purification while supplying enriched uranium for isotopic analysis. The wet purification procedure was departed in many respects, such as a peroxide precipitation step, precipitation of the acetone-sulfate complex of uranium, use of electrolytic methods, and the +4 oxalate precipitation. The success of these variations was overshadowed by considerations of large-scale production. Therefore, although the peroxide precipitation step gave excellent results, the bulkiness of the precipitate militated against using that step in large-scale operations. However, ether extraction of uranyl nitrate plus nitric acid, also studied during this work, later came into extensive use.

URANIUM METALLURGY

Hydrides

The first uranium metallurgy at Los Alamos was preparation and powder metallurgy of its hydride. Spedding's group at Ames had produced this compound. The possibility of large-scale, controlled

138

production was learned of at Los Alamos in April 1943. Use of the hydride in a bomb still was being considered seriously (Chap. 4), so metallurgical investigations of uranium hydride were in order. Early literature identified the compound as UH_4, but the chemists found that UH_3 was closer to the true formula.

The metallurgical work was modified by bomb requirements so that ways to produce high-density hydride and eliminate the pyrophoric characteristic became important. Compacting the hydride by cold-pressing and hot-pressing was attempted, as well as hydride formation under high pressures applied externally to the massive material. This work led to establishment of many control factors in the hydride formation process.

Work on the pressure bomb method of producing high-density hydride compacts was curtailed when uranium-plastic compacts were formed. Research began in February 1944 on preparing compacts in various geometric shapes in which the hydrogen-to-uranium ratio varied. This was readily done by using uranium powder and a suitable hydrogenous binding agent. It was also possible to largely eliminate use of the hydride and thus reduce the number of fires. Early in this work, a half dozen small fires a week were not unusual. The plastic bonding agents used included methyl methacrylate, polyethylene, and polystyrene. Compacts made with uranium hydrogen compositions corresponding to UH_3, UH_4, UH_6, UH_{10}, and UH_{30} were used by the physicists.

Uranium Reduction

Preparation of high-purity uranium metal was undertaken to develop small-scale methods (0.5 to 1000 g) for application to enriched uranium metal when it became available, and to use uranium as a stand-in for developing reduction techniques that might be applied to plutonium. Electrolytic and metallothermic processes were investigated. The latter was divided into the centrifuge method and the so-called stationary bomb method.

Electrolytic Process. The only successful electrolytic uranium-reduction process at this time was the Westinghouse process that used UF_4 and UO_2 dissolved in a fused mixture of sodium and calcium chlorides.

139

The product was a fine powder that contained oxide and, therefore, required washing, pressing, and melting for purification. These steps involved excessive losses, considering the value of the enriched uranium to be produced. Investigation showed that uranium could be deposited above its melting point from solutions of UF_4 in fluorides and chlorides, but that the purity was likely to be low at high temperatures. Accordingly, the most extensive investigations were limited to lower temperatures and, to simplify the container problem, to electrolytes containing no fluorides. Electrolytic methods were developed for preparing 50 mg and 200 to 300 g of uranium metal. The electrolyte was 25 to 30% uranium trichloride in a solvent containing 48 wt% $BaCl_2$, 31 wt% KCl, and 21 wt% NaCl. The operating temperature was ~630°C. The high-purity metal produced was in the form of dendrites that could be pressed and melted into one coherent piece. Recovery yields were 40 to 70% from the small-scale method and 80 to 90% from the larger.

Metallothermic Reduction Methods

The centrifuge method was to reduce uranium to 50 mg to 1 g of metal by taking advantage of the increased g-value for collecting small amounts of metal. The method reduced a uranium halide with either calcium or lithium metal in a sealed bomb. The bomb was placed in a graphite rotor that was rotated while being heated in an induction coil. Successful reductions were made using $UF_4 + Ca + I_2$, $UF_4 + Li$, $UF_3 + Li$, and UCl_3 + Li, Ca, or Ba. The metal produced using the first mixture was brittle and contained much slag. That produced using the other mixtures was malleable but usually contained some slag, increasing the impurities. Very good yields were obtained in all cases.

When work on the stationary bomb began in August 1943, only the large-scale (25 pounds of metal) reduction technique developed at Iowa State College and the possible use of iodine as a booster were known. The large-scale method was not applicable to small-scale work, where high yields and high purity were needed. Problems included developing refractory crucibles for the reaction, designing suitable bombs, investigating raw materials, and developing techniques for each scale of reduction studied. Methods for handling the very valuable enriched uranium without loss also were worked out. Successful bomb techniques were developed for reducing uranium tetrafluoride and uranium trichloride with calcium metal on the scales of 0.5, 1, 10, 25, 250, 500, and 1000 g of metal. Most of the work was on the tetrafluoride because of the hygroscopic nature and more difficult preparation of the trichloride. Experiments on the 10-g scale also showed that UI_4 could be reduced with calcium metal by the same procedures used for the fluoride and chloride. Argon was used as an inert atmosphere in the bomb. The amount of iodine used and the heating cycles varied with each reduction scale. Magnesium oxide crucibles were the most satisfactory. Methods developed for preparing the several types of MgO crucibles were later used by M.I.T. for routine preparation of the large-scale crucibles.

Uranium Alloys

Alloys were sought that would have better fabrication properties than the unalloyed metal. In November 1943, intensive work began on preparing alloys in various percentage compositions. Mixtures of uranium with molybdenum, zirconium, columbium, and rhenium indicated that many desirable properties could be produced. In particular, the uranium-molybdenum system had a much higher yield strength than pure uranium. However, in the fall of 1944 most of the alloy research was dropped because ordinary uranium was found to have adequate physical properties.

PLUTONIUM PURIFICATION

Plutonium purification, as distinguished from the more general chemistry of plutonium, was primarily the work of the Uranium and Plutonium Purification Group. Early in October 1943, the first small quantities of plutonium arrived at the Laboratory and intensive work on stand-ins was initiated to permit the Purification Group to determine and improve their techniques. The stand-ins included uranium, cerium, lanthanum, zirconium, and thorium. Uranium was used principally to investigate ether extraction methods; thorium and cerium were used to test solubilities and purification by various precipitations. In metallurgical work, cerium trichloride was used as a stand-in for plutonium trichloride, and, in other small-scale work, cerium tetrafluoride was used as a stand-in for the corresponding plutonium salt.

141

Plutonium purification developed in three parts: wet processing, dry processing, and, because of the cost of the material, recovery. The wet chemistry procedure was developed first. By August 1944, however, the dry process and recovery procedures had not been determined completely.

The Wet Process

The plutonium from Clinton and Hanford was received as a highly viscous mixture of decontaminated and partially purified nitrates consisting of about 50% +4 plutonium and 50% +6 plutonium. This material had to be dissolved out of its stainless steel shipping container and diluted, and a sample had to be removed for radioassay. This preliminary work required 3 or 4 days and permitted further processing by wet purification. Many purification procedures were investigated. Early in 1944 the first tentative one involved a double precipitation of sodium plutonyl followed by a double extraction of ether using sodium nitrate as a salting-out agent. Potassium dichromate was originally used to go from +4 to +6 plutonium ion, but it was soon superseded by sodium bromate plus nitric acid. For selective reduction from +6 to +3 plutonium, hydrogen iodide or potassium iodide in acid solution was used throughout the work.

The first major difficulty was lack of a process for separating small amounts of uranium impurity from large amounts of plutonium. Various compounds such as carbonate, peroxide, fluoride, iodate, and oxalate were investigated. The iodate $Pu(IO_3)_4$ gave 99.5% removal in two precipitations with a selective reduction step, but it was extremely difficult to convert to sodium plutonyl acetate. Finally, precipitation of the +3 oxalate solved the problem and became an important part of all future processes.

The procedure gradually changed. Two oxalate precipitation steps were incorporated, and ether extraction and a plutonyl acetate step were dropped. By July 1944, completely enclosed 1- and 8-g apparatus were being set up, and the process known as the "A" process had taken form. It involved reduction to +3 oxidation state, oxalate precipitation, oxidation to +6 oxidation state, sodium plutonyl acetate precipitation, ether extraction from nitric acid and ammonium nitrate solution, reduction to +3 oxidation state, and a final oxalate precipitation. The

142

process gave yields of about 95%. The product was turned over to the dry chemists as an oxalate slurry, and the residue supernatants were returned for recovery. The reason for developing the enclosed apparatus was primarily the plutonium health hazard (Chap. 9).

The Dry Process

Because converting the wet oxalates to the dry halide of plutonium led to a product that the metallurgists had to reduce to metal, they collaborated to find a compound suitable for reduction and selected the tetrafluoride in July 1944. The preceding investigations covered most of the halides of plutonium (for calcium bomb reduction) and PuO_2 (for carbon reduction). Plutonous chloride and bromide were rejected because they were highly hygroscopic. Carbon reduction of the oxide also was dropped.

The tetrafluoride was prepared variously from the nitrate, oxalate, and oxide by using anhydrous hydrogen fluoride. The nitrate conversion was poor, and research was concentrated on converting the oxalate and oxide. The final choice of the oxide was made early in 1945, as were the final production methods (Chap. 17).

Plutonium Recovery

Except for the peroxide recovery method (Chap. 17), the Recovery Group developed all procedures before August 1944. Recovery was necessary from the supernatants of plutonium purification and from metallurgy liners and slags. From the supernatants, the procedure involved concentrating the plutonium and subsequent purification. Reduction was with sulfur dioxide followed by a precipitation with sodium hydroxide. Treatment with aluminum hydroxide carrier brought down further plutonium. About 1 mg/liter remained in solution, and these secondary supernatants were stored. Purification originally involved iodide reduction, oxalate precipitation, oxidation ether extraction, sodium plutonyl acetate precipitation, iodide reduction, and final oxalate precipitation. The purified product went directly to the dry chemists.

Early work on liners and slag showed that complete solution would be necessary for good recovery. The major difficulty was to remove iodide ion before solution. The first method was CCl_4 or sodium sulfite extraction of the I_2. Then liner and slag were dissolved in hydrochloric or

143

nitric acid, followed by $Pu(OH)_4$ precipitation from a solution almost saturated with ammonium nitrate. This precipitation was at a high enough hydrogen ion concentration to leave most of the magnesium from the magnesia liner in solution.

PLUTONIUM METALLURGY

In March 1944, the Uranium and Plutonium Metallurgy Group undertook the first bomb reductions of PuF_4 and $PuCl_3$. This was the first direct metallurgical work with plutonium. Emphasis was on the electrolytic method and chloride reduction.

Research on the physical properties of plutonium metal began in April 1944 and proved of major importance because of the metal's unique physical properties. By May 1944, metal yields were over 80% from a number of methods, and interest shifted to the stationary bomb reduction method. Remelting as a final purification step centered interest on refractories without light-element contaminants. This eliminated the usual refractory materials. Cerium sulfide was one of the principal materials investigated. Discrepancies in the density of various metal samples first hinted at plutonium allotropes. By June, metal obtained from $PuCl_3$ using calcium as a reducing agent and subsequent remelting in cerium sulfide crucibles came within an order of magnitude of meeting the prevailing purity specifications. The alpha and beta allotropes of plutonium were established, and the PuO_2 plus carbon reduction method developed in the High-Vacuum Research Group came into temporary prominence.

Two general methods of plutonium preparation, the electrolytic and metallothermic processes, were investigated. The latter was divided into the centrifuge method and the so-called stationary bomb method. Another method was also studied in which the oxide of plutonium was reduced and the metal distilled. The process finally adopted was reduction of the tetrafluoride in the stationary bomb using calcium metal as the reductant with iodine as a booster. The reasons for this selection were the same as for uranium.

Electrolytic Process

Electrolytic reduction of 1 g of metal to 50 mg of plutonium gave ~50% recovery yields. The bath was 24% $PuCl_3$ in a solvent containing

144

48 wt% $BaCl_2$, 31 wt% KCl, and 21 wt% NaCl. The metal was obtained in droplets that usually contained small amounts of the cathode element. With the discovery of ^{240}Pu, work on the electrolytic process stopped in favor of the metallothermic process. Details of the electrolytic process are given in LA-148, "Production of Plutonium by Electrolysis" by M. Kolodney.

Centrifuge Reduction Method

As it was for uranium, the centrifuge reduction method was used to prepare 50 mg of plutonium from 1 g metal by taking advantage of increased g-value for collecting small amounts of metal. Plutonium was reduced using $PuCl_3$ or PuF_4 with lithium as a reductant. Calcium reductions of these halides using iodine as a booster were not so successful. The plutonium metal prepared by the centrifuge method was the first prepared on any scale larger than a few micrograms. With development of the 0.5-g scale stationary bomb method, the centrifuge method was abandoned; however, it had served its purpose when it was needed most.

Stationary Bomb Method

Work on the stationary bomb problem began in March 1944, after much preliminary work with uranium as a stand-in. Like that of uranium, plutonium reduction involved development of refractory crucibles, design of suitable bombs, investigation of raw materials for the reaction, and development of techniques for each reduction scale. Cerium and lanthanum were also used as stand-in elements, and techniques were developed to prepare them from their chlorides and fluorides on all the scales given below. Plutonium chloride was used for the first successful reductions by the stationary bomb method, on the 0.5-, 1-, and 10-g scales. The fluoride was later found more satisfactory because of its nonhygroscopic nature and greater ease of preparation. Techniques were developed for reducing plutonium on the 0.5-, 1-, 10-, 25-, 160-, 320-, and 480-g scales. The average yields in a single button of clean metal ranged from 95% for 0.5 g to 99% for the 320-g and 480-g scales. The bromide was also reduced on the 1-g scale, but with lower yields. The methods of PuF_4 reduction developed here were used for routine production of pure plutonium metal.

145

The oxide reduction method involved reducing plutonium oxide with carbon or silicon and distilling the resulting metal onto a cold finger. This method yielded 30 to 90% spectroscopically pure plutonium on a 5-g scale. Discovery of ^{240}Pu called a halt to this ultrahigh-purity method. Remelting techniques also were investigated. Much of this work was on the choice of crucible materials. Remelting was important because of the need for metal with uniform physical properties and because it removed further impurities such as magnesium.

Extensive work on the physical properties of plutonium metal yielded inconsistent data from different metal samples. In July 1944, these inconsistencies were partially explained by proof that alpha and beta allotropes existed, with a transition from the room-temperature alpha to the beta phase between 100 and 150°C.

MISCELLANEOUS METALLURGY

Outstanding metallurgical work outside the narrow field of uranium and plutonium was done, principally by the Miscellaneous Metallurgy Group. Compacting of boron neutron-absorbers was undertaken in August 1943, as was development of beryllia tamper material for the Water Boiler. Formation of high-density beryllia bricks for this project was completed in February 1944. In May 1943, because of difficulties in obtaining refractories, magnesia liners and cerium sulfide crucibles were developed for the plutonium metallurgy program.

Boron Compacts

The remarkable properties of ^{10}B as a neutron absorber gave it several uses in the laboratory. Its potential importance was such that procurement was undertaken very early and studies of compacting methods were begun. The oxide, the carbide, and elemental boron were used as starting materials.

Beryllia Compacts

One of the Miscellaneous Metallurgy Group's accomplishments was development and production of high-density beryllia bricks for the Water Boiler tamper and scattering experiments. Beryllium metal would have been the best tamper material, but its use at that time would have virtually exhausted the country's supply.

Various methods of obtaining high density were tried, among them impregnation with magnesium fluoride, but the fluoride was undesirable from a nuclear physicist's point of view. Impregnation with beryllium nitrate followed by ignition proved rather poor. Although unusual for a refractory material, a hot-pressing technique finally was chosen.

Experimentally, the bricks were prepressed in a steel mold and then hot-pressed in graphite at 1700° C at about 1000-psi pressure for 5 to 20 minutes. Fifty-three bricks for the Water Boiler tamper were shaped to fit around the boiler's 12-1/16-in. sphere. The density averaged 2.76 g/cm^3.

Tungsten Carbide

Tungsten carbide fabrication was not emphasized until May 1944.

Crucible and Refractory Research

This important work was done to find materials that would not introduce contaminants into purified uranium and plutonium. Wetting, sticking, and thermal sensitivity also had to be considered. The many substances investigated included cerium sulfides, calcium oxide, magnesium oxide, tantalum, graphite, a tantalum-thorium nitride mixture, zirconium nitride, thorium sulfides, beryllia, uranium nitride, thoria, tungsten carbide, tantalum carbide, and titanium nitride. Effort was concentrated on trying to improve fabricated cerium sulfide's resistance to thermal shock—its main weakness.

RADIOCHEMISTRY

Foil Preparation

Preparation of thin foils for physical experiments was a Radio-chemistry Group service that involved much arduous and delicate work and continued research on methods. Foils of many different substances were made, with emphasis on the oxides of uranium and plutonium. Among other substances used were boron, protactinium, ^{235}U, ^{237}Np, and thorium.

147

The principal methods of foil preparation were evaporation, electrodeposition, and the "lacquer" method, in which an alcohol metal salt solution was mixed with a nitrocellulose lacquer, spread in a thin film, and ignited to oxide.

Boric oxide foils were prepared by the lacquer method. Aluminum boride foils were prepared by heating aluminum foils in boron trifluoride. Boron was deposited on tantalum and tungsten foils by thermal decomposition of diborane. The boron work was difficult because it required very thin foils of accurately known mass. The method was developed by Horace Russell, Jr. Thorium, uranium, and plutonium oxides were deposited by the lacquer technique and by electrolytic methods.

The virtuosity of the chemists engaged in this work was remarkable. They produced large numbers of foils that accurately met the physicists' specifications, including unusual geometries. Often the data the chemists supplied with the foils were as important in interpreting physical experiments as any of the physical measurements made with them.

Chemistry of Initiators

It was assumed from the beginning that a neutron initiator would be used with the bomb to provide a strong neutron source that would operate at the instant of optimal assembly. Naturally, the first initiators were designed for the gun. The type of initiator, if any, to be used with the implosion was not settled until after August 1944. The principal mechanism adopted was properly timed mixing of alpha-radioactive material with a substance that would support the (α,n) reaction. In practice, polonium and beryllium were used as the mating substances because the yield and energy were high. The radiochemists' problem was purification of polonium to eliminate light elements that support the same reaction and would, if left in the polonium, increase the neutron background in the bomb above tolerance. It was not certain that sufficient purification would be possible. Early discussions of the alleged "Joliot effect," spontaneous neutrons associated with alpha emission, led the Radiochemistry and Radioactivity Groups into a program of investigation. It was guessed that the neutrons that Joliot observed came from light-element impurities in his polonium sample. Because the radiochemists did not know how to purify this material, they decided to use radon, another alpha emitter, that they could purify. A radon plant was constructed early in 1944, and 2 g of radium were obtained. Work

148

with radon stopped in August 1944, but by then the difficulties of purifying polonium had been overcome and material had been prepared.

Sensitive Counters

The radiochemists developed a neutron counter based upon the Szilard-Chalmers reaction, in which a nucleus absorbs a neutron. It then loses energy by gamma emission, and the recoil of the atom frees it from its chemical bonds. This dissociation permits chemical separation of the reaction product and measurement of its induced radioactivity. This reaction's sensitivity as a neutron counter is high because it permits neutron absorption in a large volume of material. Until early 1944, when work with potassium permanganate began, ethylene bromide was used as the basis of the procedure. A detection efficiency of about 10% eventually was obtained.

Sensitive boron trifluoride counters were developed cooperatively by the Radiochemistry and Radioactivity Groups. The radiochemists' job was to prepare extremely pure material to permit effective operation at high pressures. The first such counter, the "bucket chamber," was 1 to 2% efficient. Later, the radiochemists developed another counter (Chap. 17).

Water Boiler Chemistry

Because the Water Boiler contained active material in aqueous solution, there were several chemical problems associated with the physical ones. Matters requiring investigation were choice of a compound, original purification of the material, prevention of corrosion of the containing sphere, and methods of decontamination and analysis. Uranyl sulfate was used in the original Boiler because its solubility and solution density were higher than those of the nitrate. Also, there was some saving in critical mass because the neutron capture cross section is smaller for sulfur than for nitrogen. Purity requirements were not strenuous for two or three light elements. The requirements were calculated by the rule that no impurity in the solution should absorb more slow neutrons than the sulfur in the sulfate.

Work on corrosion determined that stainless steels were suitable for Boiler container and piping. Study of the effects of working, welding, and annealing showed that weight loss dropped to zero after a few days. Boiler parts were therefore pretreated with normal isotopic uranium

149

sulfate solution, and corrosion difficulties were substantially eliminated.

The hydrates of uranyl sulfate were investigated in order to predict volume changes from final additions to the Boiler "soup." A stable hemihydrate was found with less water than the normal precipitate from saturated solutions.

The refractive index of uranyl sulfate solutions was investigated to develop a rapid method of keeping track of amounts of uranium in solution. With monochromatic light through pure solutions, concentration would be measured to 0.1%.

When the Water Boiler was being set up in May 1944, the chemists made all additions and removals of "soup" and kept accurate records of concentration by the refractive index and gravimetric analysis methods. When Boiler activity reached the critical point, the concentration was measured by refractometry and checked by other methods. The control rod was calibrated by adding small weighed increments of sulfate and determining the critical control rod setting for each increment.

Radiolanthanum

No test shots were fired in the RaLa program before August 1944. Design of a "mechanical chemist" for remote control work with this highly radioactive material began, however, as early as May. By August, the apparatus at the Bayo Canyon site was almost complete. This work is discussed further in Chap. 17.

ANALYTICAL METHODS

The high-purity analytical program was based on theoretical considerations reported in Chap. 1. The plan to use plutonium originally demanded about 3 neutrons/minute/gram, an extremely low rate of neutron emission. Tolerance limits were calculated by polonium alpha-particle bombardment of element targets and calculation of the neutron yield.

Tolerance limits for light elements, except the rare earths, were extremely low. Furthermore, effects are additive. It was generally agreed that the sensitivity of analytical methods should be one-tenth of tolerance. Because early experimental production would be very small and analytical samples might be no larger than 1 mg, it was evident that research on new submicroanalytical methods was necessary.

150

High sensitivity, rather than high precision, was sought. The analytical chemist's greatest difficulty was to identify and determine the interfering elements approximately. Then, the purification procedure could be modified to eliminate such elements or at least cut them down.

Unusual factors entered into such "submicro" work. Reagents had to be purified unbelievably so that the presence of a particular impurity should not become the limiting factor of the method. Contamination was probably the major difficulty because most of the worst elements (atmospheric dust and fumes, floor scuffings, etc.) are prevalent in any ordinary experimental environment. The laboratories were equipped with precipitrons. Floors and walls were kept very clean. A special subgroup of the CM-1 Service Group made certain that dust in laboratory air was minimal and that personnel were not unconsciously causing significant contamination. Control tests on dust deposition were run. The humidifying system was stopped because it brought in contamination. Shoe covers were adopted to avoid floor scuffing, and floor-cleaning methods were improved.

Some of the analytical methods used at Los Alamos were developed at the Chicago Metallurgical Laboratory. There was close liaison between the two projects in this particular field. An outstanding analytical development at Los Alamos was the vacuum method for carbon and oxygen analysis.

Spectrochemical Methods

The spectrochemical methods were developed by the Analysis Group. Copper electrodes for use with spark excitation had been used at Berkeley in the first spectrochemical analysis of plutonium. The direct copper-spark method was developed by the Chicago Metallurgical Laboratory. Chicago showed that plutonium could be extracted by cupferron and chloroform. The method of making this separation before sparking was conceived at Chicago but developed at Los Alamos. The pyroelectric gallium oxide method was developed at Los Alamos.

Plutonium Analysis. The cupferron and gallic acid methods were first used in early 1944 in developing trace analysis of the light elements. The cupferron method was chosen as standard. When purity standards were relaxed in August 1944, the need for very sensitive methods disappeared and further research on their improvement ended.

151

The direct copper-spark method was used throughout the Laboratory history. Plutonium is evaporated on copper electrodes, and the spark spectrum is photographed in the 2500- to 5000-Å range. The quantities of impurities are estimated from spectral line densities. This was the only method for determining thorium and zirconium. It was used for preliminary determination of impurities in plutonium solutions from Hanford.

Uranium Analysis. Except early in 1944 when the gallic acid method was tried the pyroelectric method was used for overall purity analysis of uranium. The oxide mixed with gallium oxide is arced from a crater in a graphite electrode, and estimates are made spectrographically. Volatilization of impurities along with gallium occurs in a manner analogous to stream distillations, but the complex uranium spectrum does not appear. Volatile compounds lost in ignition to the oxide are not determinable.

Very sensitive rare earth determination was made possible by a method that removed other impurities, followed by examination of the spark spectrum. Cupferron precipitable refractories (titanium, zirconium, and iron) were separated from other impurities by this method and examined in the copper spark.

Miscellaneous. Graphite purity analysis was developed as an adjunct of the PuO_2 graphite metal reduction.

The strontium fluoride band method was the only successful one for fluorine analysis. It involved absorption of fluorine in sodium hydroxide. The sodium fluoride is arced in the presence of excess strontium oxide, and the amount of fluorine is estimated by comparing strontium fluoride band head intensities with a standard. This method seems applicable to a number of materials, but Los Alamos applied it only to uranium and calcium.

Colorimetric Methods

Phosphorus in uranium and plutonium was estimated by formation of molybdenum blue from orthophosphate. Microgram quantities of acid-soluble sulfide were estimated by conversion of hydrogen sulfide to methylene blue, determined spectrophotometrically. Iron was determined spectrophotometrically in the presence of +3 plutonium after reduction to the ferrous state with hydroxylamine. Boron in calcium, uranium tetrafluoride, and plutonium was determined by distillation as methyl borate from a special quartz still. The distillate was trapped in calcium hydroxide solution and the boron was estimated colorimetrically.

152

Gravimetric Methods

Gravimetric methods were used to determine molybdenum in uranium-molybdenum alloys and carbon in uranium tetrafluoride.

Assay Methods

Before August 1944, radioassay was used to keep track of plutonium quantities received while photometric assay was being investigated. This method, however, proved untrustworthy, so radioassay was continued.

Gasometric Analysis

Description of the procedures involved in this work is limited by the extreme complexity of the apparatus used. That for oxygen and carbon microdeterminations could be classed among the most complicated analytical setups in the history of chemistry.

The oxygen method developed by the High-Vacuum Research Group solved one of the most pressing analytical problems, development of a dependable micromethod for oxygen determination. The overall method was not new, but its application on a microscale, the accuracy obtained, and the furnace tube developed were. The procedure involved vacuum fusion of a sample in a graphite crucible and analysis of the gases evolved. Oxides react with graphite at high temperatures, giving carbon monoxide. Determination is made of this compound.

A high-vacuum system (10^{-8} cm) and a revised Prescott microgas analyzer were the main parts of the apparatus. Either could be broken from the line independently. The sample was about 50 mg, and the sensitivity was about 10 ppm. The crucible and furnace tube were put in place, the sample was put in a dumper bucket, and the tube was sealed. After the crucible was suitably degassed, the sample was dropped in. The gas was then collected for analysis.

The apparatus for carbon analysis in plutonium was simply a modification of the oxygen apparatus. The sample was burned in oxygen from mercuric oxide in a low-carbon platinum crucible. The gaseous products were then analyzed by the Prescott appratus. In this case, the sensitivity was 5 ppm or less.

153

CRYOGENY

Cryogeny was limited locally to design and construction of a Joule-Thompson deuterium liquefier patterned after W. F. Giauque's at the University of California. It consisted of ethane, liquid air, and liquid hydrogen (or deuterium) cycles. The first two were completed by the beginning of 1944; the last, in April. Although designed for 35-liters/hour capacity, at the 7300-ft elevation of Los Alamos, it produced only 25 liters, a loss that could be compensated for by additional compression, if necessary. The problems of producing and storing liquid deuterium were investigated under contract by Professor H. L. Johnston at Ohio State University. He studied the ortho/para conversion of liquid hydrogen and deuterium, hydrogen-deuterium exchange, the high-pressure low-temperature equation of state for hydrogen and deuterium, the heat of vaporization of liquid deuterium, the Joule-Thompson coefficients of hydrogen and deuterium, the properties of thermal insulators at low temperatures, and long-term operation of hydrogen liquefaction equipment. The Ohio State contract began in May 1943 and continued through the life of the project.

Cryogenic work was formally suspended in September 1944 when Long declared that he could produce about 100 liters of liquid deuterium in 2 months, and about 1000 liters in 8 months, with the existing equipment.

CHAPTER 9

General Administrative Matters
August 1944 to August 1945

REORGANIZATION

Measurement of the spontaneous fission rate of Clinton plutonium by the Radioactivity Group in the summer of 1944 (Chap. 4) ended all hope of using it in a gun assembly bomb. Work on the ^{235}U gun proceeded as before, but implosion now became an absolute necessity if the Hanford plutonium production was to be of any use. Complete Laboratory reorganization was indicated. Two entire divisions—G (Weapon Physics) and X (Explosives)—were created to study the implosion dynamics problem.

G Division, under Bacher, included several of his groups from the Experimental Physics Division and several from the Ordnance Engineering Division. X Division, under Kistiakowsky, included several groups from Ordnance Engineering. Experimental Physics, renamed R (Research) Division was organized under Wilson. Ordnance, or O, Division remained under Captain Parsons. CM (Chemistry and Metallurgy) Division and T (Theoretical) Division were unchanged administratively, although some of their work changed. In A (Administrative) Division, a new group under Long included C and V Shops and many of the miscellaneous shops. Enrico Fermi arrived from the Metallurgical Laboratory early in September and became leader of F Division, which originally included the Water Boiler and Super bomb groups. Parsons and Fermi became Associate Directors of the Laboratory, and Mitchell and Shane became Assistant Directors. Parsons was to be responsible for ordnance, assembly, delivery, and engineering; Fermi, for the research and theoretical divisions and all nuclear physics problems.

The Laboratory's reorganization involved much interlocking of responsibilities and jurisdictions. G and X Divisions had to collaborate closely because they were working on separate phases of the same problem and had to share facilities, equipment, and, occasionally, personnel. O Division was responsible for items fabricated outside the Laboratory and for the final weapon design, so it had to confer regularly with X and G Divisions. It was necessary to see that all plans and specifications of these Divisions could be incorporated into the final weapon design. If O Division proposed a plan to simplify fabrication, it had to be proposed to X and G Divisions to see that their requirements were satisfied. G Division needed close cooperation from R Division in making nuclear measurements that would assist in interpreting integral experiments and predicting the behavior of an implosion bomb. For the implosion program, T, CM, and R Divisions were considered service divisions. Perhaps the most thoroughly organized was T Division, which drew up a plan for assigning theoretical groups to service work for experimental groups. Members of T Division kept informed of the activities of the groups to which they were assigned, attended meetings of these groups, and were prepared to advise them when consulted.

Shortly before the general reorganization, Oppenheimer outlined a plan to replace the Governing Board with an Administrative Board and a Technical Board, both advisers to the Director. The Administrative Board included Lieutenant Colonel Ashbridge (Commanding Officer), Bacher, Bethe, Dow, Kennedy, Kistiakowsky, Mitchell, Parsons, and Shane. Members of the Technical Board were Alvarez, Bacher, Bainbridge, Bethe, Chadwick, Fermi, Kennedy, Kistiakowsky, McMillan, Neddermeyer, Captain Parsons, Rabi, Ramsey, Smith, Teller, and Wilson. The Administrative Board was organized informally. Members were urged to raise questions about administrative problems and could invite other members of the Laboratory to discuss specific topics. The Technical board meetings consisted of prepared discussions about immediate technical problems and brief reports on recent progress or urgent problems. Such reports were by members of the board, interdivisional committees, division leaders, or other members of the Laboratory who could make special contributions to the subject under discussion.

CONFERENCES AND COMMITTEES

As the implosion program developed and the time schedule tightened, the Technical Board's functions were taken over by various interdivisional committees and conferences. Among the most important of

156

these were the Intermediate Scheduling Conference under Captain Parsons, the Technical and Scheduling Conference, and the "Cowpuncher" Committee. The last two were chaired by S. K. Allison, former Director of the Metallurgical Laboratory, who arrived at Los Alamos in November 1944. In this shift, the Director was advised by these committees and certain senior consultants, notably Niels Bohr (Chap. 2); I. I. Rabi (Chap. 1); and C. C. Lauritsen, who served as Elder Statesmen. Another important consultant was Hartley Rowe, Chief Engineer of the United Fruit Company and former Technical Advisor to General Eisenhower. Rowe came to the Laboratory in November 1944 and assumed responsibility for the transition from "bread board" models to production. He assisted greatly in procurement of the firing unit for the implosion bomb (Chap. 16) and in procurement of machinists.

The Intermediate Scheduling Conference, an interdivisional committee, began meeting in August 1944 to coordinate the activities of groups concerned with designing and testing the implosion bomb. It was formalized in November with Captain Parsons as chairman; Ashworth (Chap. 19), Bacher, Bainbridge, Brode, Galloway, Henderson, Kistiakowsky, Lockridge, and Ramsey as permanent members; and Alvarez, Bradbury, Doll, and Warner as alternates. The conference scheduled topics in advance and invited other members of the Laboratory to its meetings when the occasion arose. Eventually, it was concerned with both the gun assembly and implosion bombs. Its meetings covered procurement of items for the final weapons, the test program carried out in cooperation with the Air Force, and packaging and assembling of bomb parts for overseas shipment. Although originally planned to handle both administrative and technical aspects of bomb design and testing, this conference became almost exclusively administrative, and the technical problems were handled by the Weapons Committee formed in March 1945. The Technical and Scheduling Conference, organized in December 1944 shortly after Allison's arrival, assumed responsibility for scheduling experiments, shop time, and use of active material. Each conference was called to discuss a particular subject, such as explosive lenses or multiplication experiments on ^{235}U metal spheres. The subjects were announced in advance, and several persons were asked to make short reports. The personnel varied according to the subject discussed. This conference really took the place of the Technical Board, and it was concerned primarily with solving technical problems. It became more a technical than a scheduling conference.

Scheduling the implosion program became the task of the Cowpuncher Committee composed of Allison, Bacher, Kistiakowski, C. C. Lauritsen, Parsons, and Rowe. It was organized "to ride herd on" the implosion program by providing overall executive direction. The first meeting was early in March 1945. This group published a semimonthly report, the "Los Alamos Implosion Program," which covered progress of experiments in each group, scheduling of work in the various shops, and progress of procurement.

In April, when the implosion bomb design was frozen, G Division became responsible for the so-called tamper assembly. They had to specify the design, obtain designs drawn to these specifications, and procure all parts of the first two complete tamper assemblies. This assignment involved consultation with many sections of the project, so Bacher appointed M. Holloway and P. Morrison as G-Division Project Engineers. They maintained close relations with the metallurgists, the various Explosives Division groups responsible for design of the outer parts of the bomb, the Weapons committee for transport and storage, and the Cowpuncher Committee.

The Weapons Committee, directly responsible to Captain Parsons, was organized with Ramsey as chairman and Warner as executive secretary. It eventually included Commander Birch, Brode, Bradbury, Fussell, G. Fowler, and Morrison. It planned all phases of the work peculiar to combat delivery and later it became part of Project A (Chap. 19).

A Detonator Committee composed of Alvarez, Bainbridge, and Lockridge was appointed in October 1944 to settle all questions about external procurement of electric detonators. Bacher, Fermi, and Wilson worked on detailed planning and scheduling of experiments with ^{235}U metal to save time and make the experimental program as fruitful and illuminating as possible. In February 1945, Oppenheimer appointed Bethe, Christy, and Fermi as advisers on the design and development of implosion initiators. Niels Bohr met with this committee, which kept in close touch with the Initiator Group and the radiochemists.

Early in March 1945, two new organizations, the Trinity Project and the Alberta Project, were given division status. One was to fire an implosion bomb at Trinity; the other, to integrate and direct combat delivery of both types of bombs. Trinity was led by Bainbridge with

158

Penney and Weisskopf as consultants; Project A was led by Captain Parsons with Ramsey and Bradbury as technical deputies (Chaps. 18 and 19).

The last division created in this period was Z Division under J. R. Zacharias, who came from the M.I.T. Radiation Laboratory in July 1945. Z Division was to carry out engineering and production, chiefly concerned with airplane and ballistic problems, to replace the program at Wendover Field, Utah. By this time, the project had acquired a small airfield near Albuquerque, formerly an Army base called Sandia; it was to be assigned its own planes; and it had use of the large Army base at Kirtland Field near Albuquerque. The new division was barely organized before the war ended.

LIAISON

Many liaison problems that had been so difficult (Chap. 3) had been solved or were no longer major at the time of the general reorganization. Liaison with the Army and Navy became increasingly important as designs were determined, actual airborne tests became necessary, and preparations were made for combat delivery (Chap. 19).

The principal liaison problem still existing was with the Camel Project at the California Institute of Technology (C.I.T.). In the fall of 1944, Oppenheimer learned that the C.I.T. rocket project had almost completed its research and development and was entering production. The C.I.T. project combined high professional scientific ability with practical wartime experience in weapon engineering and had its own procurement, laboratory, and field facilities. Because both manpower and facilities were becoming badly overstrained at Los Alamos, in November 1944 Oppenheimer discussed possible collaboration between the two projects with C. C. Lauritsen, head of the C.I.T. group. After correspondence with Bush, Conant, and Groves, and negotiations about contracts, the Camel Project was formed.

The work at Camel was determined by the facilities there, the experience of the staff, and the stage of the work at Los Alamos when the Camel Project began. The Camel staff confined their work to problems of the bomb assembly mechanism and its combat delivery, specifically implosion design and delivery. For implosion design, the Camel staff

159

researched and engineered special components of the assembly, detonators, lens mold design, impact and proximity fuses, and high-explosives components. In addition to this division of labor with Project Y, Camel had its own general implosion program, set up when the final "freeze" at Los Alamos in April 1945 forced abandonment of alternative lines of implosion development. Los Alamos adopted the multiple-lens bomb as final, and Camel was to carry out a standby program. Camel work on weapon delivery covered the production of implosion bomb mockups, of "pumpkins" (bomb mockups loaded with high explosive and intended for eventual practice bombing of enemy targets) with special impact fuses, and a special program of drop tests that paralleled the Los Alamos program at Wendover and Sandy Beach (Chap. 14) and provided data on bomb ballistics.

The main liaison was between Oppenheimer and Lauritsen, who spent part of his time at Los Alamos, where he was a member of the Cowpuncher Committee. In March and April 1945, there were extensive discussions, as a result of which a C.I.T. liaison office was established as part of the Director's office at Los Alamos with McMillan responsible for coordination. Mail service was improved, teletype connections were established, and eventually, regular airplane schedules were established for freight and passengers between "Kingman" (Wendover Field, Chap. 19), Los Angeles, Inyokern (Camel's field site), Santa Fe, Sandy Beach (Chap. 14), and Albuquerque.

ADMINISTRATION

In July 1945, Laboratory administration was organized into the following groups.

A-1	Office of Director	D. Dow
A-2	Personnel Office	C. D. Shane (Assistant Director)
A-3	Business Office	J. A. D. Muncy
A-4	Procurement Office	D. P. Mitchell (Assistant Director)
A-5	Library and Document Room	C. Serber
A-6	Health Group	Dr. L. H. Hempelmann
A-7	(absorbed in Groups A-1 and A-9)	

160

A-8	Shops	E. A. Long
A-9	Maintenance	J. H. Williams
A-10	Editor	D. R. Inglis
A-11	Patent Office	Major R. C. Smith
A-12	Safety Office	S. Kershaw

Director's Office

One of Dow's most important duties as Assistant to the Director was liaison between the using technical groups and the Post Operation Division, which handled construction (Chap. 3). During the fall of 1944, increased consumption caused several power failures, and Dow's office had to solve this problem. Eventually, power was increased by tying in with the Albuquerque line. Another responsibility, shared with the Personnel Office, was to prepare employment contracts for staff members on leave of absence from academic institutions. The first of these was prepared in September 1944, covering six months and extended to the beginning of the next academic year if the project terminated. Dow's office also cooperated with the Business Manager's office in securing insurance policies for personnel. One of these made available to University of California employees in July 1945 was an accident policy issued by the Indemnity Insurance Company of North America insuring "against bodily injuries caused by accidents and arising out of and in the course of the insured's duties in connection with war research undertaken by or on behalf of the contractor." Unlike previous Manhattan District Master Policies (Chap. 3), this one insured against certain aviation hazards, which were important because of the expanding test program.

Personnel

Abandonment of the plutonium gun program released several chemists and physicists, but they were absorbed into the implosion program. In fact, a general expansion was necessary and Shane, Bacher, and Long went on recruiting drives to the other projects of the District. Through their efforts, a group of civilian scientists came from the Metallurgical Laboratory and from Oak Ridge, and several technical military personnel came from the SAM Laboratories in New York and from Oak Ridge.

161

Personnel procurement was always hampered by the housing shortage (Chap. 3), and the situation grew worse as the Laboratory rapidly expanded. McKee completed the third section of the housing area in December 1944, and no additional multiple-unit housing was to be constructed. A policy was established that employees should be housed if they could not be hired without housing. This policy covered machinists, scientific personnel, essential administrative personnel, and 16 technical maintenance men (Chap. 3). Machinists were encouraged to come without their families in exchange for a bonus payment. Several dormitories were built, but that solution was not adequate.

Salary policy remained a principal difficulty. A working agreement provided that salary increases be limited to 15% of the minimum range per year, and that only 25% of all employees hired within a year might be hired at above the minimum applicable salary. There were to be no increases above $400 per month. In January 1945, Shane tried to remove the $400 restriction and proposed semiautomatic merit increases; eventually, the policy was changed. Shane also asked that the project be granted an exemption from the 25% hiring provision because of the special employment conditions prevailing there. The shop work, especially, required a greater proportion of highly skilled workers than an ordinary production shop, and the assignment of relatively young, inexperienced enlisted personnel made it necessary to hire highly skilled civilians to fill the responsible positions. The Contracting Officer agreed to certain exceptions, especially for shop personnel.

In July 1945, the salary situation reached another critical point, over approval of salary increases. Since March, the Personnel Office had had difficulty in agreeing with the Contracting Officer about salary changes, although they had been following the same rules since the July 1944 agreement. Shane requested a conference, which was held in July 1945. Although agreement was reached on several minor points, the main issues were not settled, and when the war ended, Shane resigned as Personnel Director.

Procurement

The Procurement Office was not directly affected by the Laboratory reorganization, except that its work increased and continued to do so until shortly before the Trinity test.

162

In October 1944, the Property Inventory Section was established with Captain W. A. Farina in charge. He was responsible for making a physical inventory of the Laboratory, for making the Procurement Office record system compatible with War Department regulations, and for advising the University on government property policy. Making an inventory and having someone at the site responsible for material accountability had been discussed since the early days but always postponed because of more urgent work.

By the end of 1944, the Ordnance Division had established its own special Procurement Group under Lockridge, and to avoid confusion, it was necessary to outline the responsibilities of each procurement group. Mitchell continued to be responsible for all stock catalog items. Lockridge was responsible for all fabrication jobs involving chemical machine shop work and mechanical assembly, and either could place orders for fabrication jobs involving chemical and metallurgical techniques, plastics, and electrical work. To avoid duplication, the office making the requisition would notify the other office. Much of Lockridge's purchasing was through special channels rather than through the University Purchasing Office in Los Angeles, but he cooperated closely with Colonel Stewart's office. A good part of Lockridge's ordering was done from the C.I.T. project and from G. Chadwick of the Detroit Office.

The polonium hazard, although parallel in many ways to the plutonium hazard, was never as serious a problem for the Health Group. No research was done on the subject at Los Alamos, but routine urine tests were done on all exposed personnel according to Manhattan District Medical Section standards. Polonium is not so dangerous as plutonium per unit of radioactivity even though it spreads around a laboratory very rapidly. Health group records indicate that only two people temporarily exceeded the tolerance limit for polonium excretion. The typical costume of a worker using plutonium or polonium included coveralls or laboratory smock, rubber gloves, cap, respirator, shoe covers, and often a face shield. All were worn only once and then laundered. In July 1945, when there were about 400 people in CM Division, 3550 rooms were monitored, 17 000 pieces of clothing were laundered, and 630 respirators were decontaminated, along with 9000 pairs of gloves, of which 60% were discarded. In June 1945, laboratory decontamination was made the responsibility of the laboratory workers themselves.

163

The hazards of external radiation, which had been negligible and confined mostly to accelerating equipment and radioactive sources, became more critical in the fall of 1944. At that time three potential sources of danger appeared: the Water Boiler and later the power boiler, the implosion studies of the RaLa Group, and critical assembly experiments. The power boiler caused several mild overexposures to radiation from leaks in the exhaust gas line and one serious exposure of several chemists during decontamination of active material. The implosion studies, involving large amounts of radioactive barium and lanthanum, brought a serious situation that the Health Group monitored constantly. A series of accidents and failures caused overexposures for about six months until remote control operation was perfected (Chap. 17). The most serious potential radiation hazard was from critical assembly experiments for which the Health Group had no responsibility except to be sure the men were aware of the dangers involved. These experiments were especially dangerous because there was no absolute way of anticipating the dangers, and the experiments seemed so safe when properly carried out that they led to overconfidence. Two serious accidents resulted from the critical assembly work during this period. One caused acute exposure of four individuals and the other killed one person.

The Health Group made extensive reports of the radiation hazards caused by the Trinity test (Chap. 18).

In the fall of 1944, the Health Group was understaffed and unable to maintain personal contact with all those engaged in technical work. Consequently, its records of external radiation dosages became less accurate, particularly where radiation hazards were not serious and changed infrequently and where experiments were performed after natural sources had been transferred from one person to another without the Health Group's knowledge. This was not true where exposure to external radiation was prolonged or severe. There were also instances where blood counts were not made, or were made less frequently than desirable, because of poor personnel cooperation. Complete radiation and hematology records are invaluable as legal evidence in case of future claims against the project. Where the Health Group could not meet all its obligations, it performed jobs in the following order: (1) procedures that actively protected personnel against industrial hazards, (2) accident reports or termination records for those leaving the project, and (3) records or reports of routine exposure, hazards, etc.

Resignations and dismissals tabulated in January 1945 showed that a third of the 219 men recruited in November and December had left by this time. Over half of these resigned because of the living conditions. These men viewed the project as one among many possible war jobs and had little reason to think otherwise. The majority, who did their work well and remained with the project, nevertheless felt, with some justification, that they were discriminated against by the Laboratory and the community.

The second recruiting drive in February 1945 led to fewer resignations because men were selected more carefully and facilities at Los Alamos had improved. To help ease the work load, shop facilities at the Metallurgical Laboratory were used to some extent, beginning in the spring of 1945.

This period presented several difficult technical problems. One was machining of full-scale explosives castings (Chap. 16). Long and Schultz assumed responsibility for designing tools and fixtures for this work in the spring of 1945. They were responsible also for accurate gauging of full-scale castings.

Another problem was construction and use of molds for high-explosive lens casting. Outside firms were unable to do this work. The shops suggested changes in design, worked out the techniques for producing molds, and sent representatives to outside producers to teach them the necessary techniques.

Several technical difficulties in uranium machining were encountered. It constituted a minor health hazard, that of normal heavy-metal poisoning. Uranium was machined from the beginning in a special shop supervised by Schultz. It was moved from a small Cryogeny Building annex to a special enclosed region in C Shop in the spring of 1944 and then to its own building in the spring of 1945.

CHAPTER 10

Technical Review
August 1944 to August 1945

THE GUN PROGRAM

During this last year of the war, the Gun Group of the Ordnance Division completed design of the ^{235}U gun assembly, tested its components at reduced and full scale, and undertook their final engineering and procurement. After an elaborate program of final field and drop tests, it produced a weapon more certain of high-order operation without testing than the more radical implosion design tested at Trinity. Earlier, the possibility of a ^{235}U implosion bomb had not been ruled out. Acquisition of accurate means of calculation and reliable cross-section data showed that such an implosion would be less efficient than the plutonium implosion, so it seemed wise to plan to use ^{235}U by the gun method alone. Toward the end of the war, the possibility of composite (^{235}U + ^{239}U) implosion bombs was considered (Chaps. 11 and 20). When the laboratory program was "frozen" in February 1945, the decision to use only ^{235}U in the gun model was final.

This quiet and efficient Gun Group continued at the center of an affiliated program in the Research Division, the Theoretical Division, other groups of the Ordnance Division, and the Alberta Project (Chap. 19). From the Research Division, Group T-2 got enough information on the nuclear properties of ^{235}U to provide accurate data for calculations of critical mass and of the amount of material that could be used safely. The sphere multiplication experiments of R Division permitted a more accurate calculation of the critical mass by extrapolation. R Division "mocked" the gun in a model experiment to provide an integral check of its performance calculations, including predetonation probability. The finished projectile and target were brought to the critical point by the Critical Assemblies Group of G Division shortly before being shipped to Tinian for combat use. This assembly was a final check of the accuracy of predictions about the point at which the system would become supercritical. A reliable efficiency calculation was made possible by theoretical and experimental estimation of the initial multiplication rate of the fully assembled bomb.

Fabrication of the projectile and target was the responsibility of Groups CM-2, -7, and -11 of the Chemistry and Metallurgy Division. Fabrication included forming the active material into pieces of proper shape and purity, and forming the steel casing that housed the target. The final design of the outer case, originally the responsibility of the Engineering Group of the Ordnance Division, was almost entirely transferred to the Gun Group. Responsibility for the fusing and detonating system remained with the Fuse Group of the Ordnance Division. The Gun and Fuse Groups collaborated in the drop tests of the Little Boy for Project Alberta.

THE PLUTONIUM BOMB

In mid-1944 hope for successful implosion was very low, so F Division was given the task of investigating the slim possibility that autocatalytic assembly using plutonium might be used. The desirability of such systems was not immediately evident, and the Weapon Physics, Explosives, and Theoretical Divisions were preparing a direct attack on the implosion problem.

On the basis of Clinton plutonium analysis, the Hanford plutonium was expected to produce many neutrons per second, in bomb amounts of plutonium, from spontaneous fission of ^{240}Pu alone. Light impurities would produce additional neutrons, but purification would keep this contribution small compared to that from spontaneous fission. Only implosion could assemble the plutonium quickly enough to avoid predetonation.

The "direct attack" on the implosion problem included continuation of small-scale implosion studies in the new X Division with particular emphasis on interpreting the causes of jets and irregularities, including careful investigation of the source of timing errors in multipoint detonation and their contribution to asymmetries. The first lens test shot was fired in November 1944. Meanwhile, G Division was starting to examine the implosion experimentally, beginning work on electric detonators, and planning the hydride critical experiments as a step to eventual critical assemblies of active metal. The Theoretical Division was

167

completing its studies of the "ideal" implosion (which began with a spherically converging shock wave) and beginning theoretical interpretation of the jets and asymmetries found in less-than-ideal experimental implosions.

The Explosives Division was searching for ways to prevent development of irregularities. The lens program in X Division showed that a converging spherical detonation wave could be approximated by a lens system if enough simultaneity could be obtained for all lenses. Although there was no sure path to success, promising directions of development had been marked out.

IBM calculations showed that efficiencies would be low, unless the nuclear reaction could be initiated at the time of maximum compression. If the reaction was initiated at this time, efficiencies would not be significantly lower than those predicted for the hollow implosion. The model was more promising than any previously investigated. There was, however, no experimental evidence that mere explosive "squeezing" of an already solid mass would give high compression.

The first experimental evidence came from the RaLa Group of G Division, which fired two shots in December 1944 using Primacord initiation and homogeneous high explosive with solid cadmium spheres. By February these shots had been repeated using electric detonators and had shown a compression not far below that predicted theoretically. Therefore, the possibility of the solid model was considered promising. As a corollary of the solid model, various types of nuclear initiators were discussed in February, but with no experimental evidence for feasibility.

At the end of February 1945, it was decided that the Laboratory's program should be frozen so that the July deadline for the first bomb test could be met. All further work would concentrate on the solid-core lens implosion with a modulated nuclear initiator.

A detailed schedule for all implosion work was established. By April 2, full-scale lens molds had to be ready for full-scale casting. Full-scale lens shots had to be ready by April 15 to test the timing of multipoint electric detonation. Hemisphere shots had to be ready by April 25. The detonator had to come into routine production between March 15 and April 15. By April 15 large-scale production of lenses for engineering tests had to start. A full-scale test, by the magnetic method, had to be made between April 15 and May 1. Full-scale plutonium spheres had to be fabricated and tested to determine their degree of criticality between May 15 and June 15. By June 4, fabrication of highest quality lenses for

168

the Trinity test had to be under way. The Trinity sphere fabrication and assembly would begin by July 4.

To meet this schedule, the Cowpuncher Committee (Chap. 9) was set up on March 1 to "ride herd." The feasibility of a modulated initiation was accepted on April 27, so the solid-core model definitely would be used. Full-scale lens molds were completed in May after innumerable procurement delays. Lens-timing measurements were successful shortly thereafter, but also delayed. By June 12, two full-scale plutonium hemispheres were tested for neutron multiplication, something over two weeks late. The delays, primarily because of the difficulty of procuring good lens molds on schedule, shortened the time for final engineering tests. But the Trinity test actually was made four days ahead of the target date—July 20, 1945.

THEORETICAL AND EXPERIMENTAL PHYSICS

The Theoretical Division was able to close its earlier investigation of techniques for solving neutron diffusion problems with accuracy, reliability, and speed. It had developed these and other techniques for treating other theoretical problems involved in the implosion and the nuclear explosion. Therefore, it could give realistic guidance to the last phase of the experimental program, final weapon design, and preparations for the Trinity Test. Investigation of the Super continued at relatively low priority in F Division until the end of the war. Men freed from other work brought this work to partial completion. The results indicated, convincingly but not decisively, that such a weapon was indeed feasible.

Experimental nuclear physics continued in R and F Divisions. R Division completed neutron number and spontaneous fission measurements, measurement of the fission spectrum and of fission cross sections, and scattering cross-section measurements. Integral experiments, including tamper measurements and integral multiplication experiments, using solid spheres of ^{235}U were increasingly emphasized. F Division made new measurements of the deuterium and tritium cross sections, which indicated a materially lower ignition temperature for tritium-deuterium mixtures than first obtained. The high-power Water Boiler was completed and operated routinely as a neutron source. Other work included cooperation with R Division on sphere multiplication experiments and preparation for radiochemical measurements at the Trinity test.

169

The G-Division Critical Assemblies Group made critical assemblies with various uranium hydrides and finally undertook the task of making metal critical assemblies with active materials to be used in the first bombs.

CHEMISTRY AND METALLURGY

The Chemistry and Metallurgy Division achieved full-scale production of active and tamper materials for both the ^{235}U gun model and the ^{239}Pu implosion model bombs. It also processed and handled highly radioactive materials for the RaLa implosion tests and the neutron initiator program. The basic research on plutonium purification had been completed by the time ^{240}Pu was discovered and purity requirements entailed by that discovery were relaxed. An efficient purification scheme was developed, as were the necessary metal-reduction and metal-forming techniques.

Chemistry and Metallurgy Division work was hampered by serious personnel hazards from plutonium and polonium toxicity. The division developed an elaborate system for monitoring buildings and examining employees. This health control work was hampered by increasingly crowded facilities until new ones became available.

THE TRINITY TEST

Planning for the Trinity test became high priority in March 1945. The Jornada del Muerto test site, called Trinity, had been selected the previous fall. Necessary road and building construction had started, and a Military Police detachment had taken up residence at "Base Camp" in December 1944. Although most of the Trinity program was under way in March, there was a vast amount of work still to be done in planning and instrumenting the great variety of mechanical, optical, and nuclear records that were necessary. Several square miles of desert became the habitat of a complex laboratory, tied together in one vast system by thousands of miles of electrical wiring.

The first aim of Trinity was the rehearsal shot of 100 tons of high explosive early in May. This tested the organization for the final shot and gave data for calibration of blast measurements.

As the July deadline approached, larger and larger contingents of personnel arrived from Los Alamos to ready the equipment for firing the bomb and for recording measurements of blast and shock, spectrographic and photographic information, and nuclear data. The two plutonium hemispheres were delivered from Los Alamos on July 11, and assembly of the high explosives began on the 13th.

The first atomic explosion was set off on the morning of July 16 after weeks of intensive preparation and hours of tense waiting to see if the weather, which had turned bad the night before, would clear. The event can hardly be summarized more concisely than it is in Chap. 18. After the Trinity test—whose yield exceeded most expectations and was soon estimated to be 15 000 to 40 000 tons of TNT—the machinery of Project Alberta, the overseas mission, began to operate.

CHAPTER 11

The Theoretical Division

INTRODUCTION

Between August 1944 and August 1945, the Theoretical Division increased the size of its groups and added three new ones. The division was not seriously involved in the general Laboratory problems of personnel, construction, and transportation, nor, except in an advisory way, with the complicated procurement and scheduling for the Trinity test and Project Alberta. It was, therefore, able to administer itself and work rather unobtrusively. It was able to handle more realistic and complex problems with increasing efficiency, to gain increased understanding of the difficult hydrodynamical questions involved in the implosion and the nuclear explosion, to refine its earlier calculations of critical masses and efficiencies, and to provide reliable interpretation of many integral experiments.

The division structure in August 1945 was as follows.

T-1	Implosion Dynamics	R. E. Peierls
T-2	Diffusion Theory	R. Serber
T-3	Efficiency Theory	V. F. Weisskopf
T-4	Diffusion Problems	R. P. Feynman
T-5	Computations	D. A. Flanders
T-6	IBM Computations	E. Nelson
T-7	Damage	J. O. Hirschfelder
T-8	Composite Weapon	G. Placzek

Group T-6, under S. Frankel and E. Nelson, was added in September 1944 to operate the IBM machines. Group T-7 formed in November 1944, when its name was changed from O-5, to complete earlier investigations of damage and of the general phenomenology of a nuclear explosion. Group T-8, added in May 1945 when G. Placzek arrived from Montreal, was to investigate fission bomb possibilities, specifically the composite core implosion intended to use ^{235}U (with ^{239}Pu) more efficiently than would be possible by gun assembly.

DIFFUSION PROBLEMS

Although the essential difficulties of the one-velocity diffusion problem had been overcome, even expansion of the neutron distribution in spherical harmonics was still rather expensive. Group T-2 greatly simplified these calculations in the fall of 1944 with an analytical expression that gave accuracies within 1 to 2%. This method used simple solutions for the shape of the neutron distribution far from boundaries (such as that between core and tamper) and then fitted these solutions discontinuously at the boundary so that the critical radius was given. From then on, solutions for many critical radius or mass problems were proliferated extensively, and even reduced to nomographic form, permitting very rapid calculation.

Various Theoretical Division groups, but particularly T-2, were concerned with special problems arising from sphere multiplication experiments done in R Division. These calculations had to account for the variation of the average cross sections after the initial and each following collision of neutrons emerging from a central source. The number of neutrons coming out of the sphere as a function of the number of source neutrons was calculated for varying sizes of spheres and various source dispositions. These calculations agreed well with the measured values. As larger spheres of ^{235}U became available, it was possible to extrapolate to the critical mass very accurately (Chap. 12).

T-4 developed critical mass calculations to predict the size of the hydride critical assemblies made by Group G-1 of the Weapon Physics Division. In the first assemblies a discrepancy of about 50% was found between the actual and predicted critical sizes. Newer fission spectrum values reduced the discrepancy materially. The discrepancy apparently arose from experimental errors in nuclear constants, not theoretical errors. However, there was doubt about the adequacy of the methods used.

THE GUN

Relatively little new work had to be done on the gun assembly. Group T-2 prepared final calculations of the expected efficiency, including the

173

predetonation probability, using integral data that Group R-1 got from the gun model experiment.

THE IMPLOSION

By August 1944 most calculations of ideal implosion (starting with a converging symmetrical shock wave at the outer edge of the tamper) were complete and this part of the subject was almost completely understood. Within the Theoretical Division, the design "freeze" occurred in February 1945.

Several new problems were set for the Division. The first was to determine how temperature affected the course of the implosion. This effect was expected to be greater for the solid model than for a hollow implosion. The second was to prove the stability of convergent shock waves. Because it would be very difficult to get experimental information on the irregularities produced inside a solid core, it would be necessary to rely entirely on indirect evidence and theory. The third major theoretical problem was to provide specifications for and to help design a modulated neutron initiator.

To determine how shock heating affected the course of the implosion, equation-of-state calculations for uranium were completed by April, and IBM calculations of the solid-core implosion thereafter included the temperature effect. Efficiencies were decreased slightly, but less than anticipated.

It was finally proven that plane shocks were stable, and the decay rate of irregularities was found. Even without a complete theory for convergent shocks, the effects of instability were roughly estimated.

In initiator design, specifications were provided for the initial neutron intensity required by the sudden mixing of α-n materials. By April 1945 the initiator design had been frozen.

As time for the Trinity test approached, Group T-1 tried to explain discrepancies between the G-Division data from implosion studies and the results of the latest and most comprehensive IBM calculations. Densities measured by the betatron and RaLa methods were lower than the theoretical values, as were the measured shock velocities. Material velocities measured by the electric pin method agreed with theory, but those measured by the magnetic method were lower than predicted. The conclusion was that the theoretical calculations should be lowered but not enough to alter the expected weapon performance significantly.

EFFICIENCY

The effect of mixing between core and tamper, expected because of the Taylor instability at the interface, was examined carefully. Another calculation improved was prediction of the initial multiplication rate for supercritical assemblies. Allowance for several groups of neutron velocities, in particular, introduced a transient time dependence of this rate in the initial stages of multiplication. Comparison of this theory with R-1 and G-1 experiments was not satisfactory and required reexamination of theory and experiments to reach good agreement.

Qualitative understanding of the radiation effect on the course of the explosion also was achieved, but reliable quantitative calculations remained impossible. Just before the Trinity test, therefore, the final efficiency prediction ignored this effect, which, it was understood, might produce much higher efficiency than predicted.

The Theoretical Division June Progress Report predicted that Trinity yield would be 5000-13 000 tons of TNT. The actual yield certainly exceeded this. One successful measurement during the test was the initial multiplication rate of the bomb as it exploded. Using this value and omitting some pessimistic assumptions gave a value near 17 000 tons. Because of the uncertainties in measuring the yield and the large theoretical uncertainty introduced by ignoring the effects of radiation, the close agreement between the figure and the "official" yield of 17 000 tons conceals several unsolved problems. The first was the effect of radiation. The second was the proportion of the energy released converted into blast energy. Uncertainty about the latter relations necessitated distinguishing between the "nuclear" efficiency (fission energy released) and the "blast" efficiency derived from measurement of the blast wave. In view of the complex efficiency problem and the unknown factors entering into it, the correspondence between theoretical and measured values was remarkably good.

Shortly after the Trinity test, there was some discussion of possible unanticipated effects that might have accounted for the unexpectedly high yield. The most familiar of these was radiation. Another was that the short-lived ^{239}U formed in the tamper by neutron capture might be a slow-neutron fissioner, like ^{233}U, ^{235}U, or ^{239}Pu. In the very high neutron flux during the explosion, double neutron captures would be common, so the tamper would act essentially as an explosive "breeder." Whether or not such a reaction is possible depends upon the fission cross section of

175

^{239}U, which is difficult to measure because of its extremely short life, and which had not been measured.

DAMAGE

The behavior and effects of a nuclear explosion were investigated much more extensively than had been possible before, tracing the history of the process from the initial expansion of the active material and tamper through the final stages. Investigations included formation of the shock wave in air, the radiation history of the early stages of the explosion, formation of the "ball of fire," attenuation of the blast wave in air at greater distances, and the effects of blast and radiations on human beings and structures. Much of this information was important in planning for the Trinity test. It was also essential to know the probable fate of plutonium and fission products in the ball of fire and the smoke cloud ascending out of it. These calculations, plus calculations of blast and radiation, were essential in planning experiments, observations, and personnel protection. Theoretical studies of damage to structures and personnel were made in anticipation of combat use. Extensive use was made of British data on high-explosive bomb damage to various kinds of structure. This work was given to Group T-7, with the advice and assistance of W. J. Penney.

EXPERIMENTS

The Theoretical Division's consulting services continued to occupy much of its time. The most varied assistance was given to the Trinity experiment, all phases of which were under Theoretical Division surveillance. The Division was responsible for preparing adequate order-of-magnitude calculations of the expected effects and for ensuring that, in terms of these calculations, the experimental program was properly planned and coordinated. Credit must be given to the computing group, T-5, for its services in obtaining the numerical solutions required in most theoretical investigations.

176

CHAPTER 12
Research Division

INTRODUCTION

By August 1944, the Laboratory had learned the difficulties of implosion dynamics and was ready to begin integral investigation of chain-reacting systems. The first development was the most important organizing influence from this time on. Several physicists went to the new G Division to assist in implosion dynamics investigation. Other experimentalists went to G Division to work with hydride and later with metal critical assemblies. The Water Boiler Group and part of the Detector Group also went into the new F Division. The remaining four groups made up the new experimental physics division (Research, or R, Division).

R-1	Cyclotron Group	R. R. Wilson
R-2	Electrostatic Generator Group	J. H. Williams
R-3	D-D Group	J. H. Manley
R-4	Radioactivity Group	E. Segrè

R. R. Wilson became the Division Leader while remaining Group Leader of the Cyclotron Group.

The Research Division was influenced by the new stage the Laboratory was entering. Differential experiments investigated the finer points of the chain fission reaction, and more semi-integral and integral experiments were made. Finally, in January 1945, R Division was asked to assist in preparing for Trinity, and it began to develop test instrumentation.

The rest of this chapter covers work in progress or that planned in August 1945.

NEUTRON NUMBER MEASUREMENTS

The Cyclotron, Van de Graaff, and D-D Groups all compared neutron numbers from fissions induced by slow and fast neutrons. The Van de

Graaff experiments were most reliable, but all showed no appreciable dependence of neutron number on energy.

In September 1944, measurements of ^{233}U sample showed that it was a good potential bomb material. The Cyclotron Group found its neutron number to be slightly greater than that of ^{235}U. The only other direct measurement of a neutron number was that made by the Radioactivity Group for spontaneous fission of ^{240}Pu. They gathered enough data to show that the number of neutrons per fission was about 2.5.

While constructing apparatus to measure the decay constant of nearly critical assemblies of ^{235}U, the Cyclotron Group measured the product of the neutron number times the fission cross section at high energies.

SPONTANEOUS FISSION MEASUREMENTS

In the fall of 1944, the Radioactivity Group moved its spontaneous fission work from Pajarito Canyon to the new East Gate laboratory. This moved them away from expanding high-explosive firing sites of the implosion program.

From August 1944 to August 1945, spontaneous fission data were taken on many heavy elements, including isotopes of thorium, protactinium, uranium, neptunium, plutonium, and element 95 (later called americium). Careful measurements were made on early plutonium samples produced by cyclotron irradiation, later samples from the Clinton pile (including a reirradiated one), and still later samples of Hanford material. This work was necessary to learn the neutron background from spontaneous fission of ^{240}Pu in the Hanford bomb material.

FISSION SPECTRUM

The Electrostatic Generator Group measured the fission spectrum from fast neutrons impinging upon a plate of enriched uranium, using the photographic emulsion technique. The results agreed qualitatively with those obtained earlier using slow-neutron initiation.

The Electrostatic Generator Group also began investigating the low-energy end of the fission spectrum using the cloud chamber technique, a mock-fission source, and a surrounding sphere of enriched uranium.

Their data agreed well with extrapolations from earlier measurements. Measurements on the bare mock-fission source showed that in the low-energy region, its spectrum was relatively close to the fission spectrum.

The Radiochemistry Group built three mock-fission sources using 2, 8, and 25 Ci of polonium, the last giving 4×10^6 neutrons/second. These were used in the various multiplication experiments.

FISSION CROSS SECTIONS

The fission cross sections of ^{235}U and ^{239}Pu as a function of energy were remeasured to obtain more accurate data in the intermediate energy region. The Cyclotron Group measured from thermal up to 1-keV neutrons, and the Electrostatic Generator Group measured at high energies, collaborating with the D-D Group at the highest energies. The Cyclotron Group measured both the fission and absorption cross sections. To obtain the latter, they measured the transmission of neutrons scattered in passing through the fissionable material. Because absorption is fission plus radiative capture, this experiment gave additional data on radiative capture. An important result was knowledge that the ratio of radiation capture to fission was a sensitive function of energy.

These same groups also found that at high energies the ^{233}U fission cross section was about twice that of ^{235}U, placing it between ^{235}U and ^{239}Pu in both cross section and neutron number.

The Electrostatic Generator Group compared the ^{237}Np and ^{235}U fission cross sections. They measured the ^{237}Np cross section from the threshold at about 350 to 400 keV up to 3 MeV.

SCATTERING EXPERIMENTS

A new directional proportional counter with a higher directionality factor than previously used was developed. The D-D Group made differential measurements of materials used in the bomb construction, such as aluminum, cobalt, copper, and uranium, which had been incompletely measured before. Scattering measurements were made using the first amount of beta-stage ^{235}U large enough to provide data. Scattering was smaller than for normal uranium by an amount comparable to the difference of their average fission cross sections. This

179

experiment was important negatively because if a large inelastic scattering cross section had been found in ^{235}U, it would have implied a longer time between fissions and less efficiency than expected.

Differential scattering measurements gave unreliable information about how inelastic scattering would actually affect operation of the bomb. Therefore, the D-D Group made several more nearly integral experiments, measuring neutrons at the inner surface of a hollow spherical tamper by using a central D-D or photoneutron source. By using source neutrons of different energies and detectors of different thresholds, they could measure the integral energy degradation for various tampers.

Another technique was to measure the decay time of tamper materials. A burst of neutrons sent into a tamper gives a reflected neutron intensity that falls off with time and that can be measured by a time analyzer. What is measured is, again, an integral effect, depending on the path and energy degradation of the scattered neutrons; but it is just this integral effect that influences the time scale of the explosion.

MULTIPLICATION EXPERIMENTS

Multiplication experiments, using larger and larger spheres of ^{235}U and ^{239}Pu as these materials became available, were among the most important integral experiments in this period. The technique was to use a mock-fission source surrounded by a sphere of active material and, in some cases, a tamper. These experiments were done by the Electrostatic Generator and D-D Groups. The latter measured mostly neutron distribution and inelastic scattering in core and tamper. Neutron multiplication was also measured in the process. More accurate measurement of neutron multiplication, measurement of the average fission cross section for fission neutrons, and data that set an upper limit on the branching ratio for fission neutron energies came from experiments by the Electrostatic Generator Group. Both groups measured the inelastic scattering in spheres of active material (by comparing the spectrum of outcoming measurements) in which not enough material had been used to produce substantial energy degradation.

The ratio of neutron intensities with and without the surrounding sphere of fissionable material depends on the average fission cross section, neutron number, and branching ratio. To set an upper limit on the branching ratio, the Electrostatic Generator Group measured the average fission cross section simultaneously with the multiplication. They placed foils or thin sheets of active material in the equatorial plane between the two hemispheres, so that the area of exposed foil at a given radius was proportional to the volume of a shell of material at the same radius. In the untamped sphere measurements, the hemispheres were separated by insulators and used as the two plates of a high-pressure ionization chamber. One hemisphere with foil attached was kept at high voltage, whereas the other was used as the collecting electrode. Thus it was possible to count the fission fragments from the foil. The total number of fissions in the sphere was this number multiplied by the ratio of sphere to foil masses.

If a surrounding tamper was used, the above technique was unwieldy. Instead, thin plates of material were placed between the hemispheres, separated by cellophane "catchers." The number of fissions was estimated by measuring the radioactivity of fission fragments deposited on the cellophane.

These experiments showed that the branching ratio averaged over fission neutron energies was small, as expected, and they measured the net increase of neutrons per neutron capture. Comparison with the net increase of neutrons per fission, known to be essentially independent of energy, showed that the high-energy branching ratio was quite small.

There was the possibility of extrapolating the sphere multiplications as a function of radius to the point where multiplication would become infinite, to the critical radius. These extrapolations did give a prediction of the critical radius extremely close to the values obtained with the first metal critical assemblies (Chap. 15).

Because the accuracy with which the mock-fission source reproduced the true fission spectrum was uncertain, the multiplication experiment was repeated by the Cyclotron Group for small spheres and by the Water Boiler Group in F Division for large spheres, using a true fission source. The results checked very closely (Chap. 13).

OTHER INTEGRAL EXPERIMENTS

In late 1944, the Cyclotron Group devised an experiment to measure the number of critical masses that could be disposed safely in the target and projectile of the gun assembly. In place of active material and

181

tamper, they used a mockup that for thermal neutrons imitated the absorbing and scattering properties of these materials. By this time they could calculate the critical mass of an untamped sphere accurately, so they made a mock sphere equivalent to a just-critical metal sphere. Its decay time was measured with pulsed thermal neutrons. Then comparison with odd-shaped tamped systems, such as the gun projectile and target, could be made. This experiment not only determined approximately the number of critical masses that would be safe in the projectile and target, but also gave information on the change in degree of criticality as the projectile moved toward the target, and hence on the predetonation probability.

Another Cyclotron Group integral experiment was measurement of the multiplication rate as a function of the mass of active material. This was done by two methods, one devised by the RaLa Group of G Division. The first was a Rossi-type experiment in which the counting of a single neutron triggered the counting of further neutrons as a function of time. The second was a fast modulation experiment in which the chain was started by a neutron pulse from the cyclotron and the decay of the burst was measured as a function of time. The change in decay time for small changes in system criticality was thus measured in the near-critical region. Theoretically, the curve so obtained could be extrapolated into the supercritical position. Both ^{235}U with a tamper and ^{239}Pu with a tamper were measured. The ^{239}Pu sphere was used in the Trinity test shortly thereafter. The ^{235}U measurement permitted extrapolation to the number of critical masses in the assembled Hiroshima bomb and a semiempirical prediction of its efficiency. The equipment used exemplified the high development of counting techniques at Los Alamos. The cyclotron beam was modulated to give 0.1-µs-long pulses. Time resolution of 0.06 µs could be obtained in the counting channels.

MISCELLANEOUS EXPERIMENTS

Capture cross-section measurements were continued. The Radioactivity Group made differential measurements at various energies, and the Electrostatic Generator Group measured average capture cross sections in spheres surrounding a mock-fission source. Both groups studied neutron capture in tantalum, which seemed to show some anomalies.

Mass spectrographic analysis and neutron assay of fissionable material continued. The ^{240}Pu content of new batches of plutonium was measured, and searches were made for the rarer isotopes of uranium and plutonium. Mass spectrographic work was attached to the Electrostatic Generator Group; neutron assay work, to the Radioactivity Group.

Incidental to investigation of ^{233}U fission properties, the Radioactivity Group measured its half-life. The same group continued its investigation of gamma radiation and alpha particles emitted with fission, measured the gamma radiation of radiolanthanum needed for the RaLa measurements of the implosion, and measured neutron background from assembled initiators as these were constructed.

CHAPTER 13

F Division

INTRODUCTION

F division was formed in September 1944. As division leader, Fermi was directed to investigate potentially fruitful lines of development outside the Laboratory's main program. This included the Super's theoretical and experimental aspects and means of fission bomb assembly other than the gun and implosion. Because of Fermi's previous work on pile development, the Water Boiler Group was placed in F Division. A final group was added in February 1945 to work with the high-power Water Boiler as a neutron source and prepare for measurement of fission fragments at the Trinity test.

Organization was as follows.

F-1	The Super and General Theory	E. Teller
F-2	The Water Boiler	L. D. P. King
F-3	Super Experimentation	E. Bretscher
F-4	Fission Studies	H. L. Anderson

THE SUPER

In June 1944, Teller's group was taken from the Theoretical Physics Division and placed in an independent position, reporting to the Director. This separation recognized that this group's work was exploratory, whereas the Theoretical Division was responsible for obtaining design data for fission bombs. In September, Teller's group became the theoretical branch of the new F Division.

Theoretical work on the Super continued without surprises as analysis of the thermonuclear reaction became more quantitative and concrete. More attention was given to the theory of detonation mechanisms. Work peaked in the spring of 1945 and continued for several months after the

184

end of the war. Various models were investigated. The result sought was a bomb that burned about a cubic meter of liquid deuterium. The energy release would be equivalent to that of about ten million tons of TNT.

Energy Release in High-Temperature Collisions

Bretscher's group made new measurements of the reactivity of deuterium with deuterons and tritons having low energies of thermal motion. The new D-D data agreed with previous theoretical extrapolations from the old, high-energy data. The low-energy T-D reactions, however, had higher reactivity than anticipated. At relevant energies, this reactivity is about 30 times as great as that of the D-D reaction. This is in addition to the fourfold energy release in the T-D, as compared with the D-D, reaction.

The theoretical investigation showed that a steep rise in the rate of energy evolution could be expected with a rise in temperature. At the critical temperatures for pure deuterium, the secondary tritium reactions provided about one-third of the energy developed. Given an initial admixture of only 0.3% tritium, the tritium reaction provided almost three-fourths of the energy. A half-and-half mixture of tritium and deuterium provided a hundred times the energy evolution of pure deuterium. The time scale of a reaction in this mixture is set by the fact that the pure deuterium can reproduce the energy with which it is heated in a little less than a tenth of a microsecond.

Energy Transfer from Product Particles to the Medium

The nuclear energy released in the reaction of a deuteron or triton with a deuteron is transferred into the kinetic energy of the product particles. This energy must, in turn, be used to heat the deuteron medium to produce more reactions. The product particles transfer their energy through collisions. In general, the energy is "deposited" in the medium through a chain of collisions, during which the product particles disperse themselves.

Damage

No account of Super development would be complete without a damage estimate. It must be emphasized that these considerations are

185

qualitative. With the contemplated energies, the effects of explosions began to enter a new range, which might necessitate some account of meteorological and geological phenomena normally beyond human control. Accurate calculation was less important than a thorough canvassing of the possibilities. The following account is highly tentative, both quantitatively and in degree of thoroughness.

The 10-million-ton Super would not be the largest explosion seen on earth. Volcanic explosions and collisions by large meteorites, such as the Arizona or Siberian, undoubtedly have produced greater blast energies, perhaps a thousand or ten thousand times greater. However, these explosions were very cool compared to a thermonuclear explosion, and correspondingly more familiar in their effects.

The blast effects from a 10-million-ton Super can be scaled up from the known damage at Hiroshima and Nagasaki. Taking the area destroyed by a 10 000-ton bomb to be 10 square miles, the Super should produce equal blast destruction over a 1000-square-mile area. This would be more than enough to saturate the largest metropolitan areas.

More widespread ground damage perhaps would result from an underground or underwater explosion near a continental shelf. Because estimates indicate that a severe earthquake produces energies of the same order as the Super, the surface effects might be comparable. To produce these effects would require ignition at a depth of several miles.

This bomb begins to reach the upper limit of blast destruction possible from detonation in air. Just as the radius of destruction of a fission bomb exploded in shallow water would be limited by the depth at which it is exploded (Chap. 14), so it would be with a Super in the atmosphere. It would "blow a hole" in the atmosphere so that the maximum radius of destruction would be comparable to the depth of the atmosphere.

Neutrons and gamma rays from the Super would not be a significant part of its damage because their intensity falls off more rapidly with distance than do the blast effects. Even at Hiroshima and Nagasaki they did not cause a large percentage of casualties. From a larger bomb, their effects would be greater, but not proportionately so.

The effects of visible radiation fall off more slowly than blast effects. This destruction can be made directly proportional to the energy release. Whereas blast damage can be increased a hundredfold, visible radiation damage can be increased a thousandfold. For the first purpose, the bomb would be detonated about ten times higher than at Hiroshima and Nagasaki; for the second, about thirty times higher. The real point is that there is no limit to the possibility of detonating larger bombs at higher altitudes. Thus a Super that burned a 10-m cube of deuterium at a height of 300 miles would equal in effect a thousand "ordinary" Supers detonated at 10-mile altitudes. The area of damage would be about a million square miles. It should, of course, be emphasized that such a high-altitude weapon is only a theoretical possibility.

It is difficult to estimate damage from visible radiation. In Hiroshima and Nagasaki, the total effect was a composite of blast, gamma radiation, and visible radiation. The last was intense enough to ignite wooden structures for about a square mile. Casualties from visible radiation alone would be smaller because of protection by clothing and walls. Effects of a Super would be comparable, and either more or less intense depending on the relative military importance for extensive versus intensive burning. The figures already given would correspond to an intensity about the same as that at Hiroshima and Nagasaki.

The most worldwide destruction could come from radioactive poisons. It has been estimated that detonation of 10 000 to 100 000 fission bombs would bring the radioactive content of Earth's atmosphere to a dangerously high level. If a Super were designed with a large amount of ^{238}U to catch its neutrons and add fission energy to that of the thermonuclear reaction, only about 10 to 100 would be needed to produce an equivalent atmospheric radioactivity. Presumably such Supers would not be used in warfare, for just this reason. Without the uranium, poisonous radioactive elements could be produced only by absorption; for example, ^{14}C could be produced in the atmosphere, but not in dangerous amounts. Moreover, poisoning would be obviated by detonation above the atmosphere, where the Super's destructive effects seem greatest.

OTHER THEORETICAL TOPICS

The gloomy prospects for implosion in the fall and winter of 1944-45 made it desirable again to investigate autocatalytic and other possible methods of weapon assembly. This whole subject had been investigated earlier and given up because of the uniformly low efficiencies indicated. Autocatalysis makes use of neutron absorbers removed in the course of the initial explosion. Therefore, one or more paraffin spheres coated with ^{10}B may be placed inside the fissionable material so that the whole assembly is just subcritical. If a chain reaction is started, heating of the material compresses boron "bubbles," reduces the neutron absorbing area, and increases the degree of criticality. Thus, in principle, the progress of the explosion creates conditions favorable to its further progress. Unfortunately, the autocatalytic effect is not large enough to compensate for the poor initial conditions of this type of explosion and the result is not impressive.

Another type of assembly mechanism examined made use of shaped charges to attain much higher velocities of a slug of active material than would be possible with conventional gun mechanisms. This method also gave low efficiency when calculated for the high-neutron background of Hanford plutonium.

A topic of continuing interest was the possibility of various types of controlled or partly controlled nuclear explosions that would bridge the gap between such experiments as the "dragon" (Chap. 15) and the final weapon.

Safety calculations were made for the K-25 diffusion plant, principally estimations of critical assemblies of enriched uranium hexafluoride under various conditions and degrees of enrichment.

A topic of interest was the formation of chemical compounds in air by the Trinity nuclear explosion. Such compounds as oxides of nitrogen and ozone are poisonous, and the quantity produced had to be estimated. It was also anticipated that they would affect the radiation history of the explosion, which was to be examined spectrographically.

DEUTERIUM AND TRITIUM REACTION CROSS SECTIONS

The low-energy cross section of the T-D reaction was higher than extrapolation from high-energy data had indicated. This discovery, which lowered the ignition temperature of T-D mixtures, was made by Group F-3, the Super Experimentation Group.

Because both the T-D and D-D cross sections at low energies were known only by rather dubious extrapolation, they were to be measured simultaneously. The first measurements were made with a 50-keV Cockcroft-Walton accelerator constructed at Los Alamos for the purpose. Experiments in the 15- to 50-keV region measured the total number of disintegrations as a function of the bombarding energy from which the reaction cross sections could be derived. The target was heavy

188

ice, cooled with liquid nitrogen. The D-D reaction was produced by a deuterium ion beam, and the protons produced in the reaction were measured. For the T-D reaction, the procedure was analogous except for precautions taken to conserve the small amount of tritium available as an ion source. In this case, the alpha particles from the reaction were counted.

These measurements showed that the extrapolated values of the D-D cross section were approximately correct. The tritium cross section, however, was much larger than anticipated at energies of interest. The measurements were later extended to the 100-keV region by using a larger accelerator.

THE WATER BOILER

When the Water Boiler experiments described in Chap. 6 were completed, plans were made to develop a 5-kW boiler for use as a strong neutron source. The new boiler was put into operation in December 1944. The 5-kW power level was attainable with the amount of enriched material available at the time, cooling would be simple, and the chance of trouble from frothing or gas evolved from electrolysis of the solution would be small. Such a boiler was calculated to give a flux of 5×10^{10} neutrons/cm^2/second.

Uranyl nitrate was used rather than uranyl sulfate because of the ease with which it could be decontaminated if necessary (Chap. 17). Additional control rods were installed for more flexible operation. Water-cooling and air-flushing systems were installed, the latter to remove gaseous fission products. Finally, the Boiler had to be shielded carefully because of gamma radiation and neutrons.

Boiler decontamination proved unnecessary, even after 2500 kilowatt-hours (kWh) of operation, partly because the air-flushing system removed 30% of the fission products, and partly because the stainless steel container did not corrode.

The tamper for the high-power Boiler was chosen on the basis of experiments with the low-power Boiler. Partly because of the difficulty of procuring enough beryllia, and partly to avoid the (γ,n) reaction in beryllium, the new tamper was only a core of beryllia bricks surrounded by a layer of graphite. The new Boiler had a graphite block for thermalizing fission neutrons.

189

NEUTRON PHYSICS EXPERIMENTS

The Electrostatic Generator Group's important sphere multiplication experiments, repeated and verified in F Division, were performed independently by the Water Boiler and F-4 groups. A source of fission neutrons was obtained by feeding a beam of thermal neutrons from the Water Boiler and graphite block onto a ^{235}U target in the center of the 3-1/2- and 4-1/2-in. ^{235}U spheres. The Water Boiler Group measured the fissions in the source and throughout the sphere by catching the fission fragments on cellophane foils. F-4 measured the fissions produced by use of a small fission chamber identical to that used to measure fissions in the sphere. Comparison of fissions in the source chamber with those in the detecting chamber at various distances gave the multiplication rate. Of these two experiments, the first gave results closer to those of the Electrostatic Generator Group and to the final empirically established values of the critical mass.

Several thermal cross-section measurements for the various elements were made using the high-neutron flux from the Boiler. One such was the absorption cross section of ^{233}U. During the thermal scattering cross-section measurements of ^{235}U and ^{239}Pu, cross sections for other elements also were measured.

To calibrate gamma-ray and neutron intensity measurements at the Trinity test, the Water Boiler Group measured delayed neutron and gamma-ray emission from ^{239}Pu as a function of the delay time (the time after irradiation). They shot a slug of material with a pneumatic gun into a pipe through the middle of the Boiler and measured the decay of activity with time by means of an ionization chamber for gamma rays and a boron trifluoride counter for neutrons.

CHAPTER 14
Ordnance Division

INTRODUCTION

In the August 1944 reorganization, the Implosion Experimentation, High Explosives Development, and S-Site Groups of the old Ordnance Division were transferred to the new Explosives Division. The Instrumentation, RaLa, and Electric Detonator Groups became part of the new Weapon Physics Division. The remaining six groups (Chap. 7) made up the new Ordnance Division, but the old Proving Ground, Projectile, Target, and Source Groups became a single Gun Group. Soon two new groups were added, one to investigate underwater explosion of the weapon and to compile bombing tables for the Little Boy and Fat Man, and one as a special ordnance procurement group. By the end of September, the organization was as follows.

O-1	The Gun Group	A. F. Birch
O-2	Delivery	N. F. Ramsey
O-3	Fuse Development	R. B. Brode
O-4	Engineering	G. Galloway
O-5	Calculations	J. O. Hirschfelder
O-6	Water Delivery and Exterior Ballistics	M. M. Shapiro
O-8	Procurement	Lieutenant Colonel R. W. Lockridge

The Ordnance Division had completed its earlier research and design activities and increased its weapon test program in preparation for final delivery. In March 1945, this second activity was formalized as Project Alberta (Chap. 19). This chapter, however, discusses other Ordnance Division work that deserves separate treatment.

The Calculations Group's computations of pressure-travel curves for the gun (Chap. 7) were essentially complete, and the group was attached to the Theoretical Division as Group T-7 (Chap. 11).

Not long after August 1944, the main design of the outer cases for Little Boy and the Fat Man was determined. The Engineering Group designed the many detailed modifications of the outer case and layout of the Fat Man, which became necessary as the design of inner components progressed and as the test program revealed weaknesses in earlier designs.

The Engineering Group had been relieved of the burdens that caused most of its earlier troubles (Chap. 7). General coordination of the weapon program was taken over by the Weapons Committee (Chap. 9). Shops administration was placed under the new Shop Group of the Administrative Division. The Ordnance engineers continued to coordinate the Fat Man design under the general supervision of the Weapons Committee. George Galloway remained in charge of the group after September 1944. This group is credited with designing the outer components of Fat Man. The corresponding elements of Little Boy were designed primarily within the Gun Group.

GUN ASSEMBLY

When the high-velocity gun project for plutonium was abandoned in August 1944, gun work shifted to making a weapon out of the gun for ^{235}U. All work on guns, targets, projectiles, initiators, and bomb assembly for the gun was consolidated under Commander F. Birch. Because of uncertainty about how plutonium could be used, the objective was to produce as reliable an assembled weapon as possible. Field operations would be as simple as possible so that the ^{235}U being produced could be used effectively.

There was still considerable uncertainty about the isotopic concentration in which the uranium would be received and, hence, about the critical mass. Experimentation with the higher velocity, however, had abolished almost all other uncertainties, so there was no essential problem in projectile, target, and initiator design or in interior ballistics. The problems were how to make this unit serviceable and how to establish proof of the overall assembly. For work that depended upon the mechanical properties of the active material, the normal uranium metal was a perfect substitute.

One lesson learned was that it required six months to procure new guns. Guns ordered in March arrived in October, and the special mount

192

arrived even later. It was December before active proof work on the new model could begin. Proof of tubes, as such, consisted of instrumented firing of each two or three times at 1000 ft/s with a 200-pound proje:tile. They were then greased and stored for future use. A few tubes were used in other experiments, notably proof of fuel-scale targets and determination of the delay between application of the firing current and emergence of the projectile. In addition to these "live" barrels, dummies were procured for use in drop tests of the assembled bombs. These guns, made mostly from discarded Naval guns, were not meant for firing and required no "proof."

The gun presented no ballistics problem, but it was far from conventional in appearance. It weighed only about one-half ton, was 6 ft long, and had a large thread on its muzzle. Two types of these guns were made: Type A, of high-alloy steel, not radially expanded, and with three primers inserted radially; and Type B, of more ordinary steel, radially expanded, and with the primers inserted in the "mushroom" (nose of the breech assembly). The primers and type of propellant were proved in an earlier gun. The Type-B gun was selected for further production because of its light weight and because the process of radial expansion was an excellent test of the forging quality.

Between August and December, proving was done on targets, projectiles, and initiators at reduced scale. When it became apparent that work had to be done at full scale before the new guns arrived, an old barrel was bored out to diameter and shipped to Site Y. The laws of dimensional scaling worked out quite well, and a large amount of design research was done on the 3-in./50 and the 20-mm Hispano guns. This was a particularly acceptable procedure for testing tampers. Many other parameters were varied, extensively in 20-mm tests and less extensively at 3-in. scale, and tested, using substitute materials. So, by the time the gun arrived, reasonably firm designs of the target and projectile components had been established.

From December on, practically all firing was at full scale to determine whether the results at small scale were misleading, as they might be because of the inability to duplicate heat treatments at different linear scales. The target cases for full-scale tests were impressive objects and hard to handle, particularly to take apart for inspection after the shot. That difficulty was alleviated by developing a tapered assembly that could be pushed apart hydraulically so that the outer cases of high-alloy steel could be reused. An amazing fact about the target cases is that the

193

first was the best ever made. This case was used four times at Anchor Ranch Range and subsequently fitted to the bomb and dropped on Hiroshima. Certain failures in subsequent target cases of the same design emphasized the importance of careful heat-treating and led to slight design modifications.

Generally, the pieces heat-treated at Site Y were far superior to those procured from industry. Many dummy targets had to be made for drop tests of the assembled bomb, and these also were not made carefully. The fact that they shattered when the projectile seated, in those drops where a live gun was used, only helped when the inner portions were recovered for study.

By far, the most extensive program was engineering and proving of assembled units, ironing out the mechanical integration of the bomb in cooperation with the Fusing and Delivery Groups. Varying completeness of assembly was required, depending on the completeness of the test. Practice drops ranged from tests of fusing and informers and bomb ballistics in which dummy guns and targets were used to drops of units that were complete except that ordinary uranium was used instead of the active variety. In the latter tests, the bomb was dug up for further study of the assembly.

Bomb design was decided in February 1945. So that only the Laboratory would possess the complete design, the heavy fabrication was divided among three independent plants. The gun and breech were made at the Naval Gun Factory; the target case, its adapter to the muzzle threads, and its suspension lug were made by the Naval Ordnance Plant at Centerline, Michigan; and the bomb tail, fairing, and various mounting brackets were made by the Expert Tool and Die Company, Detroit, Michigan. The contract with the latter firm was through the Project's Engineering Office in Detroit (Chap. 7). Smaller components, such as the projectile, target inserts, and fuse elements, were made or modified for their ultimate use at Site Y.

The component parts were assembled at Wendover Field, Utah, in preparation for drop tests. However, some assemblies were made at Site Y, both for preliminary experience and for instrumented ground tests. An assembled gun and target system was fired at Anchor Range in free recoil with complete success. At Wendover, 32 successful drops were made, and in only one did the gun fail to fire. This failure was traced to incorrect electrical connections.

The airborne tests led to one breech revision. It was desirable to be able to load the gun after takeoff or unload it before landing. The breech was modified to allow one man in the bomb bay of the plane to load or unload the powder bags.

ARMING AND FUSING

The Fuse Group was involved in the test program at Wendover Field, Utah. When the final tests began, in October 1944, overall design of the arming and fusing system, begun in April, was complete. The main component of the final fusing system was the modified APS/13 tail warning device, called "Archie." This radar device would close a relay at ₊a predetermined altitude above the target. Four such units were used in each fuse, with a network of relays so arranged that when any two units fired, the device would send a firing signal into the next stage. This stage was a bank of clock-operated switches, started by arming wires that were pulled out of the clocks when the bomb dropped from the plane's bomb bay. These clock switches were not closed until 15 seconds after the bomb was released. They were to prevent detonation in case the A units were fired by signals reflected from the plane. A second arming device was a pressure switch that did not close until subjected to a pressure corresponding to a 7000-foot altitude. In the gun weapon, the firing signal went directly to the gun primers. In the implosion weapon, this signal actuated the electronic switch that closed the high-voltage firing circuit.

The alternative to Archie, but with lower altitude range, was the PRM or "Amos" unit, another radar device developed at the University of Michigan. Amos was stand-by equipment in case lower altitude firing should be favored. It was tested and found fairly satisfactory, but the tests were not extensive. As in all parts of Project Alberta, the testing programs served to train combat crews.

BOMB BALLISTICS

Bombing tables for Little Boy and Fat Man were generated at Los Alamos. The ballistic data were from field measurements at Sandy

195

Beach, the Salton Sea Naval Air Station (a small rocket-testing station) that afforded an approach over water nearly at sea level, which simulated the conditions over Japan. Similar data were obtained for the blockbuster "pumpkin" program by the Camel project at Inyokern, and the two groups consulted about measurement techniques and the data obtained. The Los Alamos work was done by Group O-6.

SURFACE AND UNDERWATER EXPLOSIONS

A tactical use considered for the bomb was as a weapon to damage harbors, by surface or underwater detonaton. However, it became clear that surface or shallow underwater detonation would expend much of the explosion's energy in producing cavitation, and relatively little energy would go into a shock wave in water. To maximize shock-wave damage, much deeper detonation would be necessary than would be possible in harbors.

The effects of shallow explosions were investigated at a very small scale by the sudden withdrawal of an immersed cylinder that resulted in the creation, and sudden collapse, of a cylindrical cavity in water. Later experiments were made with a few ounces of explosive at a 1- to 2-ft depth. The amplitudes of gravity waves produced in these tests could be scaled up to explosions of the range of interest to the Laboratory.

The data showed, surprisingly, that a surface explosion produced larger gravity waves than a subsurface explosion of the same size. A theoretical analysis gave scaling laws that made it possible to predict the effects of surface or near-surface detonation of atomic bombs. This program, begun by McMillan, was the work of the Water Delivery and Exterior Ballistics Group, assisted by Penney and von Neumann.

196

CHAPTER 15

Weapon Physics Division

INTRODUCTION

Upon its organization in August 1944, the Weapon Physics or G Division (G for gadget, code for weapon) was directed to experiment on critical assembly of active materials and to devise and use methods for studying the implosion. In April 1945, it was directed to design and procure the implosion tamper, as well as the active core. G Division also undertook the design and testing of an implosion initiator and of electric detonators for the high explosive. The Electronics Group was transferred from the Experimental Physics Division to G Division, and the Ordnance Division Photographic Section became G Division's Photographic Group.

Division organization was as follows.

G-1	Critical Assemblies	O. R. Frisch
G-2	The X-Ray Method	L. G. Parratt
G-3	The Magnetic Method	E. M. McMillan
G-4	Electronics	W. A. Higginbotham
G-5	The Betatron Method	S. H. Neddermeyer
G-6	The RaLa Method	B. Rossi
G-7	Electric Detonators	L. W. Alvarez
G-8	The Electric Method	D. K. Froman
G-9	(Absorbed in Group G-1)	
G-10	Initiator Group	C. L. Critchfield
G-11	Optics	J. E. Mack

A large new laboratory, Gamma Building, was built for G-Division work. New firing sites with small laboratory buildings were established (Fig. 3). Most of G Division occupied new office, laboratory, and field facilities, but its work was well under way by the beginning of October 1944.

197

CRITICAL ASSEMBLIES

The Critical Assemblies Group worked at Omega Site (Chap. 6) with the Water Boiler Group. It experimented with critical amounts of active materials, including both hydrides and metals, and investigated the necessary precautions to be observed in handling and fabricating active materials—to prevent uncontrolled nuclear reactions. When G Division acquired the job of designing and preparing the implosion bomb core and tamper, called the "pit assembly," the Critical Assemblies Group was given this responsibility.

Early in this group's existence, many critical assemblies were made with various uranium hydride mixtures. These assemblies were investigated rigorously because there was not yet enough material for a metal critical assembly without hydrogen and because successively lowering the hydrogen content of the material as more ^{235}U became available gave experience with faster and faster reactions. Also, it was still possible that hydride bombs using small amounts of material might be built.

By November 1944, enough hydride-plastic cubes of composition UH_{10} had been accumulated to make a cubical reacting assembly in the beryllia tamper, if the active composition was reduced to UH_{80} by stacking seven polyethylene cubes for each cube of UH_{10} plastic. Further experiments were made with less hydrogen and other tampers. This hydride was sent back to the chemists and metallurgists for recovery and conversion to metal in February 1945, and the program of hydride critical assemblies was ended.

The most spectacular hydride experiments were those in which a slug of UH_{30} was dropped through the center of an almost critical assembly of UH_{30}, so that for a short time the assembly was supercritical for prompt neutrons alone. This experiment was called "tickling the dragon's tail," or simply the "dragon." The velocity of the falling slug was measured electrically. Before the experiment was actually performed, several tests proved that it was safe; for example, that the plastic would not expand under strong neutron irradiation, so the slug would not stick and cause an explosion. On January 18, 1945, strong neutron bursts were obtained of approximately 10^{12} neutrons.

These experiments evidenced an explosive chain reaction, and they produced energy up to 20 MW, with temperature rising in the hydride up to 2°C/ms. The strongest burst produced 10^{15} neutrons. "The dragon" is

198

historically important because it was the first controlled nuclear reaction that was supercritical with prompt neutrons alone.

Because of the burst intensity and short duration, better delayed-neutron measurements were possible than had been made previously. Several short periods of delayed emission were found, down to about 10 ms, which had not been reported before.

The Critical Assemblies Group did many safety tests for other groups. Most of these involved placing varying amounts of enriched uranium with differing geometries in water, in order to determine the conditions under which the accidental flooding of active material might be dangerous. During these tests the first accident with critical materials occurred. A large amount of enriched uranium, surrounded by poly-ethylene, had been placed in a container to which water was being slowly admitted. The critical condition was reached sooner than expected, and before the water level could be sufficiently lowered, the reaction became quite intense. No ill effects were felt by the men involved, although one lost a little cranial hair. The material was so radiative for several days that experiments planned for those days had to be postponed.

Similar safety tests were made on gun assembly models because a bomb might immerse accidentally when transporting it or jettisoning it from aircraft. Immersion could also be considered the limiting case of wetting from other possible sources.

Tests made with ^{235}U metal assemblies were much more numerous than those with ^{239}Pu, because sufficient plutonium was not available until later. The first ^{239}Pu critical assembly, which was a water solution with a beryllia tamper, was made in April 1945. By fabricating larger and larger spheres of plutonium and inserting these in a bomb mock-up, the critical mass in a bomb was determined.

In April 1945, when G Division was made responsible for designing the pit assembly of the bomb, the "G Engineers," Morrison and Holloway, were taken for this work from the Critical Assemblies Group. The G Engineers worked with the design staffs of X and O Divisions in the final detailed design of the implosion bomb.

IMPLOSION STUDIES

The X-Ray Method

The x-raying of small spherical charges was developed in the Ordnance Division and was in successful use in August 1944 (Chap. 7).

The small-scale work was placed then in X Division, and the G-Division X-Ray Group extended this implosion study method to larger scale. In addition to this development work, the G Group continued servicing and improving the x-ray tubes for the X-Division work. Other demands made upon them included the radiography of explosive charges for the RaLa Group. The final weapon tamper was also radiographically examined at this later time, but radioactive sources of gamma rays were used rather than x-rays.

Early in the program, the chief objective was to adapt x-ray techniques to large-scale implosions (up to 200 pounds of high explosives). The workers tried to follow the implosion course by using x-rays as a time function on a grid of small Geiger counters. These counters were to be either 1 or 3 mm in diameter and cross-shaped in the x-ray record. Mainly, the program aims were reliable counters and reduction of the scattered radiation that would be recorded. These objectives were pursued relentlessly, but because of the great technical difficulties involved, with little real success, so the program was dropped in March 1945. Although the difficulties were probably not insurmountable, there was insufficient justification for retaining the highly trained personnel required for this work. It had also become clear that the same essential information was obtainable by repeated betatron exposures (Chap. 15).

Although the counter and electronic circuit development and scattered radiation reduction were questionable in the new x-ray program, preparation of a field site (P Site), where the technique could be tested was completed by early fall 1944. The protection required for x-ray equipment near exploding charges was designed so that it would be useful in other x-ray experiments as well. In preparing to use the x-ray and counter technique, the problems of using the magnetic method (Chap. 15) with the x-ray method were solved at P Site. The first combined x-ray and magnetic record was obtained in late January 1945, and this was directly applicable to combining betatron and magnetic records (Chap. 15).

As emphasis on lens design increased, flash x-ray photography helped study the detonation waves. These studies paralleled those of the x-ray section of the Implosion Studies Group X-1. In April 1945, a second x-ray team began working on the initiator problem, and much experimental work on initiators was accomplished. P Site was used for the initiator program, but various recovery experiments were done out there, and later the alpha-counting experiment was begun (Chap. 15).

The Magnetic Method

By August 1944, the magnetic method was a practical way to determine the velocity of the external metal surface of an imploding sphere (Chap. 7). By integration the average compression was also obtained. The development of the method was directed along three main lines: (1) the improvement of the instrumentation and adaptation to electric detonation for large charges; (2) cooperation with RaLa, x-ray, and betatron experiments to get coordinated records; and (3) the development of new methods.

A separate proving ground in Pajarito Canyon (Fig. 3) was completed in December 1944. Meanwhile, laboratory work improved the circuits and developed shielding techniques. When field work began, the main problem was to protect the magnetic record from spurious signals caused by the electric detonators and by the static charges developed in the explosion. New results were obtained when it became possible to "purify" the magnetic records and interpret their details. Several reflected shock waves from the metal core could be recognized. The intersection of detonation waves also produced reliable signals.

The magnetic techniques were adapted to larger charges, more detonation points, and finally to electric detonation of lenses. The increase of surface velocity with number of detonation points was demonstrated in this way.

A unique property of the magnetic method was that it could be applied to a full-scale implosion assembly. A chief objective was to prepare for investigation of full-scale shots. Unfortunately, the method could not be used at Trinity because that shot had to be fired with the same type of metal case as was used with the bomb assembly, and that case made it impossible to obtain magnetic records.

As technical difficulties were surmounted, the method was applied successfully at various tamper diameters. At these scales, the method was used alone as well as in coordination with the RaLa and betatron methods. The timing results were particularly useful in these cooperative tests. By itself, its main value lay in following the dynamic sequence in the implosion and giving data on how the sequence changes with increasing implosion size.

The Betatron

Use of the betatron as an instrument to study implosion was proposed earlier but not decided upon until about the time G Division was formed.

201

The 15-MeV betatron of the University of Illinois was obtained for this work after its inventor, D. W. Kerst, made an expert analysis of its capabilities and after Neddermeyer checked on possible use of a vertical cloud chamber for recording. The early fall of 1944 saw experiments on the performance (rise and burst times) of the betatron at Illinois, and on the refinement of shadow recording by flash photography of the cloud chamber.

Construction of the betatron site (K Site) began concurrently. As at the x-ray site, the test implosion was detonated between two closely spaced bomb-proof buildings. One building contained the high-voltage gamma-ray source; the other, the cloud chamber and recording equipment. Equipment was protected from the blast by aluminum nose pieces over the exit and entry ports through which the radiation was passed, and by shock-mounting all equipment that could be damaged by the shock wave. Construction and installation were completed early in 1945. Meantime, the cloud chamber technique was improved by trapping the low-energy tracks to obtain better definition of the image in cloud chamber, and by developing an intricate and ingenious sequence circuit to correlate the firing of the explosive charge, the gamma-ray burst, and the expansion and photographing of the cloud chamber.

By the time the first test shots were fired, emphasis was mostly on the solid metal implosion assembly, and results from the betatron were crucial because of the probable abandonment of the counter x-ray method. The field installations, electronic circuits, and cloud chamber were adapted to their special purposes so rapidly that by April 1945, systematic study of compression under scaled lens shots was started. Early results showed an unexpected irregularity, but evidenced definite compression of the uranium core by an amount close to that predicted theoretically. By June, the data spread was reduced enough to provide valuable experimental information for correcting the constants used in the theory of the bomb. Data from the magnetic method indicated that the spread in experimental results might be caused by variation in the high-explosive behavior.

The betatron program was one of the few that maintained a single purpose and function from inception to production of final results. Although for this reason its history can be written briefly, the technical achievements are among the most impressive at Los Alamos.

202

The RaLa Method

The RaLa program, like the betatron effort, was a single-purpose, more or less direct adaptation of a known radiographic technique to study implosions. Both methods had to be adapted to microsecond time resolutions. For RaLa, this meant developing unprecedented performance with ionization chambers as well as unheard-of gamma radiation sources (Chap. 17). The source of radioactivity was the radiobarium from fission products of the chain-reacting pile. By the fall of 1944, radiolanthanum extraction had been put on a working basis and fast enough ionization chambers were produced.

The RaLa firing program got under way in Bayo Canyon (see Fig. 3) on October 14, 1944. Because of lack of knowledge as to how serious the contamination would be after a source equivalent to many hundred grams of radium was imploded, permanent installations were held to a minimum for these first trials. Sealed army tanks were used as observation stations. The contamination danger seemed so small, however, that permanent bomb-proofs were installed in November 1944. The early shots were fired by multipoint Primacord systems, and the results were correspondingly erratic. However, in February electric detonators were used and the quality of RaLa results immediately improved. During the spring, the ionization chambers were further refined. Protection was provided against spurious signals and provision was made for obtaining simultaneous magnetic records. This so improved the significance of the results that RaLa became the most important single experiment affecting final bomb design. The principal advantage was that analysis of the data gave an average of the compression as a function of time. This threw light on two of the most critical attributes of the system, the duration and the magnitude of maximum compression.

A second RaLa site was started in March 1945. The first high-quality lenses produced were allocated to this work in April, so that the final design could be achieved as rapidly as possible. RaLa's success was possible only because of sparing no pains in constructing and testing the equipment. Each shot involved destruction of ionization chambers and electronic equipment that could not be removed from the testing area. Each new set of equipment had to be checked completely to be sure of proper operation on the single occasion that counted.

The Electric Method

The principal technique of the electric method of investigating the implosion was to electronically record the electrical contacts formed between the imploding sphere and prearranged wires. This was a well-established method but, as always in adapting old methods to implosion work, it was necessary to sharpen the time resolution and meet more serious interference with the circuits on the part of the explosive than had been encountered previously.

The first effort in August 1944, was to make oscillographic recordings of the position of a plate as a function of time, when the plate was accelerated by a high-explosive charge. This was done by merely spacing small pins at intervals differing by about a millimeter from the surface to be accelerated. By early 1945 quantitative data on plate acceleration were being obtained. The method was then adapted to severe implosions of partial spheres, and finally used on lens systems in which only one lens was omitted. Information on velocities of material at various depths in the core, and on shock-wave velocities gave the Theoretical Division direct, quantitative data on which to test its conclusions and base its predictions for the implosion bomb.

The condenser microphone and resistance wire methods also were studied. The Initiator Group pursued use of tourmaline crystals for timing signals, as well as for possible pressure recording. Thus, they were able to determine the pressure distribution in a detonation wave, and the operation, and the location and velocities of spalls and jets. They also measured many velocities of interest for initiator design, and the velocity of shock-operated jets formed in the thin crack between tamper halves.

The Electric Method was (except for the magnetic method) the only one that could be applied at very large scale. By June 1945, it was being applied regularly at large scale and to almost complete spheres of lens charges.

INITIATORS

The Initiator Group was concerned primarily with development of a bomb component rather than with a particular research method.

A proving ground was built in Sandia Canyon (Fig. 3) for initiator work. The nature of proof was not predetermined, however, because the performance requirements could be measured only by operation in the bomb itself. Many leads were followed, accordingly, in the attempt to simulate actual conditions as closely as possible and to learn as much as possible about the mechanisms that were supposed to make the variouus designs work.

A large part of the problem was procurement and radiochemical preparation of polonium (Chap. 17). There was theoretical investigation of proposed initiator mechanisms, and much experimental corroboration was produced by the x-ray, flash photography, and electric methods. The Initiator Group's function was to coordinate these researches with its own and with trends in design.

By February 1945, it was decided that the initiator should be of an (α,n) type. Many designs were invented, and the work of selecting promising ones and testing them began. All designs involved some mechanism for mixing the α-n material, previously kept separated by some α-absorbing material under the impact of the incoming shock wave. At this time, the possibility of complete recovery of imploding spheres was discovered. Recovery of units with and without α-n materials in them was an important part of the early work. Although the results built confidence in the feasibility of initiators, this work was superseded by studies of specific mechanisms.

The deadline for a decision on initiator feasibility was met, with a favorable report, on May 1, 1945. Acceptance specifications and fabrication procedures were established, and the first service unit was finished early in June. The Initiator Group set up handling procedures in cooperation with the G Engineers and the Radiochemists for production, surveillance, and recording of the history of each unit. This work had just begun when the war ended.

ELECTRIC DETONATORS

The need for multipoint detonation of the implosion to be highly simultaneous was realized from the beginning, as was the potential usefulness of electric detonators for this purpose. Development of Primacord branching systems, investigation of timing errors associated with them, and experimental and theoretical work on the importance of such errors were undertaken (Chap. 7). Electric detonators were already being developed, and in August 1944 this work was transferred to the new G Division.

The desired perfection was, of course, not built into commercial electric detonators, and the problem of firing many such units simultanously had not been faced. This required developing an adequate high-voltage power supply and a simultaneous switching device. The high-voltage supply could be made from commercial units and was simply a bank of high-voltage condensers. The switch problem, acute at first, was met by straightforward developments that led to several designs adequate for experimental purposes.

The switch finally developed for weapon use and the particular detonators were proved in X Division (Chap. 16). But having a firing circuit adequate for experimental purposes soon showed a serious lack of simultaneity and a source of failure in the detonators themselves.

Detonators of two types were studied, the bridge wire and spark gap. Errors were reduced by use of powerful high-voltage spark discharges. Because it was assumed that even with powerful electric discharges only primary explosives, normally used for detonation of less sensitive explosives, could be used, a contract was arranged with the Hercules Powder Company for production of several types of experimental detonators, mostly loaded with lead azide. Because of manufacturing delays, only a few batches of these detonators were received. Meantime, facilities for rapidly loading experimental detonators were developed at Los Alamos, and this local supply became the primary one.

The criterion for acceptance was, of course, a small maximum time spread and few failures. Tests were made by oscillographic methods and later by photographic observation using the rotating mirror camera.

In the spring of 1945, the Detonator Group supervised preparation of thousands of detonators for field work in the implosion program. The specifications for the combat service units were completed by the end of May, and Group X-7 became responsible for putting the units into service (Chap. 16).

PHOTOGRAPHY

G Divison's Photographic and Optics Group, until August 1944 part of the Instrumentation Group of the Ordnance Division, was responsible for developing optical instruments and operating technical photographic facilities. Partly a service and partly an experimental group, it prepared photographic and spectrographic equipment for the Trinity test. Before that, it played a substantial part in the Wendover drop test.

Besides procuring photographic equipment and maintaining a photographic stockroom, the Photographic Group designed and built cathode-ray oscilloscope cameras, armored still cameras for various high-explosives test sites, an armored stereoscopic camera for flash photography of imploding hemispheres (Chap. 16), and a cloud chamber stereoscopic system for the betatron cloud chamber. They built boresights and a photovelocity system for the 20-mm gun and designed and developed the rotating-prism and rotating-mirror cameras. Much of the Laboratory's accurate, high-speed data recording can be credited to the Photographic Group.

ELECTRONICS

Early in September 1944, the Electronics Group was expanded greatly because of heavy demands upon it. This group made many standard items, such as scalers, power supplies, discriminators, and two types of amplifiers with rise times of 0.1 and 0.5 μs, and also designed and built many new pieces of equipment. They built a scaler with a resolving time of 0.5 μs per stage, amplifiers with less than 0.05-ms rise time, a 10-channel pulse-height analyzer, and a 10-channel time analyzer. They also built most of the electronic equipment used on the betatron and much auxiliary equipment for the cyclotron, Van de Graaffs, and D-D sources. They designed and built new sweep circuits, delay circuits, calibrating circuits, and a host of other circuits needed by a laboratory as large and diversified as Los Alamos had become. Membership of the group averaged about 50, ranging from electricians to designers of new equipment. The number of major constuction items ranged over a thousand, and the number of service and repair items far beyond that figure.

It is safe to say that without the Electronic Group's services many of the experiments of G Division and the Laboratory would not have been done so well, they would have required more time, and completion of the bomb itself would have been delayed.

207

CHAPTER 16

Explosives Division

ORGANIZATION AND LIAISON

The Explosives Division, organized in August 1944, consisted originally of the Explosives Experimentation, High Explosives, and S-Site Groups of the old Ordnance Division. It grew rapidly by subdividing groups into sections and adding new groups. The following organization lists are for September 1944 and August 1945.

September 1944

X-1	Implosion Research	Commander N. E. Bradbury
X-1A	Photography with Flash X Rays	K. Greisen
X-1B	Terminal Observations	H. Linschitz
X-1C	Flash Photography	W. Koski
X-1D	Rotating Prism Camera	J. Hoffman
X-1E	Charge Inspection	Technician Third Class G. H. Tenney
X-2	Development, Engineering, and Tests	K. T. Bainbridge
X-2A	Engineering	R. W. Henderson
X-2B	High Explosives	Lieutenant W. F. Schaffer
X-2C	Test Measurements	L. Fussell, Jr.
X-3	Explosives Development and Production	Captain J. O. Ackerman
X-3A	Experimental Section	Lieutenant J. D. Hopper

August 1945

X-1	Implosion Research	Commander N. E. Bradbury
X-1B	Terminal Observations	H. Linschitz
X-1C	Flash Photography	W. Koski

X-1D	Rotating Prism Camera	J. Hoffman
X-1E	Charge Inspection	Master Sergeant
		G. H. Tenney
X-2	Engineering	R. W. Henderson
X-3	Explosives Development	Major J. O. Ackerman
	and Production	
X-3A	Experimental Section	Lieutenant J. D. Hopper
X-3B	Special Research Problems	D. H. Gurinsky
X-3C	Production Section	R. A. Popham
X-3D	Engineering	B. Weidenbaum
X-3E	Maintenance and Service	Lieutenant G. C. Chappell
X-4	Model Design, Engineering	E. A. Long
	Service, and Consulting	
X-5	Detonating Circuit	L. Fussell, Jr.
X-6	Assembly and Assembly	Commander N. E. Bradbury
	Tests	
X-7	Detonator Developments	K. Greisen

The new division was headed by G. B. Kistiakowsky, previously Deputy Division Leader of the old Ordnance Division in charge of the Implosion Project. Kistiakowsky was assisted by Major W. A. Stevens for administration and construction. Group X-4 was created early in October under E. A. Long and J. W. Stout and was charged with engineering of molds for S Site, research on sintered and plastic-bonded explosives, and miscellaneous services. A section added to this group in December 1944, under W. G. Marley, was responsible for some aspects of lens research. In November 1944, a second research section under D. H. Gurinsky was added to Group X-3.

In March 1945, Group X-2, under Bainbridge, was dissolved. Bainbridge was put in charge of the new Trinity Project, preparing for the Trinity test. Three new groups were formed to continue the work of X-2: an Engineering Group, X-2, under R. W. Henderson; a Detonator Firing Circuit Group, X-5, under L. Fussell, Jr., whose section of X-2 had already acquired responsibility for designing and developing a firing unit for the electric detonators; and an Assembly and Assembly Test Group, X-6, under N. E. Bradbury. In March 1945, M. F. Roy joined the Laboratory as Assistant to the X-Division Leader. In May 1945, Section X-1A was discontinued and a new group formed under K. Greisen, X-7, to carry through the final development of detonators.

209

X Division was, in essence, responsible for developing the explosive components of the implosion bomb, as G Division was responsible for developing the active components. X Division was made responsible for

investigating methods of detonating the high-explosive components;

developing methods for improving high-explosive castings;

developing lens systems and methods for fabricating and testing them;

developing engineering designs for the explosive and detonating components of the actual weapon;

providing explosive charges for implosion studies in G and X Divisions; and

specifying and initiating design of those parts of the final weapon they had developed.

Because X and G Divisions were both responsible for the fundamental implosion development work, they cooperated closely. Both, for example, studied implosion dynamics. G Division was concerned primarily with measuring the assembly velocity and comparison of the bomb pit; X Division, with explosive techniques for achieving high-velocity assembly and compression. Methods already developed and known to be reliable remained in X Division, while G Division concentrated on developing new methods.

X Division was responsible for the explosive components and O was responsible for the case, including work necessary to convert a bomb such as was set off at the Trinity Test into a combat weapon.

The important outside Explosives Division contacts were with the Explosives Research Laboratory (ERL) at Bruceton, and later with the Camel Project of the California Institute of Technology. Work also was done at the Yorktown Naval Mine Depot and by the Hercules Powder Company. Some lens development, and investigation of possible fast- and slow-component explosives, had already been done at ERL. The Yorktown Naval Mine Depot supplied explosives and, when a new type was under investigation, supplied information on its physical and explosive properties. Hercules produced spark-gap detonators to Los Alamos specifications.

X Division applied a lot of theoretical effort to interpretation of implosion jets. Their work also included observation of imploding cylinders and hemispheres by various optical techniques, and x-ray observation of small (1-1/2 in.) spheres, by the Implosion Research Group.

Three principal techniques were used: rotating-pyramid and rotating-mirror photography, high-explosive flash photography, and flash x-ray photography. Measurement at maximum compression was possible by careful timing of the x-ray flash and the explosive detonation. The rotating-prism or mirror techniques and the high-explosive flash technique gave shadow photography of imploding cylinders. Images were obtained on the same negative at different stages of collapse by using a device that gave a succession of high-explosive flashes. Hemispherical implosions were observed by reflected high-explosive flash light and photographed stereoscopically. These observations made it possible to verify predictions about jetting based on two-dimensional cylinder results, and removed the last possible doubt that the cylinder jets might be optical illusions. Early in 1945, this hemisphere technique became standard and the cylinder work was dropped.

From so-called slab shots, the Implosion Research Group got experimental data needed to test the theory that jets were caused by shock-waves interaction. Slabs of explosive were placed on top of metal slabs and detonated at two or more points. The effects of interaction between the shock waves were observed by flash photography and terminal observation. It was thus possible to study jetting as a function of the angle between detonation waves.

The need for a special means of initiating the high explosive at several points "simultaneously" was recognized from the beginning, but the degree of simultaneity required was not known.

EXPLOSIVES DEVELOPMENT AND PRODUCTION

The size of the Los Alamos undertaking is indicated by X-Division records, which show that in an 18-month period about 20 000 castings of adequate experimental quality were produced, and a much larger number rejected through quality control tests. At its peak, S Site used over 100 000 pounds of high explosives per month. Seven or eight different explosive materials were used in castings in an enormous variety of shapes and sizes. The principal explosive was Composition B; others were Torpex, Pentolite, Baronal, and baratol. Casting methods were used wherever possible. In some of the experimental lens work, slow-component explosives were produced by hand tamping, until castable slow explosives were obtained.

211

It may be thought that in high-explosives casting there was an existing art that could have been used. This, unfortunately, was not so. Military techniques for loading explosives were crude when measured by standards needed in the implosion. There had been very little incentive to regard high explosives as possible precision means for producing phenomena outside the ordinary range of experimental physical techniques. Hence, many of the problems faced were new. Their solution was undertaken primarily by the Explosives Development and Production Group.

Machining of explosives, entailed by the use of risers and overcasting (in which its imperfections had been concentrated) gave a charge that met the necessary standards of quality. This machining program required close cooperation among the explosives groups and the Shop Group (Chap. 9). Holding jigs and cutters finally used represented several months of development and experimentation. The earliest molds were designed to minimize machining by use of small risers. This kind of machining normally is considered very dangerous and its development is considered revolutionary in explosives manufacturing. By careful design and control of operating conditions, all hazards were virtually eliminated, as is shown by the fact that more than 50 000 major machining operations were conducted without detonating the explosive.

S Site always was forced by growing demands for explosives production to grow faster than completion of facilities and arrival of personnel would easily allow. That S Site successfully fulfilled its objective is largely the result of faithful and efficient work by the soldiers who constituted more than 90% of its staff.

S Site, or Sawmill Site as it was called originally, was placed in limited operation in May 1944 after a winter of construction and difficult equipment procurement. It was subsequently enlarged and its equipment completely modified.

All early castings made in the large casting buildings at S Site were cylinders of Composition B weighing 30-500 pounds. Small castings of Pentolite were being cast at the Anchor Ranch casting room, placed in operation in October 1943. This small casting room was equipped with four 2-gallon kettles. It continued to furnish very small cylinders and special castings of Pentolite throughout the war.

Examination of the cylinder castings made by standard methods of ordnance practice showed that extensive research would be necessary to produce large castings of the required quality. Therefore, a Research and

Development Section was set up in June 1944. It started with four men but was enlarged throughout the war. Building S-28, the trimming building, was used until the laboratory buildings were completed in September 1945.

Early research was confined to determining the nature of the explosives and securing castings adequate for the testing program. Little time was available for the longer range program of making large castings of uniform quality.

Beginning in November 1944, the experimental work was divided into solving of problems encountered in making high-quality large castings with slurries and development of methods for manufacturing explosive lenses.

The adoption of lenses created a tremendous development program for S Site in research for and production of many forms of lenses. Much of the research in late 1944 and early 1945 was on production line manufacture of small-scale lens charges. Incidental research included development casting of special explosives such as Tropex and baronal.

When use of lenses became a certainty, the designs for a series of lens molds covering the range of predictable explosive rates and lens sizes, from small to full scale, were frozen. This design freezing was a wise decision, as it permitted later schedules to be met. However, it greatly increased the burden on research, to make available molds work despite their known deficiencies.

In February 1945, with adoption of cast baratol-Composition-B lenses, research moved feverishly toward developing almost foolproof methods for manufacturing full-scale lenses and inner components of a very high quality and reproducibility. As ultimately manufactured, these charges were reproducible to very close tolerances.

Late in 1944, the demand for various sizes and shapes of castings increased greatly because new methods of studying implosion were introduced. The new testing methods included the betatron, x-ray equipment, new mirror camera installations, RaLa, flash x rays, the magnetic method, and the "pin" method. These tests stimulated requirements for lenses as well as solid charges.

A new small casting line, consisting of five buildings, was constructed between December 1944 and March 1945. It was designed to meet the predictable requirements for small castings, but later was called upon to meet more varied requirements in the lens program. The quantity production of small lenses used in the testing program was conducted in

213

this line of buildings. The adoption of lenses and full machining of charges for the full-scale unit revealed that the facilities were inadequate for full-scale work. Therefore, Buildings S-25 and S-26 were built between March and June 1945.

Demand was increased for full-scale, high-quality castings not only for Trinity but for combat use, manufacturing, and research, so control and production were intermingled during June and July 1945. New problems involving major changes in process and control arose continually. The requirements for inspection, both by physical measurement and by gamma- and x-ray examination, increased so greatly as to make the success of this phase of engineering one of the primary accomplishments of the period.

Throughout the work at S Site, it was common to speak of "production casting." This was really a misnomer because all castings were tailored and produced in reasonably small quantities in a large variety of molds. Such production differed from ordnance production, not only in that its maximum monthly output of 100 000 pounds of castings would be one day's run of a standard ordnance line, but more significantly in that it consisted of thousands of high-quality 1- to 120-pound charges rather than a few large units weighing thousands of pounds each.

Because men with scientific or explosives experience and willingness to work with explosives were required, each new request for assignment of enlisted men or civilians was followed by a long delay. The site was always hampered by a shortage of manpower. However, the growth that did occur made training new men a continuous problem. Similarly, the shortage of personnel available to the construction contractor delayed construction at S Site, making it necessary to reduce building requirements to the minimum to meet time schedules.

RECOVERY PROGRAM

The recovery program (Chap. 7) was dropped, after extensive research and development, for reasons extraneous to the program itself. Originally, it was argued that the chances for a nuclear explosion at the first test might be small, and it might be necessary to recover the active material. With the weapon finally designed, the chances for *no* nuclear explosion became very small. Also, the use of a containing sphere or "water baffle" would make it difficult to obtain information that would explain a partially successful nuclear explosion, which by this time was a more realistic worry than complete failure. Hence, when the first test was made, Jumbo, which was such a magnificent piece of engineering, stood idly by half a mile from the test tower. After the test, Jumbo was unscathed, but its crumpled rigging tower was a preview of the damage to steel structures in Japan. The specifications for Jumbo had been that it must, without rupture, contain the explosion of the bomb's full complement of high explosive and permit subsequent mechanical and chemical recovery of the active material.

The final Jumbo design was the elongated elastic one described in Chap. 7. This container was ordered early in August 1944 from the Babcock and Wilcox Corporation. It was delivered to the test site in May 1945 and erected on its foundation. Because of its 25- by 12-ft size and 214-ton weight, Jumbo had to be transported to the nearest rail siding on a special car over an indirect route to provide adequate clearance. From the siding it was transported overland to the test site on a special 64-wheel trailer.

Jumbo was constructed from R. W. Carlson's design. Many tests were made with scale-model "Jumbinos" to determine their ability to contain charges without rupture.

The other recovery method investigated was the "water baffle" in which fragments were stopped by a 50:1 ratio of water to high-explosive mass. This work was dropped before it gave conclusive results, but it showed that high-percentage recovery would be difficult. For use in recovery experiments, a shallow 200-ft-diameter concrete catch basin was constructed at Los Alamos.

THE DETONATING SYSTEM

The main development work on electric detonators was by the Electric Detonator Group of G Division. However, the detonators designed there were primarily intended for experimental work. Detonators for the weapon had to meet further requirements, such as durability, ruggedness, and reliability.

The electric detonators were fired by discharging a bank of high-voltage condensers through a suitable low-inductance switch. Mechanical switches could not do this. This problem was assigned to Section X-2C, later to become Group X-5, the Detonating Circuit Group. One type of switch extensively developed and used when accurate timing was required was the so-called explosive switch. In this switch, electrical contact was made by detonating an explosive charge that broke through a thin dielectric layer between two metal disks, between which the high voltage then passed. This switch was made double by using four semicircular disks. It could not be tested before use, so an electronic device resembling the thyratron was finally adopted. This switch operated between two electrodes, the discharge being triggered by charging a third "probe" electrode to a suitable high voltage. These switches could pass very high currents in about a microsecond. The firing units incorporating them were built by the Raytheon company. Because of production delays, only a few were tested before the Nagasaki drop.

ENGINEERING, TESTING, AND ASSEMBLY

The X- and O-Division engineers cooperated in designing their assembly within the Fat Man case. Once this design was completed, it had to have assembly tests and tests for reliability and ruggedness in combat use. Assembly design was done by the Engineering Group, as were the assembly tests until Group X-6 was formed in March 1945. Group X-6 also collaborated with other interested groups in the drop tests at Wendover Field (Chap. 19).

The Engineering Group was at first responsible for designing high-explosives molds, particularly lens molds of various types and sizes. Mold development and procurement became such a serious bottleneck by October 1944 that mold engineering was made a main responsibility of the Engineering Service Group established at that time. A system was developed by which experimental molds were built in the Los Alamos shops while the same designs were sent for outside procurement. The Camel Project later helped expedite procurement in the Los Angeles area. Even so, the bottleneck remained and it was only by a matter of days that enough final full-scale lens molds were obtained for the Trinity test.

CHAPTER 17

Chemistry and Metallurgy

INTRODUCTION

In August 1945 the Chemistry and Metallurgy Division had evolved into the following groups.

CM-1	Service Group	R. H. Dunlap
CM-2	Heat Treatment and Metallography	G. L. Kehl
CM-4	Radiochemistry	L. Helmholtz
CM-5	Plutonium Purification	C. S. Garner
CM-6	High-Vacuum Research	S. I. Weissman
CM-7	Miscellaneous Metallurgy	A. U. Seybolt
CM-8	Plutonium Metallurgy	E. R. Jette
CM-9	Analysis	H. A. Potratz
CM-11	Uranium Metallurgy	S. Marshall
CM-12	Health	W. H. Hinch
CM-13	DP Site	J. E. Burke
CM-14	RaLa Chemistry	G. Friedlander
CM-15	Polonium	I. B. Johns
CM-16	Uranium Chemistry	E. Wichers

In September 1944, Group CM-3, Gas Tamper and Gas Liquefaction, was transferred to the new Explosives Divison. In April 1945, uranium and plutonium chemistry, formerly concentrated in CM-5, was shared with a new group, CM-16. CM-1 was divided into a Service Group (CM-1) and a Health Group (CM-12). The old Radiochemistry Group (CM-4) was split into three groups, with CM-14 and CM-15 taking charge of RaLa and polonium work, respectively. R. W. Dodson, former Group Leader, became Associate Division Leader in charge of radio-chemistry. C. C. Balke, former Group Leader of CM-7, left the Laboratory. Miscellaneous metallurgical services were transferred to this group, whose new Group Leader was A. U. Seybolt. CM-11 remained as the Uranium Metallurgy Group. CM-10, the Recovery Group, was absorbed into the new DP Site Group, CM-13.

217

These changes were motivated primarily by the expanding radio-chemical work associated with the expanding RaLa implosion studies and preparation of polonium for the implosion initiator program, and by the need to streamline the processing of uranium and plutonium, which was arriving in ever increasing quantities.

The transition to large-scale operation involved a constant growth of personnel in the division and a constant expansion of laboratory, shop, and plant facilities. The main steps in this physical expansion were completion of a large annex to D Building in August 1944, completion of the metallurgy building (Sigma Building) in October 1944, and construction of the RaLa Chemistry Building in Bayo Canyon in November 1944. The last and greatest addition was the chemical and metallurgical production plant (DP Site), whose first buildings were occupied in the summer of 1945.

URANIUM PURIFICATION AND RECOVERY

Beta-stage enriched uranium from the Oak Ridge Y-12 plant generally was received as a purified fluoride and reduced directly to metal. For hydride experiments, the metal was converted to hydride and formed by plastic bonding. When hydride or metal experiments were completed, the material was returned for recovery, as were crucibles, liners, and other containers used in fabrication. Recovered solutions were converted to hexanitrate, extracted with ether, and precipitated as reduced oxalate. The oxalate was ignited to oxide and converted back to the original tetrafluoride.

The essential purification step was the ether extraction method (Chap. 8), also applied by the radiochemists to decontamination of Water Boiler solutions and used by the Recovery Group in experiments on test-shot recovery methods. In April 1945, uranium recovery and purification were done by a new group (CM-16). Before that time, extensive investigations were made to determine the best reagents and ion concentrations for the extraction.

The dry purification step (reduction of UO_2 to UF_4) was developed to give fluoride yields as high as 99.9%. Fluorination of the trioxide and oxalate also was investigated.

Studies of recovery from liners and slag showed that complete dissolution of these materials was necessary before recovery. Ether

218

extraction was then used. The Recovery Group designed and built continuous extraction apparatus capable of extracting a large volume of solution per hour and giving recovery yields of better than 99.9%. The average uranium remaining in stripped solutions was not more than 60 μg/liter.

The Recovery Group studied recovery of active material in case a test shot failed. Uranium was dissipated through a large volume of sand, sawdust, and similar materials, then recovery was attempted to simulate conditions of bomb material scattering over a large area by high explosives. Recovery was poor, but recovery from sealed containing spheres was successful. A 3/4-in. uranium sphere, scattered explosively inside a 12-in. "Jumbino" containing sphere, was about 99% recovered.

URANIUM METALLURGY

By August 1944, metallic uranium could be obtained with a neutron count below tolerance. Several developments after that date were essential, however, to the final ^{235}U weapon, especially stationary bomb reduction, uranium remelting, uranium forming, cladding and protection of surfaces, use of uranium sponges, and production of final weapon parts.

After adoption of the stationary bomb reduction technique, several minor improvements were made. A better product was obtained by increasing the bulk density of the tetrafluoride to be reduced. After a wide survey of liners, a magnesia liner developed by the Miscellaneous Metallurgy Group and a magnesia plus silica liner developed at M. I. T. were chosen. Other investigations included the effect of impurities in the reductant, grain size, firing technique, and use of inert gases in the bomb. This research led to a very high yield of high-purity, well-consolidated metal.

Work on uranium remelting began in June 1944. This process drove off volatile impurities and prepared the metal for shaping. Difficulties were encountered in obtaining crucibles that would not crack when cooled. Magnesia and beryllia were finally used, with special heating methods.

Techniques for forming uranium were intensively investigated after August 1944. The main techniques used were casting, hot pressing, and rolling. Both magnesia and graphite molds were used successfully for

219

casting. Difficulty in melting large amounts of uranium was overcome by resistance heating. Investment and centrifugal casting were tried, and the latter was adopted. Hot pressing was not used in preparing parts for the full-sized gun assembly because it gave no greater accuracy than casting, but it was used for smaller pieces, including hemispheres for sphere multiplication experiments and slugs for rolling.

When it was still assumed that a liner implosion would be used, much effort went into techniques for producing hollow spheres of uranium. These were hot pressed in specially designed dies as pentagonal or triangular sphere segments. Tolerances were, of course, extremely close. The rolling technique included heating to 200°C and then cross-rolling.

Many cladding techniques were investigated, including electroplates of gold, zinc, silver, and, later, nickel and chromium. Chromium gave the best, but still inadequate, protection. Evaporated metal techniques proved better.

Low-density uranium compacts were prepared by sintering metal powder obtained by decomposing the hydride. Their densities ranged from 1 to 6 g/cm^3. Treatment with nitrogen, which formed a nitride coating on the metal, protected against spontaneous combustion and corrosion.

The culminating work of the Uranium Metallurgy Group was casting the final parts for the Hiroshima bomb. This work had been scheduled more than a year before for completion on July 26, 1945. It was, in fact, completed July 24.

PLUTONIUM PURIFICATION

The plutonium purification program was the most directly affected when the plutonium gun program was abandoned in August 1944. Efforts then concentrated on simplification, production routine, efficiency, and the plutonium health hazard.

Wet Purification

In early fall of 1944, the first completely enclosed full-scale apparatus was completed. The full run of 160 g required 24 hours with about 60 liters of supernatant remaining for recovery. Aside from minor difficulties and improvements, this represented the completed form of the process "A" wet purification described in Chap. 8.

Early in 1945, investigation and testing of "B" and "C" wet processes began. Process "B" was simpler than process "A", and it gave higher yields and a smaller volume of supernatants. Process "B" involved only an ether extraction with calcium nitrate and an oxalate precipitation. It met purity requirements and gave a product satisfactory for further processing. In July 1945, process "A" was dropped.

An even simpler process "C", involved only an oxalate precipitation, but purification was insufficient. The following chart gives the essential information on the "A" and "B" processes.

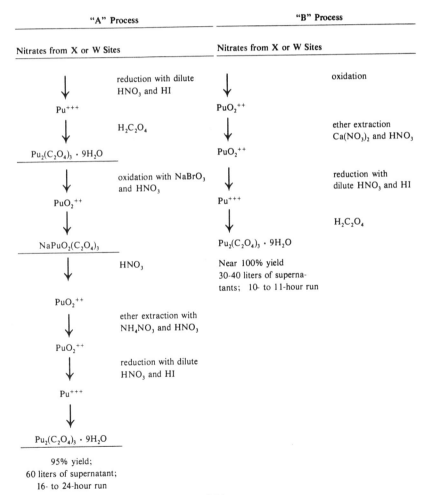

"A" Process

"B" Process

Nitrates from X or W Sites

Nitrates from X or W Sites

↓ reduction with dilute HNO₃ and HI

Pu⁺⁺⁺

↓ H₂C₂O₄

Pu₂(C₂O₄)₃ · 9H₂O

↓ oxidation with NaBrO₃ and HNO₃

PuO₂⁺⁺

↓

NaPuO₂(C₂O₄)₃

↓ HNO₃

PuO₂⁺⁺

↓ ether extraction with NH₄NO₃ and HNO₃

PuO₂⁺⁺

↓ reduction with dilute HNO₃ and HI

Pu⁺⁺⁺

↓

Pu₂(C₂O₄)₃ · 9H₂O

95% yield;
60 liters of supernatant;
16- to 24-hour run

↓ oxidation

PuO₂⁺⁺

↓ ether extraction Ca(NO₃)₂ and HNO₃

PuO₂⁺⁺

↓ reduction with dilute HNO₃ and HI

Pu⁺⁺⁺

↓ H₂C₂O₄

Pu₂(C₂O₄)₃ · 9H₂O

Near 100% yield
30-40 liters of supernatants; 10- to 11-hour run

221

Dry Conversion

After it was decided to use only fluoride metal reduction (Chap. 8), efforts concentrated on fluoride production. Nitrate, oxalate, and oxide hydrofluorination were investigated. The oxide method was chosen, which involved conversion of the oxalate from wet purification to oxide by heating in oxygen and introducing hydrogen fluoride at 325°C in the presence of oxygen. It involved a 24-hour cycle and gave 92-99% yields.

Recovery

The principal development of this period was peroxide precipitation. Of the oxalate precipitation, ether extraction, sodium plutonyl acetate precipitation, and final oxalate precipitation, the ether step was eliminated and the sodium plutonyl acetate step was used only for heavily contaminated material.

The danger of plutonium to the operator's health was greatest in the recovery operations. The need to vary procedures to fit the type of contamination involved made development of enclosed apparatus difficult. Such apparatus was, in fact, not developed until November 1945 at DP Site. The main safety effort was to monitor personnel carefully. Those who had received exposures in excess of body tolerances were removed from further exposure until counts returned to normal (Chap. 9).

PLUTONIUM METALLURGY

When the new purity tolerances were established, all metal-reduction methods were eliminated except stationary bomb reduction of the tetrafluoride. Work on crucible research for remelting continued, but it became possible to use magnesia owing to increased tolerances, in that the danger of magnesium impurities was considered less serious. Much research was done on the physical properties of plutonium metal, because more than two allotropic phases were suspected, and this primarily concerned forming operations. Work began on alloys to find one that would keep a high-temperature phase stable at room

222

temperature. The stable room-temperature phase, called the alpha phase, is brittle and difficult to work. Fabrication operations were investigated, as were methods of surface cleaning and protection. Because plutonium is highly susceptible to corrosion, these topics were more important for plutonium than for uranium.

The metal reduction and remelting techniques were well established by August 1944, but they had to be adapted to large-scale operation as more plutonium became available.

Within the limited time available, the metallurgists studied the physical properties of plutonium extensively. This first transuranic element manufactured in kilogram amounts proved to have a remarkable physical structure. It exists in five distinct allotropic forms between room temperature and the melting point, labeled in the order of temperatures at which they are stable—α, β, γ, δ, and ϵ. It is very electropositive, but has the highest electrical resistivity of any metal. It is very corrosive in water and air.

Of all the phases, the α, or room-temperature phase, is the densest. Because this phase is brittle, and the δ and ϵ phases malleable, the material was pressed at δ-phase temperatures. Hemispheres cast by this method for multiplication studies warped and cracked after standing for a day or so at room temperature. Evidently, higher temperature phases were being retained for a time at room temperature, the warping and cracking being caused by a delayed transition to the denser phase. In small spheres, this difficulty was overcome by prolonged annealing. Near the end of 1944, the solid sphere or Christy implosion was selected. Research began to find plutonium alloys that would remain in a higher temperature phase at room temperature. Alloys with a long list of elements were examined. The best was one with 3% gallium, which was apparently quite stable in the δ phase with a density of 15.9 at -75 to $450°C$. Extensive time and temperature tests showed no instability.

Although cleaning and etching plutonium surfaces caused no serious problems, protective coating did. Of many electroplated and evaporated metal coatings, electrodeposited silver was selected for the Trinity hemispheres. At the last minute, however, small pinholes were discovered in the coat, as well as blistering caused by retention of small amounts of plating solution under the coat. Because the scheduled test was only a few days away, the material was used in this condition, with the blisters polished down to restore the fit of the hemispheres.

223

MISCELLANEOUS METALLURGY

The principal metallurgy in this period, other than on uranium and plutonium, was by the Miscellaneous Metallurgy Group who made the gun tamper, beryllium crucibles and refractories, and some boron compacts. Fabrication of tungsten carbide in larger pieces than had been made before was their outstanding accomplishment. They formed pieces weighing up to 300 pounds. Compositions investigated were tungsten carbide and tungsten carbide with copper and with cobalt. The Metallurgy Group cooperated with the Initiator Group of G Division in microscopic and macroscopic examinations of flow characteristics and structure of the metal after implosion (Chap. 15).

In crucible research, cerium sulfide was used for some time after the purity standards were lowered. The material finally adopted for all plutonium and uranium crucibles and liners was a vitrified magnesia developed by the Miscellaneous Metallurgy Group and manufactured at Los Alamos, at M.I.T., and at Ames, Iowa.

RADIOCHEMISTRY

The principal developments in radiochemistry after August 1944 were as follows. The implosion initiator program was begun, and the polonium research staff was increased. Radiolanthanum work, in collaboration with the RaLa Group, was carried out at the Bayo Canyon laboratory. These two groups were formally separated from the Radiochemistry Group in April 1945. Work began with the high-power Water Boiler, with its consequent problem of decontaminating highly irradiated uranium. Foil chemistry was continued. A new sensitive neutron detector was developed. A calorimeter was built for polonium work, and a microtorsion balance was built for use in determining the mass purity of samples.

Polonium

Use of polonium for the implosion initiator was a major technical achievement that involved a lot of basic research into its chemical and "metallurgical" properties. Investigation of polonium might be said to be as novel as that of plutonium.

224

The main problems were to prepare pure enough polonium to meet the neutron background tolerance, to prepare high-density uniform foils, and to coat those foils against polonium and alpha-particle escape. This work was hazardous because of polonium's high alpha activity and extreme mobility. It is virtually impossible to work with polonium and avoid introducing it into the human body. It is eliminated rapidly, however, and does not settle in dangerous concentrations in the bone, as do radium and plutonium. The full extent of the polonium hazard can only be determined with time. Pragmatic safety rules were intended to minimize and detect polonium absorption. Persons with more than a tolerance dose were removed from possible contact with the material until their urine counts dropped below tolerance level.

After the first half-curie, which was recovered from residues from radon capsules, the polonium used at Los Alamos was obtained from bismuth irradiated in the Clinton pile. Polonium was separated from the bismuth at two plants of the Monsanto Chemical Company in Dayton, Ohio. It was deposited on platinum foils and shipped to Los Alamos in sealed containers. Migration of the polonium off these foils onto the walls of the container caused much trouble, however.

Polonium purification was primarily Monsanto's responsibility, although there was some research on chemical purification and purification by distillation at Los Alamos.

Other work of the polonium chemists was chiefly preparation of (α,n) neutron sources for the experimental physicists. Of these, the most complicated were the mock-fission sources discussed in Chap. 12.

Water Boiler Chemistry

Work on the low-power Boiler ended in August 1944, and high-power Boiler operation began in December. The first job was to purify and convert the old material from sulfate to nitrate. The main reason for choosing the nitrate was that it had to be used in decontamination (ether extraction). It also was slightly less corrosive than the sulfate.

It became necessary to build a "hot" chemistry laboratory and remote-control decontamination apparatus. Research on use of the ether extraction method was done in collaboration with the Recovery Group.

The remote-control apparatus was placed behind a thick concrete wall. The irradiated material was run directly from the Boiler into an underground tank and then pumped up into the extraction column by air pressure. After extraction, the solution could be run back into an underground vault or let out into the "hot" laboratory for concentration. Concentration was carried out behind a shield of lead bricks.

The heavy irradiation of the nitrate in the Boiler caused decomposition and loss of nitrogen, which led to precipitation of the basic nitrate. To avoid this, the Boiler solution was analyzed frequently and nitric acid was added to make up the deficit.

Radiolanthanum

Remote-control apparatus was first used after August 1944. The determining feature of the RaLa chemistry was the enormous radioactivity involved. A single batch of material could represent up to 2300 Ci of activity.

Isolation of radiobarium from the fission products of the Clinton pile was arranged in April 1944. The material as received was a mixture of ^{140}Ba and ^{140}La, and the short-lived lanthanum had to be separated from its parent barium.

The first control apparatus and associated means for separating lanthanum from barium was designed as protection against about 200 Ci of activity, and operated by the so-called phosphate method. In small-scale tests, separation of the lanthanum from about 100 times its mass of barium was found to be nearly quantitative when the phosphate was precipitated from an acid phosphate solution. But in full-scale practice, long filtration times and strong hydrogen-ion dependence caused difficulties.

As the RaLa method was developed, moreover, the source strength and dimension requirements became more stringent. A new and better method was needed to provide shorter filtration time, precipitation on a smaller filter area, and good separation with higher yield, Also, the operators would have to be protected against higher radioactivities.

In March 1945, collaboration on a new method began among the radiochemists and the Plutonium Chemistry and Recovery Groups. For greater radiation protection, the controls were removed to a distance of 90 ft. The separation method developed was a hydroxide-oxalate process. Lanthanum hydroxide was precipitated with sodium hydroxide, filtered on a platinum sponge filter, dissolved in nitric acid, and reprecipitated with oxalic acid. The oxalate was allowed to stand for about 25 minutes, and then a small quantity of hydrogen fluoride was added. The resulting

226

precipitate was crystalline, could be filtered rapidly on a small area, and was not affected by intense radiation.

Half-life measurements of carefully purified ^{140}La gave a value of 40.4 hours.

Instruments and Services

Preparation of very pure boron trifluoride for filling proportional neutron counters was investigated, as was recovery of the costly ^{10}B from counters no longer needed. Besides filling many counters for the experimental physicists, the radiochemists developed a very sensitive "quadruple proportional counter" for quick measurement of weak neutron sources.

A calorimeter for polonium half-life measurements and a microtorsion balance for weighing polonium samples were built. Comparison of sample weight and activity was found to be a reliable method of purity analysis.

ANALYSIS

Early Analysis Group work emphasized determination of very small amounts of light-element impurities. Discovery of ^{240}Pu, however, eliminated the need for strict contamination control and further research on methods for determining these elements. Tolerance limits were determined easily by existing methods, and it became possible to turn to other investigations, particularly heavy-element analysis. The Analysis Group turned increasingly to the metallurgists' problems, to improving instrumental techniques, and to developing routine methods. The principal analytical methods investigated after August 1944 are as follows.

Spectrochemical methods
  Plutonium
    The cupferron method for heavy elements
    The gallium oxide pyrolectric method
    Determination of gallium in plutonium metal
  Uranium
  Determination of zirconium in uranium
    Determination of uranium in urine
  Miscellaneous

227

Impurities in calcium and magnesium oxides
Volumetric methods
Determination of acid-soluble sulfide in uranium and plutonium
Determination of sulfate in plutonium
Assay methods
Radioassay
Photometric assay
Gravimetric assay
Microvolumetric assay

Spectrochemical Methods

The cupferron procedure for heavy-element determination was essentially the same as that for light-element trace analysis. It eventually proved applicable to 39 element impurities. In addition to its use for plutonium analysis, it was used for other elements, including uranium. The acid-insoluble cupferride of plutonium is formed, and the impurities are extracted from the compound with chloroform. The aqueous impurity solution is then evaporated on copper electrodes, which spark and give the impurity spectrum. The Health Medical Group also used this method to determine plutonium in urine samples.

The pyroelectric gallium oxide method developed for uranium analysis became important for analyzing plutonium when that became necessary. The material was arced in a dry-box to reduce health hazards.

Gallium was determined in plutonium metal by a method that illustrates the value of seemingly irrelevant research. Before such a method was needed, the spectrochemical staff studied extraction of iron and gallium complexes with di-isopropyl ether. Their results were immediately applicable when gallium became important for stabilizing delta-phase plutonium. The complex extracted with the ether was re-extracted into a small amount of very pure water and evaporated on copper electrodes. The residue was excited by sparking, and the amount of gallium was determined photometrically.

For certain elements, notably titanium, zirconium, thorium, columbium, and tantalum, no satisfactory spectrochemical method was discovered. None of these elements, fortunately, was crucially important.

Uranium analysis was further investigated and a method for determining zirconium in uranium was being studied. A process for determining 0.1 to 1 μg of uranium in urine samples was being investigated for use in health control work.

228

Analysis of impurities in calcium oxide and magnesium oxide became important when these materials were adopted as crucibles and liners in plutonium metallurgy. The method was one of direct arc spectrography. Difficulties caused by variation in oxide subdivision were overcome by grinding the samples very fine.

Useful spectrochemical techniques developed included the double spectrograph, the double-slit spectrograph, and the dry-box arc. The double spectrograph consists of two spectrographs aligned in opposition, and it passes source light through the slits of both instruments. The double-slit spectrograph produces juxtaposed spectra of two wave-length ranges on the same film. These methods halved the sample size in one case, and the time for a complete analysis in the other. They were important because they were economical of valuable material and because they reduced analysis time and operator exposure to plutonium. The dry-box arc with outside controls was developed because of the extreme danger from arcing plutonium. Laboratory contamination is serious with ordinary arcing, and the danger to operators is very great.

Volumetric Methods

Acid-soluble sulfide was determined by distillation of hydrogen sulfide, which was absorbed and determined volumetrically. The method was used for plutonium and uranium metals. Sulfate in plutonium was determined by reduction to sulfide, followed by volumetric measurement of absorbed hydrogen sulfide.

Gravimetric Methods

Another method for determining gallium in plutonium-gallium alloys, after ether extraction of gallium, indicated it gravimetrically as the 8-hydroxyquinolate.

Assay Methods

Photometric, volumetric, and gravimetric methods were investigated for routine plutonium assay.The photometric method was untrustworthy; the gravimetric method was good, but too slow. The volumetric method was finally adopted and after early 1945, used to assay all Hanford material.

229

DP SITE

At the beginning of 1945, all plutonium production work was planned and done in the Chemistry Building (Building D). This plan changed when the seriousness of the plutonium health hazard was realized. Building D was conceived to prevent plutonium contamination by light-element impurities. The building was not ideal for preventing workers' plutonium contamination, and as larger amounts of plutonium began to arrive, adequate decontamination became increasingly difficult. The expected flow rate of the Hanford material increased and it tended, at maximum production, to overstrain the resources of D Building. A bad fire occurred in C Shop on January 15, 1945 (Chap. 9), demonstrating vividly the possibility of fire in D Building. The consequences of such a fire, including the spread of contamination over a wide area of the Laboratory, indicated that it was imperative to build a new production plant, so designed that fire would be unlikely, and so located that accidents would not retard the Technical Area work.

In February, a committee was appointed to design and expedite the construction of a new plant. Plans for this plant were enlarged to accommodate polonium processing, which was inadequately housed in the Technical Area and which was also a hazardous operation.

The new plant, the so-called DP Site located on South Point, was divided into two areas. The first of these, the East Area, was designed for polonium processing and initiator production, and it was supervised by Group CM-15, the Polonium Group. The second area, the so-called West Area, was designed for plutonium processing and bomb core production, and was supervised by Group CM-13.

Building Design

When buildings were designed, the final processes were undetermined so they were made to house any finally accepted process. There were four identical working buildings and an office in the West Area, and one working building and an office in the East Area. All were entirely noncombustible with steel walls and roof, rock wool insulation, metal lathe, and plaster lining. All rooms had smooth walls and rounded corners for easy cleaning. Each of the West Area operating buildings was 40 by 200 ft and contained two 30- by 30-ft operating rooms and two 40-

230

by 50-ft operating rooms. The East Area operating building was 40 by 240 ft, broken up into small rooms.

The chief feature of the buildings was the ventilating system whereby air was withdrawn from the rooms through hoods at a rate of about 2 ft³/s. Where the hood capacity of the rooms was too small, additional exhaust ducts were furnished so that the air in every room was changed once every 2 minutes. The exhaust ducts assembled into a common duct. The exhaust air from each Area (East and West) was then passed through a bank of electrostatic filters to remove contamination, through a bank of paper emergency filters, and finally through a series of 50-ft stacks. The air was exhausted by four 50-hp blowers in the West Area, and by two 40-hp blowers in the East Area.

Serious design work was not started until about the first of March; about March 15, East Area building construction was begun, and the buildings were essentially complete by June 1. Equipment was installed by July 15, and operations started shortly thereafter. The West Area buildings were complete by July 15, and their much more extensive installation was essentially complete by October 1, but minor difficulties prevented plant operation until November 1, after the period ended.

Process Design

Operating processes were worked out by various groups in the CM Division. While procedures were being used in D Building, a new plant committee supervised redesigning the equipment for use at DP Site.

To ensure and facilitate safe operation with plutonium and polonium, all operations were carried out in closed systems and were designed to be carried out in a routine fashion. To prevent chain reactions with plutonium, all equipment was designed so that no more than a safe amount could be charged into any piece of apparatus.

Improved protective furniture (hoods and dry boxes—sealed boxes containing an inert atmosphere with inserted gloves) was designed for every operation. This equipment was all stainless steel for corrosion and fire resistance and ease of decontamination. About 20 carloads of this furniture was fabricated by the Kewaunee Manufacturing Company in about 100 days.

231

CHAPTER 18

Project Trinity

PRE-TRINITY

Preparations for an experimental nuclear explosion began in March 1944 when the Laboratory's Director and most of the group and division leaders decided that such a test was essential. Planning integral experiments that would duplicate satisfactorily the conditions of a bomb was difficult. Many questions about a practical bomb left unanswered by theory and differential and integral experiments could only be answered by an actual experiment with full instrumentation. Kistiakowsky, then Deputy Division Leader for the implosion program in the Engineering Division, formed E-9, the High-Explosives Development Group under Bainbridge, to investigate and design full-scale HE assemblies and prepare for a full-scale test with active material. Group E-9 became Group X-2 (Development, Engineering, and Tests) during the general Laboratory reorganization.

The first systematic account of the test plan was made by Fussell and Bainbridge in a memorandum early in September 1944, in which the energy release was considered equivalent to 200 to 10 000 tons of TNT. These early plans were based on the assumption that Jumbo, a large steel vessel, would enclose the bomb so that the active material could be recovered if the test failed. Among the tests planned at this time were the following.

Blast measurements - piezoelectric gauges
paper diaphragm gauges
condenser blast gauges
Barnes' Boxes (not used)
condenser gauge blast measurement from airplane
Ground shock measurements - geophones
seismographs
Neutron measurements - gold foil
fast-ion chamber (not used in this form)

Gamma-ray measurements - recording in plane, dropped "gauges"
(not used)
gamma-ray sentinels
Nuclear efficiency and
Photographic studies - Fastax cameras at 800 yards, spectrographic
studies, radiation characteristics,
photometric data, ball-of-fire studies

SCR-584 radar
Meteorology

Additional nuclear measurements were considered in three reports by P. B. Moon, who anticipated some of the experiments later adopted.

One early problem of the test group was site selection. At one time eight different sites were considered: the Tularosa Basin; Jornada del Muerto; the desert training area near Rice, California; San Nicolas Island off the coast of southern California; the lava region south of Grants, New Mexico; southwest of Cuba, New Mexico, and north of Thoreau; sand bars off the coast of south Texas; and the San Luis Valley region near the Great Sand Dunes National Monument in Colorado. There were several factors to be considered in making the selection. Scientific considerations required a flat site to minimize extraneous effects on blast and a good climate because of the large amount of optical information desired. Safety precautions required that ranches and settlements be distant to avoid possible danger from the bomb products. Tight scheduling required a minimum time loss in personnel travel and equipment transportation between Project Y and the test site, and security required complete separation of the activities at the test site from the activities at Site Y. The choice finally narrowed to either the Jornada del Muerto region in the northwest corner of the Alamogordo Bombing Range or the desert training area north of Rice, California. The Jornada del Muerto was chosen early in September 1944 with General Groves' approval, after consultation with General Ent of the Second Air Force. The project secured the use of an 18- by 24-mile area within the base, with the nearest habitation 12 miles away and the nearest town about 27 miles away.

Once a site had been selected, securing good maps of the region became extremely important. Arrangements were made with the Second Air Force to have a 6-in.-to-the-mile mosaic made of a 6- by 20- mile

233

strip that included point zero (the point of detonation) at the center. These aerial mosaics were extremely useful in early exploratory work and for final precise planning. An inadequate supply of maps caused delay, because they had to be obtained through the Security Office so as not to reveal the Laboratory's interest in these regions. The maps finally used were obtained by devious channels and included the geodetic survey maps for New Mexico and southern California, all the coastal charts of the United States, and most of the Grazing Service and county maps for the state of New Mexico.

The original construction plans for the test site base camp were drawn by Captain S. P. Davalos (Assistant Post Engineer), Bainbridge, and Fussell in October 1944 and provided for a maximum of 160 men. Kistiakowsky gave his support by outlining the plan and scope of the proposed operations and justifying construction requirements in a memorandum. General Groves approved these two documents and contracts were let early in November for the initial construction. The camp was completed in late December and a small detachment of Military Police under Lieutenant H. C. Bush took up residence. Laboratory personnel agreed that the wise and efficient supervision of the base Camp by Lieutenant Bush, under extremely primitive conditions, contributed greatly to the success of the tests.

With the Laboratory's concentration on the implosion program, beginning in August 1944, the test program lost its priority. The manpower shortage for research and development work resulted in the members of the Development, Engineering, and Tests Group devoting most of their time and effort to engineering problems and largely abandoning their work on the test. Among the few accomplishments during this period were the test site layout, shelter design and construction, earth sample collections, the procurement of meteorological and blast gauge equipment, and planning for measuring nuclear radiations.

TRINITY ORGANIZATION

By March 1945, almost all essential physics research for the bomb was complete, and Oppenheimer proposed establishing Project TR, an organization with division status, comprising personnel chiefly from the Research Division who would have full responsibility for a complete test. The original Project TR organization follows.

Head		K. T. Bainbridge
Safety Committee		S. Kershaw
TR U.S.E.D.		Captain S. P. Davalos
Security		Lieutenant R. A. Taylor
CO MP Detachment		Lieutenant H. C. Bush
Consultants		W. G. Penney
		V. F. Weisskopf
		P. B. Moon
TR-1	Services	J. H. Williams
TR-2	Shock and Blast	J. H. Manley
TR-3	Measurements	R. R. Wilson
TR-4	Meteorology	J. M. Hubbard
TR-5	Spectrographic and Photographic Measurements	J. E. Mack
TR-6	Airborne Measurements	B. Waldman

This organization expanded rapidly and by test time involved about 250 technical men. Groups R-1, R-2, R-3, R-4, F-4, G-11, O-4, T-3, and T-7 worked full time on the Trinity Project and other groups also gave some time to this work. Group G-4 manufactured most of the electronic equipment.

In June 1945, Project TR included the following.

Head	K. T. Bainbridge
Aide	F. Oppenheimer
TR U.S.E.D.	Captain S. P. Davalos
Security	Lieutenant R. A. Taylor
CO MP Detachment	Lieutenant H. C. Bush
Consultants	
Structures	R. W. Carlson
Meteorology	P. E. Church
Physics	E. Fermi
Damage	J. O. Hirschfelder
Safety	S. Kershaw
Earth Shock	L. D. Lcct
Blast and Shock	W. G. Penney
Physics	V. F. Weisskopf
TR Assembly	Commander N. E. Bradbury
	G. B. Kistiakowsky, Alternate

TR-1	Services	J. H. Williams
TR-2	Air Blast and Earth Shock	J. H. Manley
TR-3	Physics	R. R. Wilson
TR-4	Meteorology	J. M. Hubbard
TR-5	Spectrographic and Photographic Measurements	J. E. Mack
TR-6	Air Blast	B. Waldman
TR-7	Medical Group	Dr. L. H. Hempelmann

By the time Project TR was set up, all the elaborate schemes for recovering active material were abandoned, including the use of Jumbo (Chap. 16) and the use of large amounts of sand or water. When recovery methods were considered seriously, the supply of active material was extremely limited and there was a strong feeling that the bomb might fail to explode. As confidence in the bomb's ultimate success increased and adequate production of active materials seemed assured, the recovery program no longer seemed essential. Perhaps the most important deciding factor, however, was that any effective recovery program would interfere seriously with securing information on the nature of the explosion, which was, after all, the principal reason for the test. Jumbo was taken to the site and erected 2400 ft (720 m) from its originally planned location, because it was not to be used for this test.

REHEARSAL TEST

The new group's first task was preparing a rehearsal shot known as the "100-ton shot." This had been proposed in the summer of 1944, both as a full-dress administrative rehearsal and as a way of providing calibration of blast and earth shock equipment for the nuclear bomb test. The test was scheduled on May 5, 1945, but the date was extended to May 7 so that additional equipment could be installed. Several requests for additional time were refused because further delays would delay the final test, which was already tightly scheduled.

The test was made in early morning on May 7 with 100 tons of HE stacked on the platform of a 20-ft tower. Very little experimental work had been done on blast effects above a few tons of HE; therefore, it was important to obtain blast and earth shock results to determine which structures would withstand these effects in the final shot. By using

236

appropriate scale factors, the center of gravity of the 100-ton stack of HE was made 28 ft above the ground in scale with the 100-ft height for the 4000-5000 tons expected in the final test. The stack of HE was provided with tubes containing 1000 Ci of fission products derived from a Hanford slug simulating at a low activity level the radioactive products expected from the nuclear explosion. Measurements of blast effect, earth shock, and damage to apparatus and apparatus shelters were made at distances in scale with the distances proposed for the final shot. Measurements to determine "cross-talk" between circuits and photographic observations were generally done at the full distances proposed for the final shot.

The test was successful and was useful for suggesting procedural improvements. The most critical administrative needs emphasized by the test were for better transportation and communication facilities and more help on procurement. The chief test purposes were accomplished. Men who had worked in well-equipped laboratories became familiar with the difficulties of field work. Blast measurements and earth-shock data were valuable in calibrating instruments and providing standards for the design of safe shock-proof instrument shelters. Measurements of the effects from the radioactive material inserted in the stack of HE were especially informative on the probable amount and distribution of material that would be deposited on the ground. This information, secured by the Fission Studies Group of F Division, was essential in planning for equipment recovery, bomb efficiency measurements, and personnel protection for the final test. The high percentage of successful measurements in the final test may be attributed to the experience gained from the rehearsal shot.

PREPARATIONS FOR THE TRINITY TEST

When Project Trinity was established, July 4 was set as the target date for the test even though this date could probably not be met. Preparations for the test continued at an increasingly rapid pace after the rehearsal shot. The scope and intensity of the preparations necessary for the Trinity test cannot be overemphasized. To establish a complex scientific laboratory on a barren desert under extreme secrecy and great pressure was a difficult task. Too few people were available for the amount of work to be done. More than 20 miles of blacktop road were

237

laid and an area was paved in the vicinity of the tower. All personnel and equipment were transported from Los Alamos, and after considerable effort, the Trinity staff secured about 75 vehicles. About 30 more vehicles were added during the last week by the monitoring and intelligence groups. A complete communications system had to be installed, including telephone lines, public address systems in all buildings, and FM radios in 18 cars. Miles of wires were used for both the communications system and in various experiments. A complete technical stockroom was established, and all its varied contents were transported from Los Alamos. The stockroom was known officially as "Fubar." Sanitary conditions were difficult to maintain, especially in the mess hall, because of the hard water. Owing to the extremely tight test schedule, any delay in procurement or delivery of materials meant that the work pace increased when the items finally arrived. The tight schedule and personnel shortage demanded that most people at Trinity, from mess attendants to group leaders, worked at fever pitch, especially during the last month. A 10-hour workday was considered normal and often it stretched to 18 hours.

Among the most complex administrative problems were those solved by the Services Group. This group undertook providing wiring, power, transportation, communication facilities, and construction. The construction schedule was especially tight and required careful planning and hard work to complete successfully. For a month before the test, there were nightly meetings to hear reports on field construction progress and to plan worker assignments for the following day. Construction help was assigned on a priority basis for the experiments.

Considerable attention was paid to security and to the legal and safety aspects of the test. Efforts were made to dissociate the Trinity work from that at Los Alamos. Discussion about what should be done about people in surrounding towns was finally settled by having 160 enlisted men, under the command of Major T. O. Palmer, stationed north of the test area with enough vehicles to evacuate ranches and towns if this became necessary. At least 20 men associated with Military Intelligence were stationed in neighboring towns and cities up to 100 miles away, and most of these men served also by carrying recording barographs to get permanent blast and earth-shock records at remote points.

One minor source of excitement was the accidental bombing, with two dummy bombs, of the Trinity base camp by a plane from the Alamogordo Air Base early in May. The incident was reported to the base commander through the Security Officer and precautions were taken to prevent a recurrence.

Early in April, Project TR secured the services of J. M. Hubbard, meteorology supervisor for the Manhattan District. He requested information from the various experimental groups on the particular weather conditions or surveys they would find useful in their operations, and he made an effort to find a time that would meet nearly every specification of the various groups. He cooperated with the Weather Division of the Army Air Forces and was able to draw worldwide information in making his surveys. The period he selected as first choice for the final test was July 18 to 19, or 20 to 21, with 12 to 14 as second choice, and July 16 only a possible date.

One of the most difficult problems faced by Project TR was scheduling. Weekly meetings were held, with consultants and responsible group and section leaders considering new experiments and discussing detailed scheduling and progress reports. As much information was needed from the test as possible, but it was impossible to schedule every experiment because of the limitations of time and personnel. To have a new experiment considered by the weekly scheduling meeting required a detailed account of the experiment objectives, the expected accuracy, and the requirements for equipment, personnel, and machine and electronics shop time. Based on such information, the Trinity scheduling group decided whether a particular experiment was suitable and whether it could be completed successfully.

The July 4 date accepted in March was determined unrealistic. Delays in the delivery of full-scale lens molds and the consequent delay in the development and production of full-scale lenses, as well as the tight schedule in the production of active material, made it necessary to reconsider the data. The Cowpuncher Committee tried to schedule the pacing components to determine when other components or other developments would have to be completed so the test would not be delayed. By the middle of June, the Cowpuncher Committee agreed that July 13 was the earliest possible date, with the 23rd as a probable date. Because of the great pressure to have the test as early as possible, some of the experiments, tests, and improvements would not be ready, but the July 13 date was fixed so that essential components would be ready. On June 30 a review of all schedules was made by the Cowpuncher Committee and the earliest date for Trinity was changed to July 16 so as to include some important experiments. Commitments had been made in

239

Washington to have the test as soon after July 15 as possible, and these commitments were met by firing the shot early on the morning of July 16, as soon as weather conditions were suitable.

Four rehearsals were held on the 11th, 12th, 13th, and 14th with all personnel cooperating. A "dry run" of the assembly of the HE component of the bomb was held early in July after several tests to study loading methods and transportation effects. Final assembly of the HE and of the active core began on July 13. Nuclear tests and the active component assembly were done in McDonald's ranch house—a four-room frame house about 2 miles from the detonation point. The apparatus used were identical with those already shipped to Tinian, and the operation took on the character of a field test for the overseas expedition. On July 11, the active material was brought down by convoyed sedan from Los Alamos in a field carrying case designed for overseas use. The HE components were assembled at one of the outlying sites at Los Alamos and brought down by convoyed truck, arriving at Trinity on Friday, July 13.

Before the assembly started, a receipt for the active material was signed by Brigadier General T. F. Farrell, deputy for General Groves, and handed to L. Slotin, who was in charge of the nuclear assembly. The acceptance of this receipt signaled the formal transfer of the precious ^{239}Pu from the Los Alamos scientists to the Army, to be expended in a test explosion. The final assembly took place on a canvas-enclosed flooring built within the base of the tower. Active material in large quantity was put within HE for the first time. Although the people performing the operation and those watching it were outwardly calm, the tension was apparent. Only one difficulty was encountered that made the actual assembly anything more than a routine repetition of rehearsals. The desert heat, together with the heat generated by the active material, caused differential expansion among some of the parts. Some of the assembly had only been completed the night before on the high mesa of Los Alamos and was cold to the touch. A delay of a minute or two occurred while the hot material contacted the cold material, and then it cooled sufficiently to permit entry as planned.

After the HE and nuclear components were completely assembled, the bomb was still without detonators. It was hauled to the top of its 100-ft tower, where it rested in a specially constructed sheet steel house. On July 14, the Detonator Group installed the detonators and informers, and the Prompt Measurements Group and other test groups checked and

240

completed the installation of apparatus for their experiments. Visits were made to the top of the tower every 6 hours by members of the Pit Assembly Group to withdraw the manganese wire whose induced radioactivity measured the neutron background.

Elaborate plans were made for evacuating personnel if any serious difficulty arose, with the Medical Officer to be in charge. The Arming Party, a small group responsible for final operations, also assumed responsibility for guarding the bomb against possible sabotage and remained at the tower until the last possible moment. The weather seemed unfavorable early in the morning of the 16th, and not until shortly before 5:00 a.m. did the weather reports received from Hubbard begin to look satisfactory. As originally planned, the decision whether or not to run the test was to be made by Oppenheimer, General Farrell, Hubbard, and Bainbridge, with one dissenting vote sufficient to call it off. The final decision was made and announced at 5:10 a.m. and the shot was scheduled to be fired at 5:30 a.m.

Nearest observation points were set up 5.7 miles (9 km) from the tower with Base Camp located 9.7 miles from the tower. Several distinguished visitors, including Tolman, Bush, Conant, General Groves, C. Lauritsen, Rabi, E. O. Lawrence, A. Compton, Taylor, Chadwick, Thomas, and von Neumann, came for the test. All were instructed to lie on the ground, face downward, heads away from the direction of the blast. The control station, located at 5.7 miles, communicated with the various observation points by radio. From here, periodic time announcements were made beginning at minus 20 minutes until minus 45 seconds. Then automatic controls were switched on, setting off the explosion at 5:29 a.m. on Monday, July 16, 1945, just before dawn.

TRINITY

There have been many descriptions of the explosion; one of the most graphic is that of General Farrell, who saw it from one of the closest observation points. He said, in part: "The effects could well be called unprecedented, magnificent, beautiful, stupendous and terrifying. No man-made phenomenon of such tremendous power had ever occurred before. The lighting effects beggared description. The whole country was lighted by a searing light with the intensity many times that of the midday sun. It was golden, purple, violet, gray and blue. It lighted every peak,

241

crevasse and ridge of the nearby mountain range with a clarity and beauty that cannot be described but must be seen to be imagined. . . ." Several of the men stationed at Base Camp and members of the Coordinating Council of the Laboratory, who watched the explosion from the hills about 20 miles away, prepared eyewitness accounts of their experiences. All were deeply impressed by the light intensity, and also by the heat and the blue glow. Of the heat, one man said, "I felt a strong sensation of heat on the exposed skin of face and arms, lasting for several seconds and at least as intense as the direct noon sun." Of the blue glow, another reported, "Then I saw a reddish glowing smoke ball rising with a thick stem of dark brown color. This smoke ball was surrounded by a blue glow which clearly indicated a strong radioactivity and was certainly due to the gamma rays emitted by the cloud into the surrounding air. At that moment the cloud had about 1000 billions of curies of radioactivity whose radiation must have produced the blue glow." There were also many, detailed accounts of the appearance of the now familiar mushroom-shaped cloud. It was several minutes before people noticed that Jumbo's steel tower had disappeared from view. At Los Alamos, over 200 miles away, people who were not directly involved in the test and were not members of the Coordinating Council watched for a flash in the southern sky. Because the shot had been scheduled for 4:00 a.m., many watchers grew impatient and gave up. A few did see it, however, and they reported a brief blinding flash of considerable intensity.

For many of the men who watched the test at Trinity, the immediate reaction was one of elation and relief, because the successful explosion of the first nuclear bomb represented years of difficult, concentrated work. With this elation and relief came a feeling of awe and even of fear at the magnitude of what had been accomplished. For many, the successful completion of the Trinity test marked the successful completion of the major part of their work for the Los Alamos Laboratory, and there was a general letdown and relaxation after the intensive efforts of the past months. For those men who were going overseas, however, there was no rest, and their preparations for Trinity were simply a rehearsal of their duties at Tinian.

Security, which always pervaded the work of the Laboratory, was not forgotten even in the hectic hours after Trinity. As the first cars of weary, excited men stopped for food in the little town of Belen on their way back to Los Alamos, they spoke only of inconsequential things, and the

242

occupants of one car did not recognize the occupants of another. In fact, the members of the Coordinating Council were required to return directly to Los Alamos in buses, avoiding any stops in New Mexico communities. Not until they reached the guarded gates of Los Alamos did the flood of talk burst loose. The sales of the Albuquerque newspapers the following day were great because in them was an account of an "explosives blast" at the Alamogordo Air Base. The story was credited to the Associated Press, but appeared in very few papers outside of New Mexico, and then only as a brief note about an unimportant accident.

RESULTS OF TRINITY EXPERIMENTS

To give some idea of the number and scope of the experiments connected to the Trinity test, the following summary is included. There were six chief groups of experiments: implosion, energy release by nuclear measurements, damage, blast, and shock, general phenomena, radiation measurements, and meteorology.

Implosion experiments

Detonator asimultaneity measured with detonation-wave-operated switches and fast oscilloscopes. These records were fogged by gamma rays.

Shock-wave transmission time measured on a fast oscilloscope that recorded the interval from the firing of the detonators to the nuclear explosion.

The multiplication factor (α) measurement was done by three methods—with electron-multiplier chambers and a time expander; by the two-chamber method; and with a single coaxial chamber, coaxial transformers, and a direct-deflection high-speed oscillograph.

Energy release calculation by nuclear measurements

Delayed gamma rays measured by ionization chambers, multiple amplifiers, and Heiland recorders from both ground and balloon sites.

Delayed neutron measurements done in three ways—by using a cellophane catcher and ^{235}U plates both on the ground and when airborne, by using gold foil detectors to give an integrated flux, and by using sulfur threshold detectors. For the cellophane catcher method, a record was obtained from the 600-m station. With the gold foil method, the number of neutrons per square centimeter per unit

243

logarithmic energy interval was measured at seven stations ranging from 300 to 1000 m from the explosion. Of the sulfur threshold detectors, only two of the eight units used were recovered, and these gave the neutron flux for energies of 3 MeV at 200 m.

The conversion of plutonium to fission products, determined by the ratio of fission products to plutonium, resulted in an equivalent to 18 600 tons of TNT. Attempting to collect fission products and plutonium from the dust of the shot on filters carried by planes at high altitude after it circled the world gave no results, although later some particles were obtained by this method after the Hiroshima explosion.

Damage, blast, and shock experiments

Blast measurements

• Quartz piezoelectric gauges, which gave no records because the traces were thrown off scale by radiation effects.

• Condenser gauges of the California Institute of Technology type were dropped from B-29 airplanes but no records were obtained because the shot had to be fired when the planes were out of the position.

• The excess shock wave velocity in relation to sound velocity was measured with a moving-coil loudspeaker pickup, by the optical method with blast-operated switches and Torpex flash bombs, and by the Schlieren method. By the moving-coil loudspeaker method, the sound velocity was obtained for a small charge and then the excess velocity for the bomb. This measurement indicated a yield of 10 000 tons and was one of the most successful blast-measuring methods.

• Peak pressure measurements were done with spring-loaded piston gauges at an intermediate pressure range of from 2.5 to 10 psi with the same kind of gauges above ground and in slit trenches at a pressure range of 20 to 150 psi, with crusher-type gauges, and with aluminum diaphragm "box" gauges at a range of 1 to 6 psi. The first of these methods gave low blast pressure values compared to all other methods, the crusher-type gauges gave the highest pressure range, and the box gauges gave a TNT equivalent to 9900 ± 1000 tons. This last method was inexpensive and reliable.

• Remote pressure barograph recorders gave results consistent with 10 000 tons. These were necessary for legal reasons.

• Impulse gauges, mechanically recording piston liquid and orifice

244

gauges, also gave results consistent with 10 000 tons.

- Mass velocity measurements were made by Fastax cameras with suspended Primacord and magnesium flash powder.
- Shock-wave expansion measurements were made with Fastax cameras at half-mile stations and gave a total yield of 19 000 tons.

Earth-shock measurements

- Geophone measurements with velocity-type moving-coil strong-motion geophones gave 7000 tons after extrapolation from a small charge and 100-ton data.
- Seismograph measurements done with Leet three-component strong-motion displacement seismographs approximated 15 000 tons. These were also necessary for legal reasons.
- Permanent earth displacement measurements using steel stakes for level and vertical displacements compared to results of 10 000 ± 5000 tons.
- Remote seismographic observations at Tucson, El Paso, and Denver showed no effect at these distances.

Damage from thermal radiation

The ignition of structural materials was observed using roofing materials, wood, and excelsior on stakes. Observations showed that the risk is small for fire produced by radiant energy in distances greater than 3200 ft. The risk of the fire from direct radiation was likely to be much less than the risk of fire from stoves, etc., at the time of the explosion. These conclusions were confirmed at Hiroshima and Nagasaki.

The study of general phenomena consisted chiefly of photographic studies of the fireball and the column of blast cloud effects. These studies included a radar study with 2 SCR-584 radars in which two plots of the cloud were obtained. Radar reflection, however, was unfavorable. Photographic equipment used for these studies included Fastax cameras ranging from 800 to 8000 frames per second, standard 16-mm color cameras, a 24 frames/s Cine-Special, 100 frames/s Mitchell cameras, pinhole cameras, gamma-ray cameras, Fairchild 9- by 9-in. aerial view cameras at 5.7 miles and at 20 miles for stereophotos. These photographic records were extremely valuable.

The column rise was followed with searchlight equipment, and the first 18 miles of the main cloud path was obtained by triangulation. These experiments included spectrographic and photometric measurements and total radiation measurement. Spectrographic measurements were done with Hilger and with Bausch and Lomb high-time-resolution spectrographs, photometric measurements with moving film and filters and with photocells and filters recording on drum oscillographs, and total radiation measurements with thermocouples and recording equipment.

Postshot radiation measurements

Gamma-ray sentinels were ionization chambers, recording at 5.7-mile shelters, and these gave the radioactive products distribution immediately after the shot and until safe, stable conditions were assured.

Observations in the high-gamma-flux region were made from heavily shielded Army tanks using portable ionization chambers about 4 hours after the shot. Ionization data from these chambers were radioed back to the control shelter.

The Health Group made a dust-borne product survey with portable alpha and gamma ionization chambers and Geiger counters, both at the explosion site and at remote points up to 200 miles in order to measure dust-deposited fission products.

Measurement of airborne products from B-29 planes equipped with special air filters was unsuccessful.

A detailed crater survey was made with ionization chambers and amplifiers after 4 weeks and showed approximately 15 roentgens per hour at the edge of the crater and 0.2 R/h at 1500 ft (450 m).

Weather information was obtained up to 45 minutes before the shot from the detonation point to 20 000 ft and 25 minutes after the shot. Low-level smoke studies determined the spread of active material in case the nuclear explosion failed to occur. This information was vitally important for the success of the test.

246

CHAPTER 19

Project Alberta

DELIVERY GROUP

Although Project Alberta was not organized formally until March 1945, its work had been done by the Delivery Group of the Ordnance Division since June 1943 (Chap. 7). The group was responsible for delivering the bomb as a practical airborne military weapon. During the first part of the Project's history, it participated in designing the final bomb, and it acted as liaison with the Air Force in selecting aircraft and supervising field tests with mock bombs.

After the general reorganization of the Laboratory, when it was decided that the plutonium gun assembly method would not be used, three models remained—the Little Boy for the ^{235}U gun assembly (Chap. 7), the 1222 Fat Man model of the implosion assembly, and the model that became the 1561 Fat Man. By September 1, 1944, the external shapes and aircraft requirements of the three models were determined so the Air Force could begin immediately to train a combat unit to deliver the bomb. A production lot of 15 B-29s was modified at the Martin, Nebraska plant, under the guidance of S. Dike and M. Bolstad of Los Alamos. The first aircraft were available in October. Wendover Army Air Base in Utah, sometimes called by the code name, Kingman, or the symbol, W-47, was designated as the training and test center for the new Atomic Bomb Group. Colonel P. W. Tibbets was appointed commanding officer of the combat group known as the 509th Composite Group.

The first tests began at Wendover in October 1944 and continued until the first combat drop. Several groups from O, X, and G Divisions, in addition to the Delivery Group, participated in the Wendover tests, including the Fusing Group, the Gun Group, the HE Assembly Group, the Electric Detonator Group, and the Ballistics Group. In November 1944, Commander F. L. Ashworth (USN) assumed command of these field operations. The long test series began with three tentative models and ended with two final models and included tests for ballistics

247

information, for electrical fusing information, for flight performance of electrical detonators, for operation of the aircraft release mechanism, for vibration information, for assembly experience, and for temperature effects. Because the first B-29s had poor flying qualities and because the special project modifications had several weaknesses, about 15 new planes were obtained in the spring of 1945. These aircraft, which proved extremely satisfactory, had fuel injector engines, electrically controlled propellers, very rugged provisions for carrying the bomb, and all armament removed except the tail turret. In addition to the tests at Wendover, test drops were made at the Camel Project's field at Inyokern during 1945 (Chap. 9). The Ballistics Group of O Division did some research on the problem of aircraft safety in delivery. They were concerned with such problems as the shock pressure a B-29 could safely withstand, the maneuver that would carry the plane a maximum distance away from the target in a minimum time, and the use of special shock bracing for personnel.

During the fall of 1944 and winter of 1945, the Delivery Group at Los Alamos continued to design and produce mock bombs to achieve a final model. During this period, the 1561 Fat Man was adopted in place of the 1222 model. In addition to the Wendover tests, numerous physics and engineering tests on complete units were made at one of the outlying sites at Los Alamos. The Delivery Group also began plans for establishing an overseas operating base, known by its code name "Alberta."

ORGANIZATION AND TESTS

In March 1945, Project Alberta (or Project A) was established to provide a more effective means of preparing and delivering a combat bomb than the Los Alamos Delivery Group had been able to offer. Project A was independent of any existing division and was organized as a loose coordinating body, with all specific work being done by groups of other divisions. Project A provided direction only where preparations for combat delivery were concerned. Captain Parsons was the officer in charge of Project Alberta, with Ramsey and later Bradbury as deputies for scientific and technical matters. The organization included three groups—an administrative group known as the Headquarters Staff, a technical policy committee called the Weapons Committee (Chap. 9), and a working group of representatives from other divisions. Commander Ashworth was operations officer and military alternate for

Captain Parsons and served as chief of the Headquarters Staff that eventually included Alvarez, Bolstad, S. Dike, G. Fowler, and S. J. Simmons. Simmons came to the Project in June from M.I.T. Radiation Laboratory, where he had engaged in similar liaison activities with the Air Force. The Weapons Committee, of which Ramsey was Chairman until he went overseas and was succeeded by Bradbury, included Birch, Brode, G. Fowler, Fussell, Morrison, and Warner. Groups and representatives are included.

Tests at Wendover	Commander Ashworth
Tests at Wendover after June 1945	Simmons
Measurements and airborne observations	Waldman and Alvarez
General theory	Bethe
Gun assembly	Birch
Aircraft	Bolstad and Dike
HE assembly for Implosion	Bradbury and Warner
Fusing	Brode
Electrical detonator system	Fussell
Engineering	Galloway
Supply	Lieutenant Colonel Lockridge
Pit (active material and tamper of Fat Man)	Morrison and Holloway
Radiology	Captain Nolan
Damage	Penney
Ballistics	Shapiro

Project Alberta was concerned chiefly with three problems: the completion of design, procurement, and preliminary assembly of bomb units that would be complete and ready for use with active material; continuation of the Wendover test program; and preparation for overseas operations against the enemy.

Because the time schedule was becoming tighter, the major designs were continued with as few alterations as possible, although in many cases they were the result of compromises and guesses made when the problem was not well understood. The emphasis during this period was on supplying the many details necessary for successful operation and

249

correcting faults that became apparent in tests. Problems solved were such matters as the exact design of the tamper sphere, inclusion of a hypodermic tube between the HE blocks for monitoring purposes, and strengthening the Little Boy tail. Actually, the Little Boy was far ahead of the Fat Man in design and development because the Gun Assembly Group had a long time to devote to such improvements. Members of the Weapons Committee were concerned about starting work on an integrated design for the Fat Man based on current knowledge with no commitments to past production, but they realized that such a program could not interfere with the primary job of patching up the existing model as quickly as possible. Redesigning the Fat Man from a sound engineering point of view eventually became the task of Z Division, which was barely organized by the end of the war (Chap. 9). Liaison problems with bomb development were very important during this period and were handled primarily by Captain Parsons and Commander Ashworth. Among the military and semimilitary organizations and individuals involved, in addition to the U.S. Army Corps of Engineers, were the 20th Air Force, the Bureau of Ordnance, the Assistant Chief of Naval Operations for Materiel, the Commander of the Western Sea Frontier, the Commandant of the 12th Naval District, the Commandant of the Navy Yard at Mare Island, the Bureau of Yards and Docks of the Navy Department, the Naval Ordnance Test Station at Inyokern, the Naval Munitions Depot at Yorktown, and the Naval Ammunition Depot at McAlester, Oklahoma. After Parsons and Ashworth went overseas, much of this work was handled by Captain R. R. Larkin (USN) who arrived at Los Alamos in June.

The Wendover test program under the supervision of Project Alberta continued at an increasing rate. The principal difficulty in conducting this program was that the company manufacturing Fat Man firing units, known as X-Units, failed to meet its delivery schedule. Besides reducing the number of tests possible on the X-Units, this failure prevented efficient overall testing because many had to be repeated—once at an early date with all components except an X-Unit and once at a critically late date with an X-Unit. The resulting tight schedule is best illustrated by the fact that it was the end of July before sufficient X-Units had been tested to confirm their safety with HE. The first live tests with the X-Unit were not made until August 4 (Wendover) and August 8 (Tinian). Despite these difficulties, 155 test units were dropped at Wendover or Inyokern between October 1944 and the middle of August 1945. Much information was obtained from these tests and the corresponding changes were incorporated into the bomb design.

DESTINATION

Perhaps the most important function of Project Alberta was planning and preparing for overseas operations. As early as December 1944, the initial planning and procurement of some kits of tools and materials had begun, and these activities continued quickly through July. In February, Commander Ashworth was sent to Tinian to make a preliminary survey of the location and select a site for project activities. By March, the construction needs for the Tinian Base, known as Destination, were determined, and construction began in April. The buildings used by Project Alberta had all been especially constructed by the Seabees. Most of the buildings were located in the area assigned to the 1st Ordnance Squadron (Special) of the 509th Group, near the beach. These buildings included four air-conditioned Quonset huts, like those normally used for bombsight repair, in which all laboratory and instrument work was performed. A specially guarded area enclosed these buildings within the guarded working area of the group. Five warehouse buildings, a shop building, and an administration building also were located here. About a mile away were three widely spaced, barricaded and guarded, air-conditioned assembly buildings. Ten magazines and two special loading pits equipped with hydraulic lifts for loading bombs into the aircraft were also constructed. A third such pit was constructed at Iwo Jima for possible emergency use. Materials for equipping the buildings and for handling heavy equipment in assembly, tools, scientific instruments, and general supplies were all included in special kits prepared by the various groups. A kit for a central stockroom was also started, but all the materials had not been shipped by the time the war ended. Beginning in May, five batches of kit materials and of components for test and combat units were shipped by boat to Tinian, and several air shipments for critically needed items were made in five C-54 aircraft attached to the 509th Group. Project Alberta was able to beat its schedules largely because these C-54s were available for emergency shipments.

As early as June 1944, the need for field crews for bomb delivery and for later stages of experiments and testing before delivery had been considered. However, it was agreed that because the work might change

251

and because there were many people eager to volunteer, it would be wise to delay recruiting. Actually, the personnel for the project teams at Tinian were selected early in May 1945 and were organized as follows.

Officer-in-Charge	Captain Parsons
Scientific and Technical Deputy	Ramsey
Operations Officer and Military Alternate	Commander Ashworth
Fat Man Assembly Team	Warner
Little Boy Assembly Team	Birch
Fusing Team	Doll
Electrical Detonator Team	Lieutenant Commander E. Stevenson
Pit Team	Morrison and Baker
Observation Team	Alvarez and Waldman
Aircraft Ordnance Team	S. Dike
Special Consultants	Serber, Penney, and Captain J. F. Nolan

Team members included: H. Agnew, Ensign D. L. Anderson, Technician Fifth Class B. Bederson, M. Bolstad, Technical Sergeant R. Brin, Technical Sergeant V. Caleca, M. Camac, Technical Sergeant E. Carlson, Technician Fourth Class A. Collins, Technical Sergeant R. Dawson, Technical Sergeant F. Fortine, Technician Third Class W. Goodman, Technician Third Class D. Harms, Lieutenant J. D. Hopper, Technical Sergeant J. Kupferberg, L. Johnston, L. Langer, Technical Sergeant W. Larkin, H. Linschitz, A. Machen, Ensign D. Mastick, Technician Third Class R. Matthews, Lieutenant (junior grade) V. Miller, Technician Third Class L. Motichko, Technical Sergeant W. Murphy, Technical Sergeant E. Nooker, T. Olmstead, Ensign B. O'Keefe, T. Perlman, Ensign W. Prohs, Ensign G. Reynolds, H. Russ, R. Schreiber, Technical Sergeant G. Thornton, Ensign Tucker, and Technician Fourth Class F. Zimmerli.

The Los Alamos group formed part of what was known as the First Technical Service Detachment; this Army administrative organization provided housing and various services, and they established security regulations at Tinian. Also closely associated with the work of Project Alberta at Tinian were the members of the 509th Composite Group, who would deliver the bombs to the enemy.

It was decided that Laboratory employees would remain on the Laboratory's payroll. They were provided with per diem, uniform

252

allowances, and insurance policies, in addition to their regular salaries. Each civilian was required to wear a uniform and received an assimilated Army rank in accordance with his civilian salary classification.

Team leaders formed a Project Technical Committee under the chairmanship of Ramsey to coordinate technical matters and to recommend technical actions. Project personnel were responsible for providing and testing certain bomb components; for supervising and inspecting bomb assembly; for inspection before takeoff; for testing completed units; for overall coordination of project activities; and finally, for providing advice and recommendations about the weapon use.

Although preliminary construction at Tinian began in April, technical work did not begin until July. During the first half of July, all technical facilities needed for assembly and test work at Tinian were installed. After completing these technical preparations, four Little Boy tests were made with uniformly excellent results. The last test included a facilities check at Iwo Jima for emergency reloading of the bomb into another aircraft. The first of three Fat Man tests was made on August 1 and showed that essential components were operating satisfactorily. The last of these tests (on August 8) was conducted as a final rehearsal and used a unit that was complete except for active material.

The ^{235}U projectile for the Little Boy was delivered at Tinian by the cruiser *Indianapolis* on July 26, only a few days before its tragic sinking off Peleliu. The *Indianapolis* had been held at San Francisco to wait for this cargo and had then made a record run across the Pacific. The rest of the ^{235}U components arrived on July 28 and 29, as the only cargo of three Air Transport Command C-54s. Because the earliest date previously discussed for combat delivery was August 5 (at one time the official date was August 15), Parsons and Ramsey cabled General Groves for permission to drop the first active unit as early as August 1. Although the active unit was ready, the weather was not, and the first four days of August were spent in impatient waiting. Finally, on the morning of August 5, the weather report was good for the following day, and shortly afterwards, official confirmation came from Major General LeMay, Commanding General of the 20th Air Force, that the mission would take place on August 6. The Little Boy was loaded onto its transporting trailer the moment the official confirmation came through and was taken to the loading pit, where it was loaded into the B-29. Final testing of the unit was completed and all was ready early in the evening. Between then and takeoff, the aircraft was watched continuously by a

253

military guard and by representatives of the key technical groups. Final briefing was at midnight, and shortly afterward, the crews assembled at their aircraft under brilliant floodlights with swarms of photographers taking still and motion pictures. Colonel P. W. Tibbets was pilot of the *Enola Gay*, the B-29 that carried the bomb. Major Thomas Ferebee was bombardier, Captain Parsons was bomb commander, and Lieutenant Morris Jepson was electronics test officer for the bomb.

Only a few days before the scheduled drop, the technical group decided that it was not safe to take off with the bomb completely assembled, because a crash might mean tremendous destruction to men and materials on Tinian. Full safing could not be secured, but it was finally agreed that a partial safeguard would be obtained if the cartridge containing the propellant charge was inserted through the opening in the breech block during flight rather than on the ground. This scheme had been considered before (Chap. 14) but was not adopted until this time. Captain Parsons, already assigned to the crew as a weaponeer, was given the job. This decision meant that Captain Parsons had to be trained in a short time to perform the operation, and also that the bomb bay of the B-29 had to be modified to provide him with a convenient place to stand while completing the assembly. These things were done and the bomb was not completely assembled until the plane was safely in the air.

The progress of the mission is described in the log that Captain Parsons kept during the flight.

6 August 1945	0245	Take Off
	0300	Started final loading of gun
	0315	Finished loading
	0605	Headed for Empire from Iwo
	0730	Red plugs in (these plugs armed the bomb so it would detonate if released)
	0741	Started climb
		Weather report received that weather over primary and tertiary targets was good but not over secondary target
	0838	Leveled off at 32 700 feet
	0847	All Archies (electronic fuses) tested to be OK
	0904	Course west
	0909	Target (Hiroshima) in sight

254

0915-1/2	Dropped bomb (Originally scheduled time was 0915) Flash followed by two slaps on plane. Huge cloud
1000	Still in sight of cloud which must be over 40 000 feet high
1003	Fighter reported
1041	Lost sight of cloud 363 miles from Hiroshima with the aircraft being 26 000 feet high

The crews of the strike and observation aircraft reported that 5 minutes after release, a low 3-mile-diameter dark gray cloud hung over the center of Hiroshima, out of the center of which a white column of smoke rose to a 35 000-ft height with the top of the cloud being considerably enlarged. Four hours after the strike, photoreconnaissance planes found that most of the city of Hiroshima was still obscured by the cloud created by the explosion, although fires could be seen around the edges. Pictures were obtained the following day that showed 60% of the city destroyed.

The active component of the Fat Man came by special C-54 transport. The HE components of two Fat Men arrived in two B-29s attached to the 509th Group, which had been held at Albuquerque for this purpose. In all cases, the active components were accompanied by special personnel to guard against accident and loss.

The first Fat Man was scheduled for dropping on August 11. At one time, the schedule called for August 20, but by August 7, it was apparent that the date could be advanced to August 10. When Parsons and Ramsey proposed this change to Tibbets, he expressed regret that the schedule could not be advanced two days instead of only one, because good weather was forecast for August 9 and bad weather for the five succeeding days. It was finally agreed that Project Alberta would try to be ready for August 9, if all concerned understood that advancing the date by two full days introduced a large measure of uncertainty. All went well with the assembly, however, and the unit was loaded and fully checked late in the evening of August 8. The strike plane and two observer planes took off shortly before dawn on August 9. Major C. W. Sweeney piloted the strike ship *Great Artiste*, Captain K. K. Beahan was bombardier, Commander Ashworth was bomb commander, and Lieutenant Phillip Barnes was electronics test officer.

It was impossible to "safe" the Fat Man by leaving the assembly incomplete during takeoff as for Little Boy. The technical staff realized

255

that a crash during takeoff would risk contaminating a wide area on Tinian with plutonium scattered by an HE explosion, and even risk a high-order nuclear explosion that would do heavy damage to the island. These risks were pointed out to the military with the request that special guarding and evacuation precautions be taken during the takeoff. The Air Force officer in command decided that such special precautions were not necessary, and as it turned out, the takeoff was made without incident. This mission was as eventful as the Hiroshima mission was operationally routine. Commander Ashworth's log for the trip is as follows.

0347	Take off
0400	Changed green plugs to red prior to pressurizing
0500	Charged detonator condensers to test leakage. Satisfactory.
0900	Arrived rendezvous point at Yakashima and circled awaiting accompanying aircraft.
0920	One B-29 sighted and joined in formation.
0950	Departed from Yakashima proceeding to primary target Kokura having failed to rendezvous with second B-29. The weather reports received by radio indicated good weather at Kokura (3/10 low clouds, no intermediate or high clouds, and forecast of improving conditions). The weather reports for Nagasaki were good but increasing cloudiness was forecast. For this reason the primary target was selected.
1044	Arrived initial point and started bombing runs on target. Target was obscured by heavy ground haze and smoke. Two additional runs were made hoping that the target might be picked up after closer observations. However, at no time was the aiming point seen. It was then decided to proceed to

256

	Nagasaki after approximately 45 minutes spent in the target area.
1150	Arrived in Nagasaki target area. Approach to target was entirely by radar. At 1150 the bomb was dropped after a 20 second visual bombing run. The bomb functioned normally in all respects.
1205	Departed for Okinawa after having circled smoke column. Lack of available gasoline caused by an inoperative bomb bay tank booster pump forced decision to land at Okinawa before returning to Tinian.
1351	Landed at Yontan Field, Okinawa
1706	Departed Okinawa for Tinian
2245	Landed at Tinian

Because of bad weather, good reconnaissance pictures were not obtained until almost a week after the Nagasaki mission. They showed 44% of the city destroyed. The discrepancy in results between this mission and the first was explained by the unfavorable contours of the city.

Exchange of information between Tinian and Los Alamos was extremely unsatisfactory and caused considerable difficulty at each end. Necessarily tight security rules made direct communications impossible, and teletype messages were relayed from one place to the other through Washington Liaison Office using an elaborate table of codes prepared by Project Alberta. Late in July, the Laboratory sent Manley to the Washington Liaison Office to dispel any possible friction in the regular channels of information, and to see that no information was being held up in Washington that would conceivably be of interest. The first news of the Hiroshima drop came to Los Alamos in a dramatic teletype prepared by Manley summarizing the messages sent by Parsons from the plane after the drop (Fig. 4).

On the day following the Nagasaki mission, the Japanese initiated surrender negotiations, and further activity in preparing active units was suspended. The entire project was maintained at complete readiness for further assemblies in case the peace negotiations failed. It was planned to return all Project Alberta technical personnel to the United States on August 20, except for those assigned to the Farrell mission for investigating the bombing results. Because of the delays in surrender procedures, General Groves requested all key personnel to remain at Tinian until the success of the Japanese occupation was ensured. Scientific and technical personnel finally received authorization and left Tinian on September 7, except for Colonel Kirkpatrick and Commander Ashworth, who remained to make final disposition of project property. With this departure, the activities of Project Alberta were terminated.

The objective of Project Alberta was to ensure the successful combat use of an atomic bomb at the earliest possible date after a field test and after the availability of the necessary nuclear material. This objective was accomplished. The first combat bomb was ready within 17 days after the Trinity test, and almost all the intervening time was spent in accumulating additional active material for making another bomb. The first atomic bomb was prepared for combat on August 2, within four days of delivery of all the active material needed for that bomb. Actual combat use was delayed until August 6 because of bad weather over Japan. The second bomb was used in combat only three days after the first, although it was a completely different model and was more difficult to assemble.

CHAPTER 20
Conclusion

After the war ended, the Laboratory experienced a sudden relaxation of activities. Everything had been aimed at a goal, and the goal had been reached. It was a time for evaluation and taking stock. Plans for the future of Los Alamos and of nuclear research in general were widely discussed. Members of the Scientific Panel of the President's Interim Committee on Atomic Energy met at Los Alamos and prepared an account of the technical possibilities then apparent in the atomic energy field. A series of lecture courses was organized, called the "Los Alamos University," to give the younger staff members the opportunity to make up for some of the studies they had missed during the war years.

Although research projects were being completed, plans for the next period were being formulated. One task discussed was outlining and writing a Los Alamos Technical Series, under the editorship of H. A. Bethe, to set down a more systematic and polished record of the Laboratory's work than had been possible during the war. Some effort was made to complete the theoretical investigations of the Super described in Chap. 13. Weapon production had to continue and plans were made to finish the development work on the implosion bomb, including the composite sphere assemblies mentioned previously (Chap. 11).

This history records problems and their solutions. The other side of Los Alamos history, the reactions to these accomplishments by the people who made them, is present only by implication. This account ends at a time, however, when these reactions assumed a sudden importance, and it is appropriate that it should end with some description of them.

For many Laboratory members, the Trinity test marked the successful climax of years of intensive and uncertain effort. A new kind of weapon had been made, and the magnitude and qualitative features of its operation had been predicted successfully. Despite the fact, perhaps partly because of the fact, that the explosion occurred as expected, the sight of it was a stunning experience to its creators, an experience of satisfaction and of fear. A new force had been created and would henceforth lead a life of its own, independent of the will of those who made it. Perhaps only at Trinity were its magnitude and unpredictable potentialities fully grasped and appreciated.

Four days after the first bomb was dropped over Hiroshima, the Japanese began surrender negotiations. The feelings that had marked the success of the Trinity test were evident once more. But now the Laboratory staff, experiencing the sudden slackening of effort that followed the end of the war, began to speak seriously of the bomb and its consequences for the future. The thoughts expressed were not new, but there had been no time before to express them. Since 1939, when the decision was made to seek Government support for the new development, a uniformity of insight had grown among the working scientists of the Manhattan District. They came to realize that atomic warfare would prove unendurable. The Japanese learned this in the days of Hiroshima and Nagasaki, and soon all the world was saying it.

The Laboratory members who joined in these discussions saw more incisively than many other Americans that atomic bombs were offensive or retaliatory weapons, and their existence threatened the security of every nation so that it could not venture, without the gravest risk, to meet on the military plane alone. The law of counterdevelopment, which has operated so uniformly in military affairs to produce new defenses against new weapons, could operate to open collaboration channels not previously existing among nations. The wartime scientific collaboration that produced this weapon could be, by its worldwide extension, uniquely the means for eliminating it from national armaments. Scientists, who had never been overly concerned with social and political problems, felt a responsibility to tell the American public about the nature and implications of the new weapon, and to clarify the alternatives that had arisen for the future. This concern received its best and simplest expression in a speech by Oppenheimer, given on October 16, 1945, when General Groves presented the Laboratory with a certificate of appreciation from the Secretary of War.

"It is with appreciation and gratitude that I accept from you this scroll for the Los Alamos Laboratory, for the men and women whose work and whose hearts have made it. It is our hope that in years to come we may look at this scroll, and all that it signifies, with pride.

"Today that pride must be tempered with a profound concern. If atomic bombs are to be added as new weapons to the arsenals of a warring world, or to the arsenals of nations preparing for war, then the time will come when mankind will curse the names of Los Alamos and Hiroshima.

"The peoples of this world must unite, or they will perish. This war, that has ravaged so much of the earth, has written these words. The atomic bomb has spelled them out for all men to understand. Other men have spoken them, in other times, of other wars, of other weapons. They have not prevailed. There are some, misled by a false sense of human history, who hold that they will not prevail today. It is not for us to believe that. By our works we are committed, committed to a world united, before this common peril, in law, and in humanity."

Eric Jette, Charles Critchfield, Robert Oppenheimer.

E.O. Lawrence, Enrico Fermi and I.I. Rabi.

Edward Teller with Norris Bradbury.

Base Camp, Trinity test.

Jumbo was erected on a tower 800 feet from ground zero.

Trinity Bomb being hoisted to top of tower.

Jumbo, the 25-foot long, 214-ton steel vessel designed to contain the explosion of the first atomic device, arriving at railroad siding at Pope, New Mexico, Spring, 1945.

Personnel attending a technical session, one of a special series on the future of the Laboratory. Front Row, left to right: *Norris Bradbury, John Manley, Enrico Fermi, Jerry Kellog.* Second Row: *Robert Oppenheimer, Richard Feynman, Philip Porter.* Third Row: *Gregory Breit, Arthur Hemmendinger, Arthur Schelberg.*

*Recognition given to Robert Oppenheimer upon his retirement as
Director of the Laboratory. General Groves holds citation and
Gordon Sproul, President of the University of California, looks on.*

J. Robert Oppenheimer

PART II

August 1945 Through December 1946

by Edith C. Truslow
and
Ralph Carlisle Smith

ORIGINAL PREFACE TO PART II

This second part presents the activities of the Los Alamos Laboratory from August 1945 through December 1946, when the Manhattan District relinquished its control. No attempt has been made to interpret events or to forecast the importance of scientific developments. This report is merely a chronicle of the reorganization, philosophy, and achievements during this critical period after the war.

Historical material was supplied by the Director, N. E. Bradbury; the Administrative Associate Director, Lieutenant Colonel A. W. Betts; and the Technical Division Leaders: Carson Mark, J. M. B. Kellogg, Marshall Holloway, Eric Jette, Max Roy, R. W. Henderson, and Ralph Carlisle Smith. Further assistance was given by J. F. Mullaney, Report Editor, and A. E. Dyhre, Business Manager.

15 October 1947

CHAPTER 21

General and Technical Review
October 15, 1947

NECESSITY FOR POSTWAR POLICY

General

With the war ended and with no national legislation on atomic energy, the Los Alamos Laboratory and the Manhattan Engineer District faced the problem of determining an appropriate policy for a laboratory whose existence had been devoted to the atomic bomb.

Personnel Problems

The Laboratory had been staffed with almost ruthless abandon with personnel from universities, NDRC and OSRD laboratories, and industry, and with graduate students, technicians, and scientists of every degree of eminence. Not only did their backgrounds and commitments vary, but their opinions about the Laboratory's character and future differed in almost every conceivable way. Also, with the cessation of hostilities, many of the military personnel had a profound change of attitude toward the technical work. In September 1945, the Laboratory staff could be divided roughly as follows.

Academic Personnel on Leave from Colleges and Universities. Because of the sudden end of hostilities at the start of the academic year, most academic personnel were not free to return directly to their universities and take up the normal course of academic work. Moreover, the extraordinary reputation that science had gained during the war, and the achievements of physicists in particular, lcd many universities to try to build up their technical staffs, which had been seriously depleted. Consequently, many individuals were offered jobs by other universities and colleges as well as by industry. Also, the salary increases offered tempted them to "shop around."

265

Recent Recipients of Ph.D.'s. These individuals had done most of their scientific work in government laboratories from 1940 on. They had become, often unknowingly, accustomed to the speed, intensity, and particularly the technical and administrative services, provided in wartime laboratories. Therefore, many were uncertain about the type of career they wished to follow. Academic careers, industry, and government or atomic energy laboratories competed on reasonably equal ground. Again, because of enhanced prestige of scientists, there were attractive offers among which they could choose.

Graduate Students of Varying Degrees of Experience. Most of these individuals had interrupted their graduate careers one, two, or three years beyond the bachelor's degree to work as civilians during the war. Many had acquired families and corresponding financial responsibilities. Usually, they had worked in specific fields and therefore lacked the broader knowledge that their maturity might have implied. Because they had lost touch with academic procedures and channels of thought, they foresaw that returning to graduate work might be difficult. They had to decide whether to resume their graduate studies knowing that failure to do so might handicap them later.

Technical, Administrative, and Clerical Personnel. Although some of the senior technical, administrative, and clerical personnel were on leave from other organizations, they were possibly less tied by previous commitments than the scientific personnel. Nevertheless, they also had to decide what to do about their careers. Some of the junior ones, who were housed in the less satisfactory accommodations, were anxious to return to a more normal environment, now that the patriotic pressure was removed.

Military Personnel. Military personnel fell into all the above classifications, and some were on leave from industrial organizations. The higher ranking officers were usually senior personnel who had taken leave from universities or industry to accept commissions early in the war. Many enlisted personnel were products of the armed forces special training programs. Some had been drafted; others had enlisted. All were tied to previous commitments, or to return to school, and by the almost universal desire to get out of uniform. Barracks life for enlisted personnel, many of whom had had extensive technical training, had proved a serious trial. To most of the enlisted personnel who had performed so successfully and diligently during the war, peace brought complete apathy and indifference to the Laboratory's problems and activities. Their one desire was to be discharged and go home to a "normal" life.

Miscellaneous Problems

The Laboratory staff at the end of the war were undecided whether to stay at Los Alamos or leave. The Laboratory's technical future was uncertain, and its administration was equally unclear. The combination of an absentee contractor and Army administration of the community and auxiliary services had created antagonism and irritation that, for many, could be solved only by leaving Los Alamos. To these problems was added the generally indifferent attitude of enlisted personnel on whom the Laboratory had come to depend. Civilian personnel, although remaining on the Laboratory payroll, also were indifferent about its future and awaited only the best opportunity to leave. These problems, added to the basic concern about a proper philosophy for the Laboratory, presented an extremely complicated personnel picture.

PROPOSED PHILOSOPHIES

There were many opinions about what the Laboratory should do in peacetime. One group, headed by one of the most senior Laboratory members, contended that the Laboratory should become a monument—it should be abandoned and its functions, if necessary or useful in peacetime, should be taken up elsewhere. Another philosophy suggested that the Laboratory abandon production of atomic weapons and conduct only peaceful research, or basic research whose application might be in the indefinite future. Others held that the Laboratory's basic purpose was research and development of atomic weapons, and that for the present at least, their design and production might or must continue. These philosophies were further complicated by the question whether Los Alamos, on an isolated mesa top in New Mexico, was adequate or satisfactory as a peacetime location for a laboratory of any character.

TRANSITION PERIOD

Immediately after the war, technical activity practically came to a standstill. This was the result of a complete psychological deflation after the intense technical effort, which had culminated in delivery of atomic weapons at Hiroshima and Nagasaki, followed by the climax of victory. In October 1945, J. R. Oppenheimer announced his intention to resign as Laboratory Director. On October 17, N. E. Bradbury, at the request of Oppenheimer and Major General L. R. Groves, became Director for six months or until national legislation on atomic energy was established, should that occur sooner.

In a talk on October 1, 1945, Bradbury presented to the Coordinating Council of the Laboratory his conception of a philosophy for the Laboratory (Appendix A) between the end of the war and establishment of national legislation on atomic energy. His philosophy was based on the assumption that the Laboratory would remain in operation in Los Alamos; that its problems and goals would be those pertinent to research and development of atomic weapons or related matters; that such weapon problems would be considered on both short- and long-range bases; and that the Laboratory staff would decrease to approximately one-third of its wartime level owing to the scarcity of housing and departure of military personnel previously housed in barracks who would have to be replaced by civilians with families requiring houses.

General Groves in a letter dated January 4, 1946 (Appendix B) outlined general plans for the future, which agreed in part with Bradbury's philosophy. The paramount problem was to establish the Laboratory's internal technical and administrative structure and to determine the composition of its technical staff.

During the spring of 1946, morale throughout the Laboratory was seriously harmed by the indecisiveness of many personnel. It became evident that planning an intelligent and vigorous program required a staff committed to, and enthusiastic about, the Laboratory's future. A new policy, announced in May 1946 to become effective on September 1, stated that, except in special cases, the Laboratory would no longer pay an individual's way home whenever he decided to terminate. The effect was exactly as expected—those who had been undecided announced their intentions to stay or leave Los Alamos. Those who stayed, by that act, committed themselves to work toward the success of Los Alamos.

PEACETIME ACTIVITIES INTRODUCED

The Los Alamos University

The early tendency to think of the Laboratory's task as finished led to

268

establishment of the Los Alamos "University" in September 1945. "University" was a facetious title for a program of technical lectures given as semitraditional academic courses. The intent was to encourage enlisted personnel and junior civilian personnel to complete their academic training. Unfortunately, these courses were conducted during normal working hours and conflicted with operation of the project. Of the approximately 678 persons enrolled in this program, 134 received college credit for their work. The schedule of courses is given in Appendix C.

Program for Consultants

Although many of the senior personnel who left Los Alamos after the war had strong convictions about their future association with classified work and atomic weapon research in particular, more recognized that this type of work had to continue and expressed some willingness to participate. The Laboratory also recognized the desirability of being able to consult with previous members of the Laboratory about Laboratory problems. Accordingly, a program was set up whereby these people could periodically work with the Laboratory's present staff on various problems, many of which had originated during the Laboratory's wartime existence.

April Conference

Of various conferences on highly classified matters during 1946, one of the more important was the April 17-23 conference on the "Super" led by Edward Teller (Chap. 23).

University Affiliation Conference

It became apparent that not only could the Laboratory's techniques (where declassifiable) add to the advance of science in the country, but that the Laboratory could profit by cordial relations with universities. Although it was recognized that the Laboratory could not become a regional laboratory as was proposed for Brookhaven and Argonne and suggested for Clinton, it appeared that certain aspects of this procedure could be applied with profit to Los Alamos. In July 1946 representatives from universities west of the Mississippi River were invited to a conference to explore the possibilities of cooperation with Los Alamos.

The conference centered on the possibilities that Los Alamos offered for training graduate students in physics and chemistry, particularly for doing thesis work for the doctoral degree. The Laboratory could offer responsibility for the direction of thesis work only in fields of interest to the Laboratory. Such work could be evaluated by the university granting the degree and conducting the examinations. The report of this conference is Appendix D.

August Conference

The consultant and conference program climaxed with a large conference on August 19-24 attended by the Laboratory staff, members of other Manhattan District laboratories, and consultants to Los Alamos. A total of 57 participants from outside the Laboratory attended the Nuclear Physics Conference as well as the informal conferences on Laboratory activities. The formal program of the conference is given in Appendix E.

HEALTH AND SAFETY PROGRAM

Of primary concern was the nature of the medical care provided for the Laboratory staff and residents of the community. During the war, most of the medical personnel were in uniform; as their discharges became imminent, it became necessary to hire civilian doctors. It also became evident that in peacetime the Laboratory could not afford to take any chances with human life. It was therefore necessary to increase emphasis on medical and industrial health research, as well as general safety practices. The Laboratory Director ordered that any unsafe practice be stopped, regardless of its priority and importance. In the spring of 1946, the Laboratory explored the possibility of establishing a link with a well-known medical school to provide a board from whom medical advice might be obtained. An agreement was reached with Washington University at St. Louis, Missouri, where such a board was established.

Radiation Fatalities

Despite increased emphasis on safety, there were two serious accidents during experiments with critical assemblies. The first, on August 21,

1945, caused the death of Harry K. Daghlian on September 15; the second on May 21, 1946, caused the death of Louis Slotin on May 30. The second accident, which emphasized that such accidents could occur even with the most senior personnel in charge, led to establishment of a remote control system for the necessary experimentation in this field (Chap. 25).

WATER SHORTAGE

Although no other Laboratory disasters occurred during this period, the severity and duration of the water shortage in the winter of 1945-46 requires special mention. This event climaxed the bitter resentment against the system of Los Alamos community operation and undoubtedly hastened or inspired the exodus of many already unhappy people.

TECHNICAL ORGANIZATION

Before the end of the war, the Laboratory, in addition to certain technical and administrative staff groups, was composed of the following technical divisions.

Technical Divisions	Division Leaders	Terminated
Chemistry and Metallurgy (CM)	Joseph Kennedy	December 20, 1945
	Cyril S. Smith	January 2, 1946
Physics (R)	Robert R. Wilson	February 8, 1946
Physics (F)	Enrico Fermi	December 31, 1945
Theoretical (T)	Hans Bethe	January 2, 1946
Explosives (X)	George Kistiakowsky	January 17, 1946
Gadgets (G)	Robert Bacher	January 15, 1946
Engineering (Ordnance) (Z)	J. R. Zacharias	November 1, 1945

Before these people left, the Laboratory made the following administrative changes and reorganizations.

Chemistry and Metallurgy (CMR)	Eric Jette
Physics (P) (Division combined)	John Manley
Theoretical (T)	George Placzek

271

Explosives (X)	Max Roy
Experimental (old Gadget) (M)	Darol Froman
Engineering (Ordnance) (Z)	Roger Warner
Documentary (D) (August 1946)	Ralph Carlisle Smith

In July 1946, J. M. B. Kellogg became head of the Physics Division to free Manley, at his request, for experimental work. In November 1946, R. D. Richtmyer took over the Theoretical Division. The Laboratory was fortunate in its Division Leaders. Most had had extensive experience in the Laboratory, and many had been in positions of high responsibility in the divisions they were now directing.

The Laboratory eventually was organized into seven technical divisions, several technical staff groups, and an Administrative Division. The divisions, in turn, were divided into groups that had particular but closely interrelated responsibilities. The detailed technical history of the Laboratory is given in the monthly progress reports of each division. The interaction of these reports and the corresponding general technical advance of the Laboratory are discussed in the following chapters. Specific identification of the groups and their individual responsibilities is also made.

Technical Advisory Boards

In November 1946, two technical committees were organized to help the Director formulate the Laboratory's technical program. The Technical Board consisted of all Division Leaders, whereas only the leaders of Ordnance Engineering (Z), Chemistry and Metallurgy (CMR), Explosives (X), and Experimental (M) Divisions made up the Weapons Panel. The functions of these groups soon became indistinguishable and their separate meetings gradually merged into one. The Coordinating Council, composed of Group and Division Leaders and certain senior scientific personnel, continued as the interim council. However, the colloquium, for all staff members, was discontinued.

CONTINUANCE OF THE WEAPON ENGINEERING PROGRAM

Although the Laboratory's technical program took some time to evolve, there remained a considerable program of pure ordnance engineering of the weapon. This phase of bomb development had been neglected or hurried to provide a weapon as soon as possible. The need for overall design improvement was apparent to all. During the war, all the engineering had been conducted at Los Alamos, but most of the field testing had been at Wendover Field, Utah. Selection of this site, although advantageous for security reasons, created practically insurmountable difficulties in peacetime because of its distance and relative inaccessibility from Los Alamos. In the fall of 1945, this activity was transferred to Oxnard Field, now known as Sandia Base, near Albuquerque, New Mexico. Several buildings were already available together with facilities for troops. The nearby Kirtland Field was used to base the B-29 squadron needed for the test program. Some of the explosive materials stored at Wendover were transferred to the Fort Wingate depot near Gallup, New Mexico (Chap. 28).

At this time, the Salton Sea Navy Base, which had been of some use during the spring and summer of 1945, became the main location of the Ordnance Engineering (Z) Division drop test program.

The Los Alamos facilities for full-scale high-explosive production were closed down owing to lack of personnel and deterioration of the temporary buildings in which these lines were housed (see Chap. 27). The Inyokern (California) project, initiated during the war to build experimental half-scale castings, was now overdesigned for this purpose. Arrangements were made with the Navy Bureau of Ordnance to carry out full-scale casting operations there, and the plant was revised accordingly. Delivery of reasonably satisfactory full-scale castings began in the winter of 1946.

The basic design of a "composite" implosion weapon using both plutonium and ^{235}U was developed for use in implosion conditions like those previously used for pure plutonium.

OPERATION CROSSROADS

In December 1945, the Laboratory was informed of a proposed test of atomic weapons against naval vessels. The Laboratory was asked to undertake technical direction of this test and to supply the atomic weapons. In preliminary meetings in December and January, it was decided to

(a) set up a joint Task Force Operation with the Army, Navy, and Manhattan Engineer District,

273

(b) use the type of atomic weapon used at Nagasaki, and

(c) recommend that three types of tests be conducted in the following order.

Able: Air burst over an array of representative ships.

Baker: Shallow water burst under an array of representative ships.

Charlie: Deep water burst under an array of representative ships.

The Navy selected Bikini atoll in the Marshall Islands as a satisfactory location for "Able" and "Baker," and suggested that "Charlie" might be conducted on the deep-water side of the atoll.

Ralph A. Sawyer was selected as Technical Director for this operation, with Marshall Holloway of B Division and Roger Warner of Z Division under him as heads of the Los Alamos experimental and weapon preparation groups, respectively.

Which weapon to use in these tests was the subject of much discussion. The Laboratory recommended the Nagasaki-type weapon, although a different type of weapon was urgently in need of testing. The Laboratory's recommendation came from the following arguments.

(a) The purpose of the test was strictly military. It was not even certain that nuclear efficiency and equivalent high-explosive yield would or could be measured accurately. Therefore, it was important from the point of military strategy and tactics to use a weapon that had been used before so that the same weapon could be compared in different circumstances.

(b) If a new and untried weapon gave poor efficiency, the Laboratory and Joint Task Force would be criticized for not using a "proven" weapon.

(c) Because the effect of an atomic weapon was to be studied, rather than the weapon itself, the Laboratory was unwilling to use a new weapon whose behavior might be difficult or impossible to ascertain.

The responsibilities the Laboratory agreed to accept, in addition to supplying technical direction, were as follows.

(a) Preparation of recommendations for the overall character of the test—including disposition of ships, height (or depth) of burst, and types of weapons to be used.

(b) Preparation of a "handbook" of phenomena to be expected so that other participants might have a technical guess what to expect and thereby make more reasonable technical preparations.

(c) Making at least one definite estimate of the weapon's equivalent high-explosive yield. It was decided to attempt this by radio-chemical methods, photographic methods (for the air burst), and fast neutron measurements (for the air burst). The technique of estimating yield from blast measurements was assigned to the Navy Bureau of Ordnance. The underwater burst presented the greatest measurement problem because of possible failure of the radiochemical technique. This problem was particularly serious because no weapon had ever been detonated using water as the surrounding medium. Also, it was not known whether an atomic weapon would transfer energy to a denser medium exactly as it would transfer energy to air. Accordingly, one timing measurement was to measure the time from first current to the detonators to first appearance of ionization due to the gamma rays emitted by the bomb. A proper value of this time would indicate "normal" performance, whereas too small or too great a value would suggest "subnormal" performance. Fortunately, the radiochemical technique was successful, and the time measurement merely supported the radiochemical observations.

(d) Preparation of firing circuits for the underwater test, as well as the timing system, with radio links, both transmitting and receiving, to be associated with both tests.

(e) Preparation of weapons for both tests, including an appropriately engineered barge from which to suspend the weapon for the subsurface test. Included in this responsibility was supervision of the training for practice flights before the actual drop.

The original target date for the tests, May 15, 1946, was such that no equipment could be developed or engineered to appropriate specifications. It was necessary to use equipment already on hand or that which could be modified easily. This posed a particular problem for the radio links to be used in transmission and reception of radio timing signals at definite time intervals before detonation. The haste in procuring and modifying stock items, intended for another purpose, created various difficulties. The tests were subsequently delayed by Presidential announcement on March 22, 1946. This additional time, however, did not permit any changes in equipment already ordered and in process. Although the delay may actually have done more harm than good to certain phases of the operation where the psychological letdown was

275

severe, the additional time probably benefitted some of the engineering and logistic phases.

Two ships, the *Cumberland Sound* and the *Albermarle*, were assigned to Los Alamos personnel for their operations. The former was Holloway's headquarters and was appropriately modified as a laboratory ship; the latter was headquarters for Warner and the bomb assembly groups and was suitably reworked for the assembly operations. A weekly C-54 run between Santa Fe and Washington, DC, permitted the necessary close liaison between the Laboratory and technical personnel in Washington.

The cost to the Laboratory for additional procurement was about one million dollars. About 150 Laboratory personnel (approximately one-eighth of the staff) worked on this operation for almost nine months. Although the technical divisions at Los Alamos were harmed by the Crossroads operation, the effect on the Ordnance Engineering (Z) Division, split between Los Alamos and Albuquerque, was enormous; their development and engineering programs were almost totally stopped because of their senior personnel's preoccupation with the Bikini effort.

Although Crossroads involved a high cost to the Laboratory, it was not without its gains. Because it occurred when the Laboratory was seriously engaged in postwar reconstruction, it gave the Laboratory an objective, and although it was not entirely welcome, it did provide a psychological advantage. Furthermore, the Laboratory's ability to produce satisfactory weapons without the services of many of its senior and experienced staff undoubtedly had favorable effects both within and without the Laboratory. Another advantage was the Laboratory's ability to participate in an operation that was definitely in the public eye.

The operation also provided technical gains. It gave the Laboratory staff additional experience in practical conduct of atomic weapon tests that would be useful in planning subsequent operations and it supplied technical data that ultimately were of great significance. It was somewhat of a surprise that the Bikini weapons had almost the same efficiency as those of the same type detonated previously. The rise of the vapor column in the air shot suggested further problems to be worked on, besides demonstrating the absence of fallout where dust is not present. The underwater shot proved the basic assumption that an atomic weapon behaves satisfactorily with water surrounding it; it was also an excellent large-scale demonstration of the water column effects to be observed, showing new phenomena. More penetrating radiation than can easily be accounted for was observed, substantiating certain physiological observations relative to the Nagasaki detonation.

CHAPTER 22

Administrative Organization

INTRODUCTION

The Reviewing Committee recommended appointment of an Associate Director and an Administrative Assistant to relieve the Director of some of the problems that were brought to him personally (Chap. 3). As a first step, David Dow was appointed Assistant to the Director. However, a further strengthening of the entire administrative organization was indicated.

ADMINISTRATIVE STRUCTURE

Associate Director Appointed

In the fall of 1945, Oppenheimer appointed Colonel L. E. Seeman (General Groves' liaison officer at Project Y) as his administrator. On October 31, Colonel Seeman was appointed Associate Director to Bradbury (who became Laboratory Director after Oppenheimer's resignation in October). In addition to his administrative and technical functions at the Laboratory, Seeman was also officer-in-charge of the Santa Fe Area, Manhattan District. In this capacity he was responsible for the military divisions, including Security, Post Command, Area Engineer, and the hospital.

Administration and Services Division Formed

An Administration and Services Division was formed in December 1945 with Colonel Seeman as Division Leader and Colonel A. W. Betts as his Deputy. It contained the following groups.

278

A-1	Administrative Group	David Dow
A-2	Personnel Group	E. J. Demson
A-3	Shops Group	Gus Schultz
A-4	Procurement Group	Harry S. Allen
A-5	Tech Area Maintenance	Lieutenant Robert C. Hill (USNR)
A-6	Safety Group	Eldon E. Beck

In April 1946, E. J. Demson was made Assistant Director and John V. Young was made Personnel Group Leader. The arrangement seemed to function very well as evidenced by a letter from Bradbury to Robert M. Underhill on January 28, 1946, which said in part: "We have found that to have Colonel Seeman acting as Associate Director as well as Area Engineer has been of the greatest assistance in smoothing and expediting the relationships between the Technical Area, the Post Military organization, Lt. Col. Stewart's office, and the District Engineer. I quite concur that this arrangement can serve as an additional protection to the University."

Technical Staff Groups

The following technical staff groups remained unaffected by this change and were directly responsible to the Director.

A-7	Editorial Group	David R. Inglis
A-8	History Group	David Hawkins
A-9	Patent Group	Major Ralph Carlisle Smith
A-10	Health Group	Dr. Louis Hempelmann
A-14	Library Group	Inez O'Brien
A-15	Declassification Group	Frederic de Hoffman

The technical staff groups reporting directly to the Director were reduced when D Division was formed in August 1946 under Major Ralph Carlisle Smith. This new division absorbed A-7 Editorial Group, A-8 History Group, A-9 Patent Group, A-14 Library and Document Room, A-15 Declassification Group, the Technical Series Group previously under Hans Bethe of the Theoretical Division, and the duties of Demson, the Assistant Director (Chap. 29). The A-10 Health Group, under Dr. Louis Hempelmann, remained as the only independent staff group.

279

Business Office

The Business Office, A-11, continued its administrative functions for the University of California under J. A. D. Muncy, Business Manager, until June 1, 1946, when A. E. Dyhre assumed that position.

British Mission

The British Mission personnel started leaving the Laboratory shortly after the Trinity test at Alamogordo Air Base on July 16, 1945, leaving only Ernest Titterton as their representative at the end of 1946. During the spring and summer of 1946, W. G. Penney, J. L. Tuck, and Titterton served with B Division in Operation Crossroads.

A and S Division Organization in December 1946

With time, organizational changes were made in the A and S Division to keep it streamlined and functioning as efficiently as possible. The organizational setup at the end of 1946 was as follows.

Associate Director for Administration and Services		Colonel Austin W. Betts
Assistant Associate Director for Administration and Services		Henry Hoyt
A-1	Photographic Group	Loris Gardner
A-2	Personnel Group	John V. Young
A-3	Shops Group	Gus Schultz
A-4	Supply and Property Group	Harry S. Allen
A-5	Tech Area Maintenance	Robert C. Hill
A-6	Safety	Conrad W. Thomas
A-13	Central Mail & Records	George S. Challis

Personnel Group

Beginning in the summer of 1944, C. D. Shane directed an attempt to organize the Personnel Office so that it might carry out its responsibilities in a large laboratory. Specific functions were assigned various members of the office, and the record-keeping system was revised and completed.

280

The demands for new employees continued to be the major concern, and recruiting absorbed most of the attention of the small staff. Enlarging the staff could not be considered because of the housing shortage. Administrative needs for personnel were subordinate to the needs of the technical staff.

Laboratory personnel policy, finally approved in February 1944, plus Shane's understanding with the Area Engineer and Contracting Officer, Lieutenant Colonel S. L. Stewart, that local decisions about salaries would be reviewed only in terms of stabilization rules, broke the salary deadlock that had plagued the Laboratory for more than a year. From mid-1944 until the end of the war, routine reclassifications, and salary adjustments were processed with minimal delay and disagreement, but cases not in strict conformance to the rules as interpreted by the Area Engineer continued to be troublesome. For example, proposed salary increases above the "OSRD scale" usually were disapproved even though government stabilization regulations were not violated. The OSRD scale was a rule-of-thumb based on education and experience and the Laboratory considered it appropriate for a norm or average; but the Area Engineer office considered it a maximum.

Three major personnel problems that faced the Laboratory at war's end were, in order of importance,

(a) maintenance of an adequate scientific staff,
(b) replacement of the approximately 1600 young scientists in the Special Engineering Detachment (SED) who would be discharged shortly, and
(c) staffing of a special division for technical operations of the Bikini tests scheduled for the summer of 1946.

Holding a scientific staff together was complicated by the uncertainties about the Laboratory program and by the rigid salary control system left over from wartime operation. Most scientists who considered remaining were willing to gamble on the future of the program but were not so sympathetic with the perpetuation of the salary stabilization program. There was ample evidence in the form of eager offers from other installations that wartime controls were not inhibiting our competitors for their services. This crisis was met by waiving the stabilization rules in case after case on an individual basis. Meantime, a more realistic set of classifications was hastily drafted and finally approved on February 1, 1946. The new classifications and salaries provided a necessary inducement, although scientists often persuaded themselves to stay here in the

281

face of even higher salary offers elsewhere. Intangibles such as anticipation and productive work, a belief in the necessity of the work in the national interest, loyalty to the Laboratory where often their careers had started, and the fascination of the surrounding country all were important in keeping a competent group of scientists at Los Alamos.

The impending loss of personnel because of anticipated discharges in the SED was met by two recruitment programs. These who had proved competent were offered civilian jobs and a formal request for military discharge, for the convenience of the government, if discharge was not imminent, if they would agree to stay for at least six months. A considerable number accepted, and often the offer was used to obtain an accelerated discharge under War Department regulations. A vigorous program of outside recruiting also was organized, and schools throughout the country were visited to interest promising students in employment here.

The Bikini test program was not generally popular with the scientists. Only a few were interested in the ordnance aspects, and many could not see that scientific knowledge would be advanced proportionately to the effort expended. To interest Los Alamos scientists and facilitate outside recruitment for Bikini, the Laboratory had to offer a premium wage in anticipation of probable overseas duty. Although the Laboratory ultimately was successful in recruiting additional personnel for the Bikini operations, problems inevitably arose. For example, the additional personnel created a minor crisis in the already acute housing situation. To ease the housing shortage, prefabricated units were transferred by truck from Hanford. However, it was also necessary to convert barracks into makeshift apartments.

In the fall of the 1945, C. D. Shane and R. E. Clausen returned to their university posts, and the Personnel Office was reorganized under E. J. Demson. The staff was expanded and sections were set up with more specific responsibilities, such as job evaluation, employment, and wage administration. The Files and Records section also was enlarged to make personnel information more readily available.

The end of wartime salary and wage stabilization saw a nationwide upward trend of salaries and wages. The Laboratory found it difficult to keep up with this trend because any increase allowed was subject to such prolonged scrutiny that by the time it became effective it had also become insufficient.

It became apparent that the Laboratory needed a new salary

adjustment system for long-term operation. As a first step, a job evaluation program was proposed shortly after the war ended. In the expansion of the Personnel Group, Demson hired John V. Young as a general assistant. In April 1946 when Demson was appointed Assistant Director, Young was named Personnel Director. Until the end of 1946, the personnel staff were busy handling the transition from war to peacetime operation, so permanent policies and operating procedures received little attention, although their importance was recognized.

The Laboratory had kept on its payroll certain employees of the school, nursery school, hospital, and housing office because there was no other agency to employ them. However, the Laboratory maintained no supervisory control over these agencies.

The Housing Office staffed the maid service and express service, housing maintenance (in cooperation with Post Operations), and a public laundry facility, as well as allocating all project housing, until February 1946, when the Army took over these functions. The Zia Company later accepted this responsibility as well as the hiring of personnel for the hospital and schools.

In January 1946, the Personnel Services section of the Personnel Group, under George Challis, undertook an allocation plan for providing new cars at Los Alamos. In a January 28, 1946, letter, Bradbury proposed that cars be allocated to this community for priority distribution to residents. Buick, Cadillac, Chevrolet, and Ford agreed to participate and delivered the first cars in May 1946.

The Los Alamos Times, a weekly project newspaper established under the supervision of the Personnel Services section, printed its first issue on March 15, 1946.

Personnel Services Group

A new Administrative and Service Group (A-13, Personnel Services) was formed in May 1946 to administer certain functions formerly handled in the Personnel Group. It was in charge of the technical area mail room and messenger service, stationery stockroom, car allocation plan, community radio station, and newspaper until the three latter functions were transferred to the Zia Company and the Army. Later the stationery stockroom was returned to the Procurement Group, A-4.

The radio station, KRS, was originally a true community operation with volunteer service, but with the advent of the Personnel Services organization, financial assistance for the station was obtained under the operating contract, and it was put on a regular, full-time operating schedule with hired staff. The contractor, however, did not exercise supervisory authority.

Personnel Services was dissolved on November 1, 1946, with the transfer of the newspaper, radio station, and car allocation plan to the Army and transfer of personnel employment to the Zia Company.

Group A-13 became the Central Mail and Records Group and concentrated on establishing a central mail records function and on improving mail and messenger services. All incoming Laboratory mail was distributed to offices, mail records procedures were established, and the Director and Associate Director's files were consolidated and recatalogued.

Shops

The Machine Shops Group went through the inevitable transition at the end of the war. Stress on production was replaced by emphasis on experimental work. A directive from Bradbury reduced personnel by about 33%. The night shift was discontinued on October 1, 1945, and a controlled redistribution used personnel to better advantage.

More adequate shop space was provided. HT Building was completed in July 1946, and removal of shops from other buildings began. The first was the Uranium Shop for C Building. Meanwhile, machine tools, benches, and other equipment were arriving for the new Sheet Metal and Heat Treat Shops. Final installation of the Heat Treat Shop was delayed until April 1946 because of difficulty in obtaining bus-bars and electrical equipment for the furnaces. In November 1945, the new Foundry-Pattern Shop Building was put into operation. The Graphite Shop in Sigma Building was transferred from CMR to the Shops Group on November 19, 1945; in June 1946 the machine shop at S Site was placed under the Shops Group.

In December 1945, X-4 (the Engineering Group of X Division) and A-8 (Shops Group) were combined into Group A-3 under the Administration and Services Division. On January 16, 1946, Earl Long was hired by the University of Chicago, and Gus Schultz took over the group.

In addition to the personnel reduction, in October 1945 the project changed from a six- to a five-day work week, and the Shops Group went from a 54- to a 48-hour week (9 hours, 36 minutes per day). An effort to

284

increase wages to offset the loss of take-home pay resulted in an average raise of ten cents an hour, six nonwork paid holidays, and vacation time on a daily instead of weekly basis, effective February 1946.

Although these new regulations boosted morale momentarily, they were offset almost immediately by other policies. In the fall of 1945, a new housing control plan reduced A-3's housing allotment. This completely halted the hiring of additional machinists. Further, on May 1, 1946, the Laboratory discontinued the $100.00 monthly incentive pay (Chap. 9) given to men whose families lived away from the project. This was followed by an announcement that return-travel reimbursement would be terminated on September 30.

Personnel were further reduced when enlisted men in the Shops Group were discharged. In January 1946, 90 of the 120 GI machinists were left in the Shops Group, but by August all military personnel had left the group. Although approximately 30 former members of the Special Engineer Detachment returned to their jobs as civilians, the group was still shorthanded and unable to handle the work load. Approximately 20% of the plastic and metal machine jobs were contracted to West Coast shops through the Area Engineer's Los Angeles Office. Although this plan relieved Group A-3 of an overwhelming production schedule, it did not solve the problem completely. For example, preparation of proper drawings and descriptions of research items became more involved, and production was delayed by the time required for shipment to and from the coast.

To simplify control and records of metal stock procurement and issue, Group A-3 took over the metal stockrooms in the various shops from the Procurement Group, and all general issue of metal was transferred from S Warehouse to the Metal stockroom in V Shop in January 1946.

The shops encountered technical problems, in uranium machining (Chap. 9). Warpage was solved by heat treating. Further developments in heat treating reduced the time for certain processes from 3 days to 22 hours. Changing steels and altering the heat treatment extended the life of the dies for molding ^{235}U at DP Site. This saved much time and expense.

Work begun in 1944 on molds, cutters, and machining fixtures for S Site continued constantly. Some of the new designs evolved were fabricated aluminum molds to replace Cerrobase molds, sintered and bonded casting molds, and special mixtures and gauges for machining high explosive. Design and fabrication of spheres for radiolanthanum

285

implosion tests, begun in October 1944, continued as a constant design load.

In December 1945, Philip Morrison asked the engineering section of the Shops Group to design and construct the central part of a fast reactor to be built at Omega Site. After the first assembly in the fall of 1946, David Hall asked the group to continue the design and to supervise fabrication of the shielding and parts necessary to complete the reactor. This work was still in process at the end of 1946 (Chap. 24).

Supply and Property Group

After V-J Day, Procurement Office operations were easier because many security restrictions had been removed and it was no longer necessary to conceal the fact that the University of California was operating the Los Alamos Laboratory. Requisitions formerly sent to one office to be relayed to another went directly to the supplying office, thus eliminating duplication of messages and increasing efficiency. Even though channels became simpler, requests for the almost impossible were still as frequent and as diversified as in the early days.

D. P. Mitchell, Assistant Director, who led the Procurement Section during the war, returned to Columbia University, and Harry S. Allen took charge on November 1, 1946. At about the same time the Laboratory administration was reorganized and the Procurement Group (A-4) became part of the A and S Division.

In March 1946, the Procurement Group also suffered when the military personnel began to leave. Replacements were slow in coming, and the new personnel had to be trained, all of which hampered the group in initiating new procedures.

In April 1946, the Procurement Group was changed to Supply and Property. The responsibilities of property, warehousing, stockrooms, and receiving and shipping were added to their purchasing activities. Harry Allen was Group Leader, Robert J. Van Gemert was Procurement Section Leader, and A. R. Johnson was Property Section Leader. A heavy load was placed upon the Procurement Section early in 1946, when planning for the Bikini tests began. The Procurement Group had to purchase the needed scientific equipment and apparatus and to arrange for its specialized packing and transportation.

The group built stockrooms on both the *Albemarle* and *Cumberland Sound*, which they completely equipped and stocked. These were

286

manned by A-4 personnel to ensure efficient handling of records and supplies. The Procurement Group purchased and moved about 300 tons of equipment for the Crossroads Operation.

The Property Section probably suffered most from the early severe restrictions of housing shortages, security measures, and lack of experienced personnel. All precluded a tight property system.

The first Accountable Property Officer for contractor-held property was Lieutenant Willam A. Farina, who was sent here on July 1, 1945, under the Area Engineer, Lieutenant Colonel S. L. Stewart. He advised Stewart on July 23, 1945, that the account was not operating according to TM 14-910 (the War Department Manual applying to a cost-plus-a-fixed-fee contractor regarding property accounting). Because operations could not be changed at that time, the group continued as it had been established. Captain (formerly Lieutenant) Farina was replaced on April 25, 1946, by Captain Albert C. Hull, Jr., who accepted accountability subject to correction of discrepancies revealed by a physical inventory when accomplished. No such inventory was taken when accountability was transferred to Major (formerly Captain) Hull, and he immediately started corrective measures.

An audit by the Manhattan District Property auditor (the first ever conducted) was started May 16, 1946, on the basis of a directive from General L. R. Groves. The account was found to be in an unsuitable condition for audit, and a physical inventory was begun in September 1946, under the direction of A. R. Johnson, a staff of fifteen persons, and representatives of Major Hull. All Class B and C property was inventoried. Class A property records were not dealt with, as they were maintained for the entire Los Alamos project either by the Army or by the service contractor, the Zia Company. The inventory was still in progress at the end of 1946.

In November 1946, the Operating Contractor's Property Section published a Property Control Manual, which after approval by the Army Accountable Property Officer, was distributed to all division leaders, group leaders, and group property personnel. The manual covered property issues, turn-ins, salvage procedures, disposal of contaminated property, responsibility for property, relief from responsibility, and accounting procedures for certain special items. It provided a procedure for those not associated with the Property Group and materially aided the Contractor in complying with TM 14-910.

Also in November 1946, the General Service and Warehouse Section (part of A-4) was placed under Clyde Reum, and steps were taken for closer control of stock. Rearrangement of stock rooms and warehouse space was accelerated to provide better storage, shipment, and receipt of goods. Also, proper accounting records for Class C property were established. By the end of 1946, a completely controlled property system was well under way.

Technical Area Maintenance

On December 6, 1945, Group A-9 became Group A-5 and joined the A and S Division under the Associate Director. Its function still was to provide maintenance and construction services for physical operation of the Laboratory, in collaboration with the Army Maintenance Group (Chap. 3). The Army-supervised craft shops were discontinued on February 4, 1946, and Group A-5 became responsible for the entire Technical Area maintenance program, except for power line, sheet metal, and masonry work, linoleum installation, sprinkler maintenance, and placard and sign work. A-5 also absorbed some personnel from the discontinued Post group, bringing its total to 244 employees.

After April 28, 1946, Group A-5 was reorganized. The Zia Company assumed the Technical Area craft shop and all but 15 of the original personnel. The group now became a planning body with the following authorized functions.

(a) Overall planning of Laboratory installations.

(b) Preparation of preliminary plans for new construction in consultation with representatives of the requesting group or division.

(c) Submission to the Associate Director for forwarding to the Post Operations Office of all requests for building alteration or new construction.

(d) Approval of all contract drawings and their revisions prepared by the Post Operations Office.

(e) Approval of all job orders for Zia Company work from authorized Laboratory personnel. This included followup and coordination of such orders.

(f) General engineering advice and inspection in collaboration with maintenance representatives of the various groups and divisions.

(g) Initiation of work orders for routine maintenance of buildings and grounds in Technical Areas not under the jurisdiction of group or division maintenance representatives.

288

(h) General coordination of effort of all maintenance representatives, including standardization of procedures, interchange of maintenance information, and calling of regular meetings.

(i) Certain of these responsibilities could be delegated to other competent Laboratory or consultant personnel at the discretion of the A-5 Group Leader.

Safety

Stanley H. Kershaw continued to supervise the Laboratory Safety Group, which had been made independent of the Post Safety Section. His group, primarily concerned with the technical aspects of safety, had devised a closely coordinated safety program with the safety engineer of the CMR Division and committees set up in other divisions.

On September 1, 1945, Kershaw left and E. E. Beck took over. There had always been some agitation to centralize all the accident prevention activities of the project. This finally culminated in an inspection report dated August 28, 1945, from Lloyd E. Blanchard, Consulting Engineer for the Office of Chief of Engineers, to General Groves. Blanchard strongly recommended that there be only one Project Safety Engineer responsible for both Post and Technical Area planning. The recommendation was adopted promptly, and unifying the separate organizations became Beck's primary objective. However, he left the project in December 1945 before the unification was complete.

In January 1946, the Post Safety Engineer, Sydney Ingham, became responsible for the Laboratory safety program as well as Post safety, and was to extend all possible aid to the Technical Areas. He was accountable to the Commanding Officer and the Office of the Chief of Engineers for all safety and accident prevention at Los Alamos. The position of Laboratory Safety Engineer ceased to exist. The various Technical Area safety groups and committees organized in the divisions were to continue operation as Ingham directed. At the end of November 1946, Ingham resigned and C. M. Francis became Project Safety Engineer.

Centralization proved very effective, the only apparent fault being a slight tendency on the part of certain division safety committees to lose interest and disintegrate. Although the CMR Division (with a full-time Safety Engineer, Herbert W. Drager, since April 1945), X Division, and

289

the Shops Group continued their safety programs, other parts of the Laboratory indicated the need for a Laboratory Safety Engineer. The Laboratory accident rates in 1946 were higher than in 1945. In 1945, the Frequency Rate was 6.59 lost-time injuries per 1 000 000 man-hours worked and the Severity Rate was 2.49 days lost per 1000 man-hours worked. In 1946, the Frequency Rate was 6.96 per 1 000 000 man-hours, and the Severity Rate was 8.23 per 1000 man-hours.

In December 1946, Conrad W. Thomas became Laboratory Safety Engineer directly responsible to the Associate Director. He was to set up a safety program, to improve accident prevention, and to provide adequate accident investigations, reports, and records.

Photographic Group

The photographic services, except for highly specialized technical photography handled by the technical divisions, were in the Drafting Section, Shops Group, until December 1945. At that time, a separate photographic group was formed in A and S Division, with John Keller as group leader. He was transferred to Sandia in February 1946, and Loris Gardner assumed leadership.

An expansive program was necessary if the group was to take its proper place as a widely applied service unit. With that in mind, Laboratory needs for photography and reproduction methods such as photostat and ozalid were surveyed. This survey made it possible to order supplies intelligently and use available machines more efficiently. Plans were made to replace the military personnel who had so ably staffed the photo laboratory. Arrangements were made for more space, and additional equipment was installed throughout 1946, which increased and improved the work that Group A-1 produced.

A third darkroom was added. Experiments with new materials were organized: Ozalid Transparent line, Dryphoto, chart film, Heccolith and Heccovel, and color prints by Printon, most of which became part of the group's reproduction functions. Coordination of machines to accomplish specific reproduction results gave good results. For example, the photostat was used to reduce tracings, and negatives and positives were made on the reflex printer. The resulting foils or transparencies were produced in volume on the ozalid. The photographic and microfile facilities were used later as a beginning function in this chain, with selection based on the fineness of line needed or on economic factors.

Mimeograph and ditto processes were limited as a means of reproduction for a laboratory of this type, so in November 1946, a small offset machine was ordered with necessary materials to test its feasibility and economy. It was a salvaged machine and quite old, but it demonstrated that such a method was flexible and fast, and it provided permanent master copies for further use.

Intensive efforts had been made to improve black and white prints and to speed up production so that all technical groups could profit from the facilities. Another successful project was assignment of a photographer from A-1 to different groups for full picture coverage of experiments. Series of pictures were made of the Little Boy assembly, Fast Reactor, DP Site, Bayo Canyon, and other operations.

Since April 1946, all health records have been microfilmed in triplicate and various medical experiments conducted at Los Alamos have been recorded on film. Perhaps the most spectacular Health Group pictures were the complete photographic coverage, in black and white and in color, of the effects of the tragic radiation accidents that killed Harry Daghlian in September 1945 and Louis Slotin in May 1946.

Another tremendous project that the Photo Group accomplished in 1946 concerned Operation Crossroads. B Division (in charge of the Bikini planning for Los Alamos) had countless drawings, tracings, and other engineering data to be reproduced, which severely taxed the facilities of A-1.

HEALTH GROUP

General

The Health Group, A-10, was always a Technical Staff Group, and after the reorganization (Chap. 29), it remained the only staff group directly responsible to the Laboratory Director.

Administration

Several men held the position of Group Leader during the period in question. From August to October 29, 1945, Dr. Louis H. Hempelmann was in charge. Captain James F. Nolan, M.D., succeeded Hempelmann, who was appointed medical consultant to the project. In March 1946, Nolan relinquished his position and, with Hempelmann, transferred to

291

activities connected with Operation Crossroads. Nolan was succeeded by Captain Harry O. Whipple, M.D., until August 14, 1946, when Hempelmann returned from Bikini and again assumed duties as Group Leader.

Functions

In addition to its routine duties (Chap. 3), the group was particularly involved in termination procedures and reorganization of its functions.

Termination Procedures. Because of the many people leaving the Laboratory, the Health Group had an enormous workload. Often other functions were curtailed because of termination interviews and physiological tests. To compound the problem, personnel in the hematological and clinical laboratories also were terminating. It was believed, however, that the termination procedure was most important to complete records and to establish some record for the many workers whose names were not even listed in the health files. These files were for the protection of individuals and the project, and much effort was spent to reorganize them. Photostatic copies were made to minimize harm if the originals were lost or destroyed.

Dr. Wright Langham, head of the Biochemistry Section, asked for urine assays of plutonium workers who were terminating. Often employees left very soon after notification of discharge and they were uncooperative about testing.

The hematological laboratory had been so pushed that counting techniques were not rigorous (200 cells instead of 500 on differentials, one pipette for white and red blood counts instead of two); some workers did not obtain counts every 4 to 6 weeks, as desired by group leaders and the Health Group; and development work on the hematological picture in radiation exposures of human beings (Dickie granules) was suspended.

Reorganization of the Health Group's Functions. With security lifted and pressure eased by the war's end, the Laboratory's employees were concerned about the special hazards. It became necessary, to protect the contractor and for worker morale, to do things that were desirable but not absolutely necessary for the workers' health. Nurses instead of G. I. first-aid men were employed in the first-aid rooms of outlying sites. They conducted routine health procedures and kept records of minor accidents and toxic disturbances. Another nurse and aid room was established in the Technical Area to care for minor shop accidents, although the Station Hospital was nearby.

292

Dr. J. G. Hoffman's Health Physics Section made the distribution, collection, development, and recording of film badges used in certain areas more routine. Equipment and techniques were being developed that made this function more automatic. The work was arranged so that more records could be obtained by less-skilled technicians. Recommendations were made according to actual film results instead of the unsupported opinions of the workers. These routines made enforcement methods easier.

The Health Group asked for reporting of critical assembly activities. Accidents in this field (particularly the radiation accident that killed Harry K. Daghlian) convinced those in charge that the Health Group should be represented in these operations. Louis Slotin, the group leader in charge of critical assemblies, agreed to assign one man from his group to report to J. G. Hoffman on monitoring activities. However, failure to obey the established regulations resulted in an accident on May 21, 1946, at Pajarito Site, during a critical assembly of fissile material. Slotin and seven others were present; Slotin died soon after and three others required intensive follow-up care. Dr. Paul Hageman directed the clinical care of all these men, and Dr. Hempelmann, then at Washington University, St. Louis, and lent to Operation Crossroads, was a special consultant. The Health Group followed closely the hematology and the induced activities in the blood and urine of these men (Chap. 25).

Work with radiolanthanum at Bayo Canyon was limited by a policy that tended to forego the experiment if exposures became high and limited the amount of material with which personnel could work.

The Biochemistry Section under Dr. Wright Langham attempted to establish a routine urinalysis method for plutonium in all CMR personnel (Chap. 26). This involved decontamination of his laboratory, allocation of hospital space, and development of supervision for collecting samples. It was possible that the sensitivity of this assay might make the physical setup impractical for routine use. It was anticipated that this phase of the work should be moved out of the main Technical Area, especially if a new hospital were built. A "milk route" was established to get specimens from the homes of polonium workers, who were most uncooperative about furnishing them without supervision. Secretarial aid was obtained to help the hematological group notify personnel to report for blood counts. Also, routine trips were taken to outlying sites to obtain blood specimens so that individual responsibility would be minimized.

Film badge monitoring increased during this period, and facilities for

293

routine film processing had to be expanded. The darkroom in Q-Building was inadequate, and considerable time and effort were spent in design and construction of a new darkroom in which personnel-monitoring films could be processed efficiently.

Most external radiation hazards were well controlled. However, unsatisfactory arrangements for discharging fission products from the Water Boiler represented a potential health hazard. Recommendations were made to fence the immediate area and search for a way to dispose of this material without creating a hazard or nuisance. It was thought that release of this material through a stack high enough to avoid ground turbulence might be a solution.

The possibility of a major catastrophe involving radioactive materials was being considered. Although the hazard was admittedly slight, it was felt that plans should be made for handling a situation in which much of this area might be contaminated with radioactive material. Supplied-atmosphere masks and monitoring instruments were being assembled to equip an emergency vehicle.

Although this group had no direct involvement in preparations for the Crossroads test, Doctors Hempelmann, Nolan, Hoffman, and Langham spent time on arrangements and conferences. Hempelmann, with Langham, Hoffman, and Captain Large, gave a two-week course in basic physics and radiation problems of atom bomb tests to a group of Army, Navy, and Public Health doctors on the staff of Colonel Stafford Warren. Hempelmann and Hoffman went to Washington and Rochester for conferences. Colonel Warren came to the project several times for conferences and planning. Hempelmann, Langham, and Nolan actually took part in these tests as monitors.

Hoffman's group made several trips to the Alamogordo area to obtain data for decay curves on radiation from the July 16, 1945, test. Hempelmann continued studies of the physiological effects of radiation on cows bought in that area and brought to Los Alamos. The hematology group spent a lot of time doing hematological studies on them. These activities were practically terminated at the end of 1946.

BUSINESS OFFICE

Introduction

The Laboratory Business Office was the direct representative of R. M. Underhill, Secretary and Treasurer to the Regents of the University of

California (Chap. 3). As such, it had a special interest in all phases of Laboratory operations. However, this special interest, based on the contract between the University of California and the Manhattan Project, gave a substantially different viewpoint of the Laboratory's operation. It seems only proper to review certain phases of the history from this vantage, even though there will be some repetition.

Laboratory Payroll

The average payroll of monthly employees increased from approximately $200 000 in July 1945 to a high of $437 750 in April 1946 (a sharp jump of about $160 000 occasioned by Operation Crossroads) and then slowly decreased to $300 000 in December 1946.

The payroll of hourly employees was about $125 000 in July 1945 because of the Alamogordo test of the atomic weapon. It dropped to a low of approximately $36 000 in December 1945, and then slowly rose to $63 000 in December 1946.

The entire payroll of both monthly and hourly employees seemed to remain relatively close to a monthly average $360 000 throughout this period, with a peak of over $500 000 in April 1946.

Time and Attendance Reports

Time reporting by individuals as handled by the Personnel Department was discontinued on March 4, 1946. Thereafter, the Director approved the attendance form signed by a responsible person in each group, which was forwarded to the payroll office weekly. The information from these reports was posted weekly to Kardex records. On the 10th of each month, a schedule of days to be deducted from the checks of monthly employees was sent to the Los Angeles office. The Los Angeles payroll office then wrote the checks for the monthly employees. For hourly employees, the Laboratory made payments based on the recording of actual hours worked.

Travel Disbursement

From July 1945 through December 1946, $683 312 was disbursed for official travel, an average of $37 962 per month. The peak, in March 1946, was $83 450 when many employees were sent overseas for the

295

Crossroads Operation. The second highest amount, $73 500 in September 1946, was caused by the last rush of terminees who left the project before the September 30 deadline on return travel payment. The Los Angeles office stopped issuing checks on travel expense accounts in August 1945, except for a few issued each month through May 1946. Because of the tremendous load of outstanding travel accounts from April through September 1946, the travel office faced a several hundred thousand dollar backlog, which became a major problem. Consequently, by December 1946, this section was strengthened with individuals actually experienced in travel and transportation matters.

Local Expenditures Other Than Payroll and Travel

Purchasing. Monthly local purchases averaged $4301.03. In July 1945, the peak ($14 061.37) was occasioned by purchase of materials for the Alamogordo test. The low was $1538.37 in November 1945. Purchases were nearer average during the last half of 1946, when they included material for the Technical Library, the KRS radio station, and the paper and printing for the Los Alamos Times. Local purchases also included gases, stationery and supplies, dry ice, commissary items, and miscellaneous emergency items.

Consultant Fees. The Business Office began paying consultant fees in July 1946. Average payment in the last months of 1946 was $3048.19. The peak, in October 1946, was $8003.22.

Advances (not including travel). After six months' continuous service, all employees were automatically included in the California State Employees Retirement System (SERS). Monthly payroll deductions were deposited to this fund. A wartime emergency provision under this system allowed terminating employees to fill out a power-of-attorney to the University of California so that the University could make immediate payment of the amount withheld from their salaries and deposited in their account under SERS. The University of California, in turn, presented the power-of-attorney to the State of California Retirement System to recover the money that the University of California had advanced as severance pay.

The monthly advance payments from the payroll office averaged $598.62. The peak, $2663.28, was in September 1946, chiefly for SERS advances.

Services. Payments for miscellaneous services included utilities, rent, freight, laundry, telephone, and telegraph. The monthly average was $3966.17. The peak, $29 303.84, in August 1945, was mostly freight charges. Payments for freight services decreased steadily from August 1945 because of the change from cash payment to Government Bill of Lading. After May 1946, freight payments were negligible.

Miscellaneous. Other local expenditures included hospital subsistence for employees on health passes or being treated for job-incurred injuries, contract nurse subsistence, payments to blood donors for injured employees, and other emergency payments including one $10 000 Welfare Fund payment made to Harry Krikor Daghlian's mother from the Business Manager's Revolving Fund. The average monthly miscellaneous payment was $680.10. September 1945 was the peak month with payment of the $10 000 mentioned above.

Additional Activities

Check-Cashing Facility. This office also was the check-cashing facility (Chap. 3). The following statistics indicate the volume of business.

Monthly Record—

Average month: $101 610.20
Peak month: $121 024.31 - October 1945
Low month: $ 78 265.51 - February 1946

Daily Record—

Average day: $ 4467.58
Peak day: $ 21 799.55 - August 2, 1946
Low day: $ 887.85 - October 16, 1946

Trend—

There was a slight but steady increase during the last half of 1946, probably caused by the relaxing of security regulations that required paychecks to be mailed directly to the individual's bank.

The heaviest days were usually the 1st and 2nd of the month.

The lightest day was approximately the 20th of the month.

Revolving Fund. The Revolving Fund of $50 000 maintained by the Business Office included the following constant figures.

$25 000.00 check-cashing service
$ 15.00 for petty cash, Santa Fe office
$ 25.00 petty cash for nursery school

$ 500.00 war bond account
$ 125.00 US Postal stamps—mail room
$ 15.00 petty cash in Housing Office for small express bills. (This was increased to $25.00 in August 1946.)
Because of the added expense for the Crossroads Operations, the total fund was increased to $65 000 for September 1946.

The McNierney Cattle Case. On August 25, 1946, W. L. McNierney was paid $1350 in compensation for loss he suffered in selling 140 head of cattle at reduced market value during September 1945. McNierney alleged that the hair of his cattle was discolored from exposure during the July 16, 1945, Alamogordo test.

Telephone and Telegraph. For security reasons, all bills from Western Union, Railway Express (which handled much of the telegraph service to help maintain security), and the telephone company were billed to the Business Office, which collected the amounts due from the various employees. This practice was discontinued at the end of May 1946 after security regulations were relaxed.

Nursery School. Nursery school financing was handled through the Business Office. It collected tuition and lunch fees from the childrens' parents and, in turn, paid the teachers, the food bills, maid service, dietician, and other bills. Such services as nurse, janitor, utilities, buildings, and laundry were not chargeable to the budget but were subsidized by the Government. The school was a responsibility of the Business Office until December 1, 1946, when Colonel A. C. Nauman, (in October 30 and November 25, 1946, letters) directed that the Zia Company would assume responsibility for operating the nursery school (a part of Community Services).

Newspaper. As of June 21, 1946, the Laboratory handled the local newspaper, the Los Alamos Times. The New Mexico Publishing Company in Santa Fe printed the weekly issues. The printing bills were paid through the Business Office. By letter of October 30, 1946, Colonel Nauman, representative of the Contracting Officer for the Zia Company, directed that the Zia Company would assume responsibility on November 1, 1946, for operating Community Services, including the Los Alamos Times.

Radio Station. Similar to the maintenance of the newspaper was the establishment and operation of the local radio station, KRS. Expenses, paid by the Business Office, included such items as news service, records, equipment, and special directors' salaries. Operation of the radio station began in February 1946 and was handled by the Laboratory until November 1, when the contract was transferred to the Zia Company by letter of October 30, from Colonel Nauman.

United Press Associations—News Service. By contract agreement dated June 8, 1946, between the United Press Associations and the University of California Los Alamos Laboratory, the United Press furnished leased radio wire and news reports for broadcast over Radio Station KRS and printing in the Los Alamos Times. Payment was by the Business Office through November 30, 1946. On December 10, The University of California Regents formally assigned this contract agreement to the Zia Company pursuant to the October 30, 1946, directive from Colonel Nauman.

Hospital Employees. While the local hospital was directed by the US Engineers, many of the nurses were under Laboratory contract. It was felt that, in accordance with practice in nearby hospitals, the nurses should be furnished one free meal for each eight-hour duty period. The Laboratory consequently reimbursed the US Engineers for daily subsistence of one meal for each eight hours of duty by each nurse. A like responsibility was daily subsistence for employees ordered to the hospital for one-day health passes. An allowance of $1.00 per day was made for these charges. By letter of May 29, 1946, Lieutenant Colonel W. A. Stevens, Authorized Representative, Contracting Officer, for the Zia Company, directed that effective June 16, 1946, the Zia Company would be responsible for operating the Post Hospital and that the hospital nurses and technicians should transfer to the Zia Company payroll under Contract W-17-028-Eng-90. The nurses' contracts were also transferred to Zia, and the Business Office was released from payment of such meal subsistence bills. The Business Office continued paying subsistence for those on health passes, however, because those individuals were Laboratory employees.

School. The superintendent and teachers for the grammar and high schools were on the Laboratory payroll until June 1946, when they were transferred to the Zia Company, which was operating under Contract W-17-028-Eng-90, in accordance with a directive dated May 25, 1946, from Lieutenant Colonel Stanley L. Stewart, Area Engineer.

Library. For obvious security reasons, all orders for the Library had been made through Los Angeles, and the publications were routed to the Los Angeles warehouse, where they were processed and forwarded to Los Alamos. In April 1946, it was determined that all monographic and

299

serial publications ordered by the Berkeley Accessions Department for the Project Librarian should be mailed directly from the publisher or vendor to the Library. It was believed that this change would expedite arrival of publications at the Library, promote more efficient and economical operation subsequent to lifting of security regulations, relieve the Los Angeles warehouse of the processing of Library receipts and shipments, and relieve the Los Angeles Purchasing Department of handling Library purchase orders placed outside the local campus.

This change in procedure was so successful that in May, Library control was moved to the project from Los Angeles. Plans were made to have the Business Office control Library purchasing and invoice payment. Of course, the transition was slow, taking several months. By the end of June 1946, all Library requests were made by local purchase orders. The Librarian prepared the orders and sent them to the Business Office for approval. There they were processed carefully to ascertain that all such purchases were made within the Laboratory rules. Local files were set up by vendor name to provide records and materials for payment and follow-up. Vendors invoiced the Business Office, which paid from the Revolving Fund. Until arrangements could be made with the Santa Fe bank to handle foreign drafts, invoices of foreign companies were forwarded to Los Angeles for payment. By December 1946, the bank was prepared to issue foreign drafts, and the Business Office assumed payment of these bills.

Because the Business Office maintained files by vendor name as well as purchase order number, it was decided in December 1946 that purchases could be expedited better in the Business Office than in the Library. A 60-day follow-up system was instituted on most orders, 90 days for unusual circumstances.

By late 1946 the project was handling locally all work and problems connected with the Technical Library. Most "kinks" resulting from the transition (such as missent mail, duplications, or erroneous cancellations) had been ironed out, and the Library and the Business Office agreed that the new system was satisfactory.

Compensable Industrial Accidents—July 1945 to January 1946

Workmen's Compensation and Welfare Benefits. At 9:55 p.m. on August 21, 1945, Harry Krikor Daghlian was fatally injured while working at Omega Laboratory. He lived for 25 days after the accident,

300

and died at 4:30 p.m. on Saturday, September 15, 1945. He was survived by his mother, a sister, and a brother. His mother and sister were at the project at the time of his death. On that day, the Area Manager authorized the Board of Regents of the University of California to make immediate payment from the Welfare Fund of $10 000.00 in the case of Harry Krikor Daghlian. The Business Manager at the project delivered Daghlian's mother a $10 000.00 check from his Revolving Fund before her departure from the project on September 16. No death benefits under the Workmen's Compensation Act were due in this case because Daghlian left no dependents.

During 1946 there were 195 occupational accidents involving Laboratory employees, of which 15 were determined to be compensable. Five of these accidents, although not considered "minor" by those injured were not termed "critical," nor did they cause permanent disability. These five cases were compensated by a total of $149.27 under the Workmen's Compensation Laws of the State of New Mexico.

Another accident involved 7 of these 15 compensable cases. All these men were injured and one, Louis Slotin, died. Slotin and the other six men were conducting an experiment at Pajarito Laboratory on May 21, 1946, when the accident occurred. Slotin died on May 30, 9 days later, and was survived by his parents, a brother, and a sister. After the accident, his parents were called to visit him. Their travel to Los Alamos and return to Winnipeg, Manitoba, Canada, was paid from Government funds, and this matter was turned over to Major Sidney Newburger, Post Intelligence Officer. Their lodging and meals while at the project were paid for in the same manner. Medical and shipping expenses in connection with Slotin's death were borne by the Laboratory. A maximum settlement of $10 000 under the Welfare Fund was made to his mother. Slotin left no dependents, so his case was excluded under the Workmen's Compensation laws. No final medical opinion could be made in the cases of the other six men, pending further observation of this type of injury. Two of them may have some permanent disability.

Three men were involved in a smoke-bomb explosion at Omega on August 2, 1946. One, Joshua I. Schwartz, was fatally injured. The other two received permanent injuries: one was totally blinded; the other was declared industrially and economically blind, with total loss of sight in one eye and only light perception in the other. For Schwartz, the maximum Welfare Fund settlement of $10 000.00 was made. No compensation was awarded by the Workmen's Compensation laws because he had no dependents. For one of the injured men, the maximum Workmen's Compensation settlement was in the form of $18 per week for 550 weeks (a total of $9 900). A similar decision was expected in the case of the second man. Welfare payments in both cases had not been determined in December 1946.

Public Liability. On January 31, 1946, Philip Lawson, a Laboratory employee on official duty, was driving between two housing units on the project. Two children, who were playing in the yard, moved near the driveway behind Lawson's car as it backed toward the road, and the side of the car hit them. One was unhurt; the other, Dennis Roth, age 3, received minor cuts and bruises and a fractured foot. Lawson immediately took the child and his mother to the hospital, where examinations and x rays were made. Dennis Roth was hospitalized overnight at the project and taken to Bruns General Hospital in Santa Fe the next morning, at the request of his father, Captain Lloyd Joseph Roth (US Army). The Bruns doctor recommended that the boy wear a short leg cast for about three weeks and stated that no permanent disability was likely. Within a few weeks the child's foot had healed completely. Captain Roth and his wife made formal request that some compensation settlement be made for their care to the child and the inconvenience suffered on account of the accident. By letter of June 10, 1946, Underhill stated that this matter was covered by Globe Indemnity Company Policy No. 500350, and that the case should be handled expeditiously. Accordingly, a Globe Indemnity Company check for $275.00 was tendered to the Roths and the proper indemnifying release was secured for final settlement. At that time, this was the only public liability case that the University of California had had to settle at the project.

Insurance

Master Policy FD-502. In June 1945, it was announced that the Contractor's Representative under Contract W-7405-Eng-36 could take applications for Indemnity Insurance Company of North America Policy FD-502 for Contractor's employees regularly assigned to work on the project. Certificates of insurance issued under this policy were for $10 000, which principal, however, could be increased to a maximum of $20 000 or decreased to a minimum of $5000.

The certificates of insurance were purchased by employees individually, and were frequently called "Crossroads insurance" because 80

302

employees detailed to the Bikini test program in May 1946 purchased policies.

Group Hospitalization Insurance. On June 30, 1946, free hospital and medical service for project employees was discontinued. As a result, the Director appointed a committee to study and recommend a plan of group insurance to assist the employee in meeting his hospital, medical, and surgical expenses. The committee recommended a plan offered by the Business Men's Assurance Company as offering the most complete coverage at the least cost and being generally best suited to the particular needs of the project employees.

After a careful study of the plan, the Business Manager, A. E. Dyhre, approved it and agreed that a payroll deduction plan would be set up to handle the premiums for all participating employees. This meant setting up and keeping about 1000 new records in the Payroll Section. This plan was then presented to the project employees for their approval. The plan went into effect July 15, 1946, and most employees signed up as participants. The payroll deduction plan worked smoothly, and there were no reported difficulties with the company.

General

When Dyhre (Business Manager) and L. G. Hawkins (Assistant Business Manager) assumed management on June 1, 1946, the Business Office had 17 employees. At the end of December there were 23. The increase was in the Travel Section because of the heavy Crossroads load.

CHAPTER 23

Theoretical Physics Division

INTRODUCTION

From August 1945 through the end of 1946, the Theoretical Division continued its war-time program on a reduced scale, and added the theoretical physics research on thermonuclear systems formerly carried on in F Division (Chap. 13). In the fall of 1945, much attention was given to the complex hydrodynamical problems involved in interpreting the blast measurements made at Trinity, Hiroshima, and Nagasaki. Also, much effort was spent on the radiation hydrodynamics of the implosion fission bomb, largely because of its unexpectedly high efficiency.

ORGANIZATION

The difficulties of the early interim period seriously affected the Theoretical Division. The division leader and the eight group leaders serving in August 1945 had all left by September 1946; and early in this period most of the junior members of the division returned to the universities to resume their interrupted studies. The group structure of the division in August 1945, and the group leaders' termination dates are as follows.

Group		Group Leader	Termination
T-1	Implosion Dynamics	R. E. Peierls	January 1946
T-2	Diffusion Theory	R. Serber	November 1945
T-3	Efficiency Theory	V. F. Weisskopf	February 1946
T-4	Diffusion Problems	R. P. Feynman	October 1945
T-5	Computation	D. A. Flanders	September 1946
T-6	IBM Computations	E. Nelson	January 1946
T-7	Damage	J. O. Hirschfelder	August 1946
T-8	Composite Weapon	G. Placzek	May 1946

Many of the group leaders were retained as consultants.

On November 14, 1945, Group F-1, which had been part of the Theoretical Division until June 1944 (Chap. 13), returned as Group T-7. As before, its concern was the theory of thermonuclear systems.

On December 1, 1945, Hans A. Bethe resigned as Theoretical Division Leader and returned to his position at Cornell University. His successor was George Placzek, who had joined the Theoretical Division in May after working for several years with the Montreal group.

After a general reorganization, the group structure as of January 8, 1946, was as follows.

Group		Group Leader	Termination
T-1	Hydrodynamics	Joseph M. Keller	June 27, 1946
T-2	Group (dissolved; personnel transferred to T-8)		
T-3	Efficiency Theory	V. F. Weisskopf	February 28, 1946
T-4	Diffusion Problems	R. Ehrlich	January 9, 1946
T-5	Computations	D. A. Flanders	September 4, 1946
T-6	IBM Computations	R. W. Hamming	June 24, 1946
T-7	Super	E. Teller	January 21, 1946
T-8	Diffusion Theory	C. Mark	Did not terminate

The return of key personnel to the universities for the spring semester in 1946 necessitated considerable change. Groups T-2, T-3, and T-4 were dissolved, and their personnel transferred to the remaining groups.

Radiation hydrodynamics formerly worked on by Group T-3 became the concern of R. Landshoff, of Group T-7, who in May 1946 visited Ithaca, Rochester, and Boston to discuss this work with former members of Group T-3. Group T-1 was dissolved in June and some of its problems were distributed among the remaining groups. A new group, T-1, under F. Reines, was formed to consider the theory of the dragon (Chap. 3) and to work on blast wave and damage problems that might arise. All computations, both manual and machine (IBM), became the concern of a group under D. A. Flanders. In anticipation of Flanders' leaving, Bengt Carlson became leader of a new group, T-2, on August 14, 1946.

Operation Crossroads, which seriously affected some other divisions, did not add to T-Division personnel problems. J. O. Hirschfelder and J.

L. Magee, of Group T-7 (Damage) until their resignations in the fall of 1945, returned to devote full-time attention to Operation Crossroads problems in the spring and summer of 1946.

On May 20, 1946, Placzek became ill and left Los Alamos on a leave of absence. He could not return to the project because of the high altitude, but he was not terminated until July 9. Robert D. Richtmyer transferred from the Patent Group and acted as T-Division Leader until November 1946 when he became Division Leader.

In the summer of 1946, a group of consultants helped the Theoretical Division. They included E. Fermi, F. Hoyt, and E. Teller of the University of Chicago; Lothar and Gertrud Nordheim of Clinton Laboratories; R. Marshak of the University of Rochester; R. P. Feynman of Cornell University; J. O. Hirschfelder, University of Wisconsin; J. von Neumann of the Institute of Advanced Studies, Princeton University; and V. Weisskopf and T. Welton, of the Massachusetts institute of Technology. Except for the Nordheims and Hoyt, all were members of the wartime Los Alamos staff.

In the fall of 1946, S. Ulam, who left Group F-1 in 1945 to join the staff of the University of Southern California, returned to become leader of a new group whose concern was general mathematical methods. L. Goldstein, formerly of the College of the City of New York, and of the Division of War Research, Columbia University, joined the T-Division staff at about the same time. Another group was set up in the fall of 1946 under R. D. Richtmyer to study the new thermonuclear system proposed by E. Teller.

Theoretical Division structure in December 1946 was as follows.

T-1	Theory of Dragon	F. Reines
T-2	Computations	B. Carlson
T-3	Super and Radiation Hydrodynamics	R. Landshoff
T-4	Diffusion Theory	C. Mark
T-5	IBM Computations	B. Carlson (Acting)
T-8	Mathematical Methods	S. Ulam
T-9	Advanced Designs	R. D. Richtmyer
T-10	Fundamental Nuclear Physics	L. Goldstein

EFFECTS OF TEST AND COMBAT NUCLEAR EXPLOSIONS

Interpreting the observations of the July 16, 1945, test explosion at Trinity and the combat explosions at Hiroshima and Nagasaki occupied the Theoretical Division in the early fall of 1945. Group T-1 developed a theory for estimating the energy release of a fission bomb explosion by considering the expansion velocity of the ball of fire in its early stages. This theoretical treatment differed from earlier ones in that it considered how the bomb's high-density material (and at Trinity, its supporting platform and housing) affected the pressure distribution. In an attempt to determine (Chap. 11) what proportion of energy released by a fission bomb is converted into blast energy, a calculation (Problem M) using the IBM machines was made. The blast energy from explosion of a fission bomb was found to be about two-thirds that from explosion of a corresponding amount of TNT. After further study of blast and optical data, Group T-1 gave as the most probable nuclear energy release values the TNT equivalents of 20 000 tons for Trinity, 15 000 tons for Hiroshima, and 50 000 tons for Nagasaki. The value for Nagasaki was least certain.

RADIATION HYDRODYNAMICS

The unexpectedly high yield of the Trinity explosion (Chap. 11) led to renewal of the early speculations (Chap. 5) about the simplifying assumption of neglect of radiation made in the original efficiency calculations.

THE SUPER

When the war ended, Group F-1 (which became Group T-7 on November 14, 1945) was relieved of its responsibilities for design and testing of fission bombs and could work full-time on bombs based on thermonuclear processes. E. Fermi summed up existing knowledge on relevant thermonuclear processes in six lectures given in July, August, and September 1946. Intensive activity was ended by June 1946 by serious loss of personnel.

307

SUPER CONFERENCE

In April 1946, a conference of Group T-7 members and other staff members and consultants concerned with development of thermonuclear bombs took place. The work on thermonuclear processes was reviewed and a specific model of a thermonuclear bomb was considered.

An account of the fundamental physical processes important to the thermonuclear bomb is given in Part I and not repeated here. The conference centered on the feasibility of the specific model presented by Group T-7. Proposals and suggestions for basic theoretical investigations grew from examination and consideration of this model. The model proposed was chosen for amenability to theoretical treatment rather than for engineering practicability or efficient use of precious material. The purpose of the conference was to study the feasibility of thermonuclear bombs in principle, not to propose designs for actual weapons.

POWER-PRODUCING DEVICES

The Theoretical Division was considering two power-producing devices.

The Repetitive Dragon

The first device was proposed by F. Reines of Group T-1, and because of features in common with the mechanism used in the "dragon" experiment (Chap. 15), it was called the "repetitive dragon." It was a machine that periodically positioned pieces of active metal or hydride so that for a short time they formed an assembly critical with respect to prompt neutrons. The device would be a useful research tool and could be used to produce power. Extreme accuracy was demanded of the mechanical assembling device to keep the bursts of power, varied exponentially with the length of assembly, fairly uniform. Use of the hydride, in which the mean time between fissions was about one hundred times that in the metal, would permit greater tolerance in the mechanism.

The Fast Reactor

Carson Mark of Group T-8 worked on a second power-producing "fast" reactor with the computation groups T-2 and T-5. Although R. P.

308

Feynman first suggested this type of reactor in 1943, this specific reactor was proposed by P. Morrison (Chap. 24). It was a fairly complex arrangement of plutonium rods, coolant, and tamper, which formed a critical assembly for neutrons with fission-spectrum energy. Using a modification of the diffusion-theory methods developed for determining the criticality of the simpler assemblies considered in bomb design. Group T-8 was able to determine accurate critical dimensions for the assembly.

CHAPTER 24

Physics Division

FORMATION OF P DIVISION

The Research Division under R. R. Wilson (Chap. 12) and F Division under Enrico Fermi (Chap. 13) continued experimental work throughout August and September 1945. Both divisions were greatly concerned over the problems of establishing a postwar program, and both suffered the same unrest, so no especial advances in research resulted. Before Wilson left the Laboratory he recommended that the Research and F Divisions be combined into a single Physics Division. On October 31, 1945, he wrote that the "work performed by these two divisions plus the inclusion of the fast chain reactions were particularly important to give the new Research Division a solid foundation around which to orient some of its work."

In November 1945 R and F Divisions were consolidated into the Physics Division with John H. Manley as Division Leader. The only group not absorbed was the Super and General Theory Group (F-1) under Edward Teller. This group was transferred to the Theoretical Division on November 14 (Chap. 23). These changes resulted in the following organization at the end of 1945.

P-2	Water Boiler	L. D. P. King
P-3	Cockcroft-Walton Accelerator	H. H. Barschall
P-4	Thermonuclear Reaction	E. Bretscher and H. Staub
P-5	Reactor	P. Morrison
P-6	Van de Graaff Research	R. Taschek
P-7	Cyclotron	R. R. Wilson
P-8	Radioactivity	R. Segrè
P-10	Fission Studies	H. L. Anderson

P Division was rounded out during 1946 by adding P-1, Electronics, under Wilbur Hane (formerly Group G-4 in the old G-Division) on

January 1, and P-3, Cockcroft-Walton accelerator, or D-D source, which remained under Barschall until July 1946. Jorgenson became Group Leader until the group was consolidated with Group P-6 in September 1946. The combined group retained the designation P-3, under Richard Taschek, and continued work on both the Van de Graaff (short tank) and the Cockcroft-Walton accelerators. P-9, Van de Graaff construction, under J. L. McKibben, was a new group formed in January 1946 to complete a new Van de Graaff. Late in January 1946, P-13, the Cosmic Rays group, was formed under Darol Froman for research on atmospheric radiations, particularly the neutron component.

Further reorganization in February 1946 discontinued P-7, P-8, and P-10 and left the following structure, which remained (with some changes in group leaders) throughout 1946.

P-1	Electronics	Ernest Titterton
P-2	Water Boiler	L. D. P. King
P-3	Van de Graaff (short tank) Cockcroft-Walton Accelerator (formerly P-6)	Richard Taschek
P-5	Fast Reactor	David B. Hall
P-9	Van de Graaff construction	J. L. McKibben
P-11	Betatron	William Ogle
P-12	Cyclotron	J. A. Fowler
P-13	Cosmic Rays	Darol Froman (group dissolved June 1946)

Manley resigned as Division Leader in July to do further research work, and J. M. B. Kellogg took over. Manley remained Associate Division Leader.

PHYSICS DIVISION ACTIVITIES

The division's main efforts centered around problems of the fission process, the D-D and D-T reactions, the scattering process, and the fast reactor.

311

The Fast Reactor

In the fall of 1945, Philip Morrison suggested construction of a new nuclear reactor at Los Alamos. This was the first to be undertaken in the Manhattan District since the war, and it was to be different in principle from all other reactors.

It was proposed that the Laboratory build a reactor using plutonium and operating on fast neutrons at about a 10-kW power level with mercury as a coolant. The philosophy underlying this proposal was based on the following assumptions.

- The Laboratory needed more information on the properties of near-critical systems operating on fast neutrons because the bomb itself was a supercritical system based on fast neutrons.
- The spectrum of fast neutrons produced by such a reactor would be nearly the same as that from the bomb itself, and the reactor would thus be useful for exploring problems associated with nuclear reactions in weapons.
- No plutonium reactor or fast reactor had yet been attempted although such systems were potentially of great interest for producing useful power and for breeding or converting fissionable materials.
- The desirability of such an objective, having both weapon goals and application to peaceful uses of atomic power, was apparent in a laboratory striving to establish itself on a useful and effective postwar plane.

Major General L. R. Groves granted approval for this construction and allotted the necessary plutonium from material on hand but not completely satisfactory for weapon use. Ground was broken for the new laboratory building in Los Alamos Canyon (adjacent to the Omega Laboratory housing the Water Boiler) on May 15, 1946. This was the first building on the mesa planned for permanence.

R. D. Baker's Group, CMR-5, began experiments on rods of the active material in March 1946 (Chap. 26). Extrusion of plutonium δ-phase rods at low temperature led to partial transformation to α phase, presumably because of the great amount of working. High-temperature extrusion, all in the δ-phase stable region, gave unsatisfactory dimensions, but the density remained low. Machining of extruded oversize rods in this manner was undertaken in April and was successful enough to permit starting production in June.

The canning process also began in June. One rod was experimentally canned in a special air-filled disposable can and exposed in high flux in the Water Boiler. The complete operation was moved to DP Site for general health reasons late in the year.

Morrison accepted a position at Cornell University, and David B. Hall continued as leader of this work.

Final assembly of the reactor progressed rapidly during the summer of 1946. By August the uranium tamper blocks had been electroplated with the adopted 3-mil-thick silver and were assembled. The active material had been introduced into the reactor can and the tamper had been closed by using the safety block and installed hoist under remote control.

Critical assemblies of the fast reactor started on September 12 and continued for three days until mechanical failure of the bottom tamper mechanism stopped operations for almost 60 days while the apparatus was taken down and completely reassembled. Some of the associated reactor equipment was redesigned at this time, particularly the safety block mechanism.

Early in November, reassembly was completed including the aluminum envelope enclosing the reactor pot, uranium tamper, tamper cooling jacket, steel tamper, and 4 in. of lead shielding. Progress also had been made on the electromagnetic mercury pump, the mercury flowmeter, heat exchanger, and supply and sump tanks, but this auxiliary equipment was not finished completely. Critical assembly measurements started again on November 19 (without coolant or the external radiation shield), and critical conditions were reached two days later.

The following figures give the loading and observed multiplication when all tamper materials (safety block, top tamper, control and safety rods) were in maximum reactivity position. The reactor cage had 55 holes for insertion of active material rods. The plutonium was loaded in a central array and all remaining holes were filled with rods of natural uranium. These first assemblies were made without mercury in the reactor pot.

No. of Plutonium Rods Loaded	M_{obs}	$1/M_{obs}$
10	2.20	0.455
15	3.91	0.256
17	5.41	0.185
19	8.55	0.117
21	14.8	0.068
22	23.1	0.043
23	48.5	0.021
24[a]	∞ (critical)	0

[a]The critical mass was 23.8 rods as estimated from the control rod calibration. The critical condition was found by bringing the pile to a slow period by means of a control rod.

The Electronics Group

P-1, with Ernest Titterton as Group Leader, functioned efficiently as a service unit, not only for P Division, but for all technical groups. This was particularly true during the summer of 1946 when loss of trained electronic men in the technical groups threw an extra burden of repair and design work on P-1. The group also completed and tested engineering models of oscillators, gauges, analyzers, and amplifiers.

Technical problems that arose in October 1946 received priority from the Electronics Group for the rest of the year. The most important was cyclotron arc modulation. It was necessary to redesign frequency dividers and modify certain other features of McDaniel's circuits. Research began on a degenerative stabilizer circuit for the D-D source to prevent the beam from wandering off the target. A third objective was research on McKibben's informer problem on the 8-MeV generator. The fourth major item was an order for heavy-duty power supply and control circuits for pulsing a spark-gap light source in synchronism with a fast camera. There were serious difficulties in extinguishing the spark at frequencies as high as 1500 cps, and it was estimated that it would take three months to solve the problem.

314

The Water Boiler

The Water Boiler Group, P-2 (formerly F-2), under L. D. P. King, successfully operated the "Boiler" for 3783 kWh from August 1945 to December 20, 1946, for a total of 4809 kWh. Research continued and changes were made to make the Water Boiler a permanent research tool.

During July 1945, the pile reactivity decreased greatly owing to excessive nitrogen loss and formation of a precipitate. To avoid future trouble of this type, minor design changes were made before reassembly, and the nitrogen concentration was kept at a normal level. This required adding acid and water in the ratio of 1.4 to 1 instead of the previous 2.8 to 1.

The overall radiation in the building was decreased substantially by improving the shielding in the gas outlet line, thermal column, and ports. The controls were completely rewired. The safety circuits were simplified and installation of a new large fission chamber in April permitted linear calibration of the automatic control over the entire operating range. The accuracy of this control was increased by incorporating a complete commercial potentiometer in the balancing circuit.

An analysis of the gas from the Boiler indicated that about 30% of the total fission activity was carried off in this manner. An intensity of 40 r/hour 1 foot from the outlet line was observed during full-power operation. The gas evolution due to electrolysis was about 2.3 cm^3/s/kW. Background activity from the radioactive gas caused trouble. On several occasions it was necessary to stop Boiler operation because of health hazards. A permanent solution for this problem was planned. A shielded concrete pit was constructed to house the safety liquid trap for the gas outlet, and the outlet line near Omega was all of buried stainless steel pipe. The gas outlet was removed to a wider part of South Mesa, about 1500 feet from Omega. The gas was released from a point about 60 feet above ground. In December the gas backgrounds seemed improved but not yet solved. A high stack and gas dilution might be more satisfactory.

Sample irradiation continued throughout 1946 for the health group, radiochemistry group, and other groups in P Division.

Research continued along several lines. Experiments on short delayed neutrons and gamma rays from ^{235}U and plutonium were continued. A long series of measurements was begun on the ranges and fission yields of plutonium fragments; the heavy group was essentially completed and work on the light group was begun. The cross sections of 40-h ^{140}La, ^{129}I, and ^{153}Sm were determined. Construction of a thin-lens beta-ray spectrometer for energies up to 15 MeV was completed in July 1946. A preliminary run with ^{182}Ta indicated poor resolution. Collimating slits, however, had not been adjusted and no current stabilizer was used on the generator. A large 30-in.-diameter fission chamber for measuring angular distribution in scattered neutrons was begun. It consisted of 15 1-in. concentric rings coated with ^{235}U.

Particle Accelerators

After the war, the Los Alamos Laboratory found itself with a variety of equipment borrowed or leased from various institutions. Chief among this equipment were the following accelerators.

The 40-in. Harvard Cyclotron
The University of Illinois Cockcroft-Walton accelerator
The University of Wisconsin "Short Tank" Van de Graaff
The University of Wisconsin "Long Tank" Van de Graaff

Although Los Alamos recognized the owners' need and desire to repossess this equipment so that nuclear research might be recommended in academic laboratories, it was understandably reluctant to lose equipment just as its long program of reconstruction began. The health of the Laboratory demanded that active research be pushed with as much enthusiasm as possible, and the ability to proceed with an immediate research program was one of the Laboratory's few attractions. Accordingly, efforts began to purchase outright as much of this equipment as the universities might be willing to sell. In some respects, there was more enthusiasm for these proposals than might have been anticipated. The rapid advances of the war years had filled all physicists with a desire for bigger and better accelerators—and the prospect of disposing of older and smaller equipment at a reasonable price and using the proceeds to finance larger devices was attractive.

Ultimately, arrangements were made to purchase all the above equipment except the "long tank" Van de Graaff accelerator. The success of this program in providing the Laboratory with efficient research tools when it desperately needed such attractions cannot be overestimated. It is dubious that an operating division in nuclear physics could have been maintained had a long program of accelerator construction been required, and the cooperation of the universities in selling their equipment indicated a real willingness to assist the Laboratory.

316

COCKCROFT-WALTON ACCELERATOR
AND SHORT TANK

Cockcroft-Walton Accelerator

This group's experimentation was greatly curtailed by the water shortage in late 1945. The D-D source could not be operated, DP Site suspended operations and could not deliver source, and work on ion sources ceased.

By the end of February 1946, water was more abundant and experiments with the accelerator, ion sources, disk scattering, D-D cross-section measurements, and scattering of high-energy neutrons by hydrogen and deuterium went forward.

In March 1946, the results of measurements for D-D neutrons were obtained. The scattering of neutrons by hydrogen still showed a slight ($<5\%$) preference for the backward scattering of neutrons.

The scattering of D-D neutrons by deuterons showed a strong anisotropy. In the center-of-mass system, the differential scattering cross section per unit solid angle is 2.2 times greater for neutrons scattered through 180° than for those scattered through 90°. The area under this peak in the differential scattering cross section is about 15% of the total cross section.

Tests were made in April and May to establish that the yield of neutrons and protons from the D-D reaction is constant for a given set of conditions. Discrepancies were traced to a too-low pumping speed at the target chamber. In June, a by-passing pumping lead and smaller diaphragms were installed to restrict the beam. This decreased the yield variations somewhat but not completely.

The Cockcroft-Walton accelerator group continued runs on D-D yields after the Van de Graaff was changed over to the tritium program. The main reason for the data variation seemed to be unstable operating conditions of high-voltage output. An electronic high-voltage stabilizer was installed, but experimentation was still in process at the end of 1946, and evidence that the trouble had been eliminated entirely was not conclusive.

As previously stated, the Cockcroft-Walton Group was an independent unit of the Physics Division (P-3) until September 1946 when it was combined with the short tank group, because their experiments were often the same, except on the different types of accelerator.

317

Short Tank Van de Graaff

This group's activities were completely halted by the lack of water during the winter of 1945, and did not continue until early spring. By March 1946, they had completed evaluation and analysis of measurements on saturation behavior and the characteristics of the Frisch grid.

The short tank program changed in August 1946 when various aspects of the tritium source were placed under observation. Assembly and construction of the D-T experiments called for tank modifications. The diffusion pump was set up with silicone oil. All graphite and carbon shutters, liners, slits, and diaphragms were replaced with tantalum. A glass-uranium pumping system was installed with forepumps sealed. Safety circuits were made with large burettes for the forepumps to permit the tritium to go to the forepumps in case of a double accident.

A tritium sample of rated 18% concentration was admitted to the uranium pump on November 29, 1946. Data from bombardments indicated that the sample was weak, probably not containing more than 5% tritium. Before experimentation could proceed, a richer sample was needed for target material.

Van de Graaff Construction

Loss of the "Long Tank" created a serious gap in the range of neutron energies the Laboratory could investigate, and in experiments devoted to use of thermonuclear reactions. Because such long-range problems were considered fundamental to Laboratory philosophy (Appendix A), remedying the situation was considered.

A new machine had to be constructed, and because no adequate accelerator construction was planned by any major electrical companies, the Laboratory was able to consider the most desirable type for its purposes. After preliminary small-scale experimentation, it was decided that a Van de Graaff accelerator in the 8- to 12-MeV range would permit investigation of neutrons from the lowest energies up to more than 20 MeV. The cost was estimated at $500 000 and, if constructed and successful, this would be the largest generator of its kind in the world. Permission for this program was requested of General Groves, and granted on December 12, 1946.

J. L. McKibben, Group Leader of P-9, assumed responsibility for the machine's design. J. H. Williams of the University of Minnesota, R. G.

318

Herb of the University of Wisconsin, and Van de Graaff and Trump of the Massachusetts Institute of Technology consulted.

It was specifically proposed that this accelerator be used to continue the scattering and cross-section experiments, particularly with fundamental particles at higher voltages; to produce and study neutrons up to energies of approximately 25 MeV; to bridge the gap with experiments using neutrons between 6 and 12 MeV, a range that could not be reached with existing equipment; to study the ^{14}N-^{14}N reaction (important in possible ignition of the atmosphere); and to study new (p-n), (p-2n), and (p-d) reactions that might occur at these higher energies.

Almost as important as the experimental field this instrument offered was the opportunity it presented to attract and keep good nuclear physicists. Such as a tool would be invaluable in maintaining a competent staff interested in basic nuclear physics and its potential application to military weapons.

McKibben outlined the proposed design and progress made on the generator in an October 11, 1946, report to Bradbury. The proposed design had a separation column that permitted high pressures of special gases around the high-potential electrode that could be operated without the charging belt in a high-windage, corrosive atmosphere. This would give the machine flexibility, keep it freer from dirt, and make servicing simpler, because the critical parts would be made accessible by raising the vessel.

A small test generator using standard parts of the instrument under construction was operational by July 1, 1946. It had a separation column 1-1/2 ft tall by 1 ft in diameter. The high-voltage electrode was charged with a 6-in.-wide belt. The tank had a 30-in. outside diameter and operated up to 500 psi. This generator was to be a source of experimental information as well as a training mechanism for new personnel.

Plans for the large instrument took form as a divided pressure device with the belts running in a lower pressure region inside the separation column. Three evacuated tubes were to be provided in the low-pressure region. Large shells in the high-pressure region outside the separation column divided the potential between the high-potential head and the tank. The insulating column, or separation column, was a series of insulator "Mykroy" rings and steel rings. Mykroy, a lead-glass-bonded mica, has high compressive strength, good dimensional stability, and high puncture voltage, and it is fireproof and malleable. Therefore, it was selected over other possible plastics and phenolic-bound paper.

Small-scale experiments under final conditions gave voltage gradients of 1.5 MeV/ft. Original plans called for a control room containing two laboratories, shielded by a concrete wall, and a small shop. The best design seemed to be a 150-ft-tall steel tower with a windbreak, with a 125-ton hoist at the top for dissassembly to accomplish internal repairs.

By the end of 1946, the generator tank design had reached the point where the manufacturers were considering many orders. The separation column had been ordered. The 256 steel rings were being manufactured by the Consolidated Steel Company. The die for making the Mykroy rings was completed in Los Angeles, and a press had been lent to Electronics Mechanics for molding these rings. The tank was being designed by the Consolidated Steel Company with an estimated five month's completion date. Designs for the inner column and buildings were not complete, and the site location had not been approved.

The Betatron Group

P-11, under S. H. Neddermeyer, became part of P Division in January 1946, after M Division ceased using the betatron for research (Chap. 25).

Throughout 1946 experiments were conducted on photo fission thresholds in normal uranium, ^{235}U, ^{238}U, and plutonium. The most dependable results came from using a shielded paraffin geometry about 5 feet from the betatron. These findings indicated that all fissionable materials (such as plutonium, normal uranium, and ^{238}U) have thresholds at about 5.2 MeV. However, this work was not conclusive because the thresholds for production of neutrons from the betatron were also 5.2 MeV. Therefore, there was doubt whether the fission observed was neutron-produced fission or gamma fission.

An independent section of this group worked on chronotron and counter development. The chronotron is a transmission line and a detector, or an array of detectors, coupled to the line at evenly spaced intervals, whose function is to determine the region of superposition for two transient pulses traveling along the line in opposite directions. The term "detector" is reserved for this specific meaning and the term "counter" is used to mean a device that produces a pulse when traversed by a charged particle. The ultimate purpose was to produce a system of chronotron and counters for measuring velocities of charged particles by comparison with the propagation rate along the line.

320

The first chronotron model was completed in the summer of 1945. Tests with a single detector showed that, with pulses generated by a condenser discharge with 50- to 150-V amplitude and time constant of approximately 10^{-10} s, time differences could be measured to an accuracy of about 3×10^{-11}s.

A second model finished in March 1946 gave cleaner operation, but the pulser gave serious difficulties. Experiments continued on this model until the last of June when Neddermeyer left for Washington State University. Shortly afterward, this equipment was transferred to him at Seattle. William Ogle became group leader of betatron activities on July 1.

Cyclotron

After the P-Division reorganization, the Cyclotron Group (P-12), under J. A. Fowler, had to indoctrinate a new crew and repair the cyclotron.

Exploratory experiments on distribution of fission fragment energy as a function of incident neutron energy were conducted. A suitable fission fragment energy counter was constructed in March 1946. This work was tabled temporarily in May to allow investigating the apparent fine structure of the fission fragment energy spectrum but was resumed the following month. Experiments continued, but high radiation near the cyclotron caused difficulty with the counter.

In October 1946, the Electronics Group started building circuits for modulating the cyclotron arc with variable pulse widths. The detecting equipment was modulated so that the neutrons arising at the target due to the arc pulse were separated into energy groups by the time of flight over a fixed distance. Twelve successive time intervals (corresponding to twelve neutron energies) were used. This equipment was not complete when the year ended.

Cosmic Radiation Studies

A temporary group under Darol Froman was set up in January 1946 to further scientific data on cosmic radiation. Four BF_3 counters built for Operation Crossroads were adapted for measuring the neutron components of the cosmic rays. These counters had about 2% efficiency for thermal neutrons.

321

A B-29 airplane from Z Division was used for this work. It was modified by removing all gunnery equipment, armor glass, and hydrogenous material from the tail section, and by installing the four counters: a bare (unshielded) enriched counter 3 ft from the other three, an enriched counter covered with a 0.030-in. cadmium sheet, an enriched counter shielded with 1 in. of B_4C (normal boron—density 1.3 g/cm^3), and a normal BF_3 counter shielded with 1 in. of B_4C. These three counters were 2 ft from each other. The high-voltage box and preamps for these counters were also in the tail section. The filament supply and coaxial leads for the signals were strung from this section to the radar room, which contained the amplifiers, scalers, recorders, and power supply. A cadmium-paraffin-covered counter to monitor total neutron intensity also was mounted in the radar room.

Initially, it was feared that the gasoline the plane carried (about 6500 gallons) would affect the bare and cadmium data. However, measurements taken at the same altitude at the start and end of a flight after the gas load had shrunk 3500 gallons were identical. Apparently, the counters could be considered as being in free space.

It was thought that, as altitude was gained, counting effects from showers might start, even with high bias settings. This was checked by running an enriched BF_3 counter and a normal BF_3 counter, bare and with a B_4C shield, at various altitudes to discover whether there was a decrease or any other trend. The ratio stayed constant, which showed that the count was, at all times, either of neutrons or natural background.

Several flights were made before June 1946 when the group activities were dissolved. The first experiment showed that the cadmium ratio remained constant above 7000 ft and had a value of 2.18. Also, the counting rates in all counters varied in the same way with altitude. Bad weather and faulty mechanical performance during two flights curtailed the time aloft, and good statistics were not obtained. One flight was attempted during cloudy weather, forcing the plane to remain below the main cloud body. Directly under the large cloud mass, the count was high except in the paraffin monitor. As the plane approached openings in the clouds, the count dropped.

Lack of personnel to staff this group and urgency of other experimental work in the division finally caused the program to be abandoned in June 1946. H. M. Agnew, W. C. Bright, and Darol Froman recorded full details of the operation.

CHAPTER 25

M Division

FORMATION OF M DIVISION

In the fall of 1945 reorganization of the Laboratory, M Division was formed under Darol Froman, with the responsibilities previously assigned to the Weapon Physics Division or G Division (Chap. 15). The following list gives the various groups in the new division and their relationship with groups in the previous organization.

M-1	Design and Production	R. E. Schreiber	In part, G-Eng. and G-1
M-2	Critical Assemblies	L. Slotin	In part, G-Eng. and G-1
M-3	Initiator	H. W. Fulbright	G-10
M-4	Electric Method	A. Graves	G-8
M-5	RaLa	D. Hall	G-6
M-6	Flash Photography	W. Koski	X-1C
M-7	Super Mechanics	J. Tuck	In part, G-2
M-8	Optics	B. Brixner	G-11
M-9	Magnetic Method	E. Creutz	G-3
M-10	Betatron	S. Neddermeyer	G-5
M-11	Consulting Engineering-Physics	J. T. Serduke	In part, G-Eng. and G-10

Later, H. W. Fullbright returned to Princeton and D. P. McMillan took over M-3. Alvin Graves was the Group Leader of M-4 until November 1, 1946, when he was appointed Associate Division Leader of M Division. At this time, Stanley Barris took over M-4.

323

GENERAL RESPONSIBILITIES

Conforming to long-range Laboratory policy, M Division was assigned to the following work, including both peacetime applications of nuclear energy and continuation of weapon development.

Maintenance of the Weapon

This included engineering design, production, inspection, and surveillance of all parts of the gadget inside the HE charge for weapons stockpiling. It also included maintenance of field equipment, written instructions, and personnel with adequate knowledge and experience for assembly and tests. Of course, there was division of responsibility with the CMR Division in production of active cores and initiators.

Critical Assemblies

This included not only routine measurements of the multiplication of fabricated cores, but also experiments on safing, on new models, and on spectral and intensity distributions of neutrons throughout the pit. Also, there were extensive measurements to help design fast reactors and miscellaneous measurements relative to safety problems for the Los Alamos Project and other Manhattan District Projects.

Bomb Improvement

This experimental work included study of the induced motion and compression of parts of the pit by HE, design and testing of new models, measurements of the improvement effected by new explosive arrangements, a detailed study of initiators, fundamental studies of shock waves and the associated hydrodynamics, and measurements of explosives effects in collaboration with X Division.

Super Mechanics

This phase included experimental studies of proposed mechanical methods of initiating thermonuclear reactions.

324

Optical and Engineering Physics Service

This responsibility embraced the design, procurement, etc., of special optical and photographic equipment and of physical measuring instruments for other groups and divisions as well as for M Division.

In September 1946, M-1 was reorganized to undertake weapons control, covering the pit in all its parts. The first plan was to transfer all completed pits, apart from the initiators and active cores and plugs, to the stockpile at Sandia (Chap. 28). M-1 was to maintain surveillance, records, and stockpiling of the plugs, fabricated active material, and initiators. La Roy Thompson became Acting Group Leader of M-1, and R. E. Schreiber took over Group M-2.

A committee consisting of M. G. Holloway, R. E. Schreiber, La Roy Thompson, and William C. Bright (chairman), was appointed to advise M-1 on preparation of manuals and kits for the Armed Forces at both storage and advance bases.

Design of cases for carrying and storing hot plugs was well under way in November 1946. This work was necessitated by the decision to deliver hot plugs to the Armed Forces as a simplification of procedures at advanced bases.

The Design and Production Group conducted a training program for Army Officers under the direction of M. G. Holloway. This training gave these men (chosen because of their excellent qualifications) enough background in routine mechanical assembly operations and in safety problems so that they might intelligently cope with unpredictable occurrences or accidents. On November 1, 1946, Bright assumed leadership of group M-1.

Critical Assembly

The G-Division critical assembly group continued critical assembly of active material in various tampers after the war. Harry K. Daghlian, a staff member of this group, on the evening of August 21, 1945, performed the experiments and accidentally obtained a supercritical arrangement. He quickly dispersed the assembly, but in doing so received lethal exposure from radiation and neutrons. He died on September 15, 1945.

Louis Slotin's group, M-2, continued experimentation on composite cores, critical masses in various tampers, nuclear safety, and the Fast Reactor (on a temporary basis until Philip Morrison returned to Los

Alamos in April 1946 and his Group P-5 resumed operation) (Chap. 24).

Plans were made in the fall of 1945 to transfer critical assembly work from Omega Site to Pajarito Site, and construction of a laboratory building began. The new site was in operation by April 1946.

Most of the experiments were conducted as a service to other groups. For example, several measurements on critical masses of ^{235}U biscuits in iron, in iron and uranium, and in iron and sodium tampers were made in April and May at the request of Walter Zinn of the Argonne Laboratory. In the measurement of the composite cores, the fit of the tamper to the core was not always good. There were cases in which the cavity was known to leave a gap of definite size around the active core. Qualitatively, it was found that a small gap reduced the observed multiplication by a large factor.

An experiment with a critical assembly of a combat-type plutonium core and a beryllium tamper on May 21, 1946, resulted in another serious radiation accident that caused the death of the group leader, Louis Slotin, nine days later. This occurrence practically halted all work in critical assemblies. The Laboratory devoted serious thought to a means of continuing this essential work without danger to those involved. It was apparent that such studies were essential to Laboratory progress, were required in weapon production work, and were expected of the Laboratory by many other parts of the Manhattan District. Although it was clear that no machine could think as well as a trained man, it was also clear that a machine could do what it was prescribed to do and could be provided with a variety of automatic controls and safeguards.

Because this work would cease unless something drastic was done, the Laboratory decided to forbid (they had been stopped anyway) all manual critical assemblies, to provide a critical assembly laboratory operated by remote control and provided with every practical safety device that could be devised, and to locate that laboratory at a distance from the control room and separated from it by earth embankments. Then should an accident occur, there would be the protection of both inverse square law and absorption. Finally, all critical assembly experiments were made subject to certain procedures and required both the presence of specific individuals and detailed approval at high levels.

After much deliberation, it was decided to construct an assembly and instrument building in Pajarito Canyon about 1250 ft from the main laboratory, which was to become a control room. The instrument room was thoroughly shielded from the assembly room. The assemblies were to be photographed through periscopes, and television and telephoto equipment was to be incorporated. Special remote control equipment was designed and constructed. One such piece called "Topsy" ("I just growed," from *Uncle Tom's Cabin*) was to stack various cubes of ^{235}U into an assembly by remote control. At the end of 1946, approximately 60% of the equipment had been constructed and installed.

After Slotin's death, R. E. Schreiber became Group Leader and M-2 worked on storage safety problems and weapons measurements while critical assembly work was suspended. The group also prepared the Pajarito Safety Manual for use by all M-2 personnel.

A new guarding system with the following three phases was initiated at Pajarito in December 1946.

Plan 1, the normal operating condition with no active material present, involved no special restrictions on persons entering and leaving the area.

Plan 2 went into effect when active material was present but no experiments were in progress. Access to the laboratory was only by means of an exchange badge.

Plan 3 went into effect when an experiment with active material was in progress. Access to the area was then controlled by the person in charge of the experiment. Everyone entering at such a time was to have a film badge.

Special Photography

The Photographic and Optics Group continued its role as a service and experimental group. It began preparing for the Crossroads tests in late 1945. All cameras and photographic equipment had to be thoroughly tested before shipment. Supplies were stockpiled for the tests and Kardex files were set up for all equipment involved. Special studies were made to determine the variation of image size as a function of density in order to get corrected measurements of the expanding ball of fire. Preliminary studies of ways to calibrate effective focal lengths of lenses also were made. Part of the group was sent to Bikini to construct camera installations and to act as advisors.

After Able and Baker Tests, the group was deluged with the work of correlating the photographic results. All the returned equipment had to be catalogued, cleaned, and repaired. Some movie and still film had to be processed. Great quantities of photographs required sorting and filing with proper explanatory data. The group edited two complete films: one (16-mm) showing Los Alamos activities in the Crossroads Operations; the other (35-mm) showing the two explosions.

The Photographic Group also prepared various reports supported by photographs and graphs. One, completed April 2, 1946, was "Time-Space Relationships" by Julian Mack. Another, completed December 10, 1945, was by Donald C. Livingston on "Gamma Radiation at Hiroshima."

At Sandia Base, the group also studied airborne camera installations for combat use.

Four M-Division groups were ultimately discontinued because their purpose no longer existed or because their function could be achieved more efficiently by another group or division.

- The Super Mechanics Group, M-7, devoted largely to the Crossroads tests, was discontinued entirely in February 1946.
- The Magnetic Method Group, M-9, was dissolved January 1, 1946, when the Magnetic Method (Chap. 15) was discarded except as an auxiliary in the RaLa methods.
- The Betatron Group. On January 1, 1946, it was decided not to use the Betatron Group, M-10, to study implosion (Chap. 15), but to retain all facilities at K Site so that work could be resumed on short notice. This group was transferred to P Division (Chap. 24) so the accelerator could be used in physics experiments.
- The Consulting Engineer Physics Group, M-11, was appointed to M-6 and the CMR Division. It became entirely inactive on November 1, 1946.

328

CHAPTER 26

Chemistry and Metallurgy Research Division

DIVISION ORGANIZATION

Late in October 1945, the Chemistry Metallurgy Division became CMR Division under Eric R. Jette, who became Division Leader when the co-Division Leaders, Joseph Kennedy and C. Smith, left the project. CM-1, the Service Group, dissolved about this time, and the designation was not reassigned until December 1. Then CMR-1, Analytical Chemistry (formerly CM-9), under H. A. Potratz, and CMR-3, Polonium Chemistry, with D. S. Martin, Group Leader, were added to the division. The organization at the end of 1945, quite different in structure and personnel from that established in August 1945 (Chap. 17), was as follows.

CMR-1	Analytical Chemistry	H. A. Potratz
CMR-2	Chemical Research and Development	C. S. Garner
CMR-3	Polonium Chemistry	D. S. Martin
CMR-4	Radiochemistry	G. F. Friedlander
CMR-5	Heat Treatment and Metallography	G. L. Kehl
CMR-6	Metal Fabrication	J. M. Taub
CMR-7	Corrosion Protection	D. Lipkin
CMR-8	Metal Production	R. D. Baker
CMR-9	Metal Physics	E. F. Hammel
CMR-10	^{235}U Chemistry	K. M. Harmon
CMR-11	Plutonium Production	J. E. Burke
CMR-12	Health Instruments	W. H. Hinch

In February 1946, CMR-10 was dissolved and its functions were transferred to CMR-8 under R. D. Baker, and in July 1946, CMR-7 was discontinued.

329

A new Process Development group, CMR-13, under R. B. Duffield, was established March 1, 1946, to develop a new plutonium purification procedure for DP Site. After fulfilling its mission, the group was dissolved in July 1946 and all personnel were transferred to CMR-11 at DP Site.

Further redesignations of groups and group leaders resulted in the following organization at the end of 1946.

CMR-1	Analytical Chemistry	C. F. Metz
CMR-2	Chemical Research and Development	J. F. Lemons
CMR-3	Initiator Chemistry	D. I. Vier
CMR-4	Radiochemistry	R. W. Spence
CMR-5	Physical Metallurgy	F. M. Walters, Jr.
CMR-6	Metal Fabrication	J. M. Taub
CMR-8	Metal Production	R. D. Baker
CMR-9	Metal Physics	E. F. Hammel, Jr.
CMR-11	Plutonium Production	F. K. Pittman
CMR-12	Health Instruments	J. Tribby

GENERAL POLICY OF CMR RESEARCH

Chemical and Metallurgical research for the Los Alamos project dealt with problems of all fissionable materials (with slight attention to uranium). It also dealt with highly radioactive materials, especially where large quantities of such materials were desirable or necessary; in other words, with problems that could not be handled adequately on micro- or milligram scale.

This policy was based primarily on the following. The division was concerned mainly with production, isolation, and use of large quantities of fissionable or radioactive materials. Very small-scale investigations had proved unsatisfactory for the Laboratory's purpose.

Protection of those working on the larger amounts of such materials required elaborate equipment, special techniques, medical inspection, and auxiliary service developed to a high degree during the early years of Laboratory operation. The health hazards involved in working with large quantities, whether expressed in mass or radiation energy, cannot be appreciated by inexperienced persons. Another grave danger in large operations was the possibility of contaminating the surrounding community. Here again, Los Alamos was experienced in providing protective measures.

330

Specific Division Program

Early in the spring of 1946, Jette outlined the various phases of work for his division.

Metallurgical and Physical Problems

Studies in the physical and mechanical properties of plutonium, other transuranic elements as they became available, and polonium

Phase diagrams of plutonium and other transuranics and their alloys

Studies of alloys including transformation rates and mechanisms, and precipitation hardening

Development of methods for fabricating plutonium and its alloys

Corrosion rates and methods of retardation

Diffusion rates involving plutonium, uranium, and polonium

Plutonium Chemistry

Metal preparation

Dry chemistry

Wet chemistry

Polonium Chemistry

Tritium Research

Intense Radiation Effects

Transuranics Research

Classical Radiochemistry

Analytical Chemistry Research; usually specific problems that were not of great interest outside this project had to be solved; methods for plutonium and uranium analysis included spectroscopic, fluorimetric, polarigraphic, etc.

Miscellaneous Chemical Problems; included study of the solubility of various metals in mercury and preparation of anhydrous halides of various metals

CHEMISTRY ACTIVITIES

Analytical

Group CMR-1 leadership changed twice between August 1945 and March 1946. H. A. Potratz remained Group Leader until January 1946, when he was succeeded by L. P. Pepkowitz, who was followed by C. F. Metz in March 1946.

Analytical research for CMR practically came to a standstill in late 1945. The only work that continued was routine analysis having to do with production. This quiescence continued until the spring of 1946, when Group CMR-1 reflected the new interest born in the Laboratory.

After that routine analysis progressed, analytical procedures were refined and standardized, and research went forward. Research covered improvement of existing analytical methods and development of new ones to solve new problems. An example was the investigation to determine alloying constitutents in both plutonium and uranium, and procedures for determining the plutonium and uranium content of waste solutions. The group cooperated closely with CMR-8 in the analytical work on recovery and purification of ^{235}U. CMR-1 entered practically every phase of new and old experimentation, not only for the CMR division, but as a service group for other divisions.

Chemical Research and Development

Group CMR-2 was another service unit, which performed routine radioassays for plutonium and maintained the electronic equipment for radioassays. It performed certain phases of research on solubility determinations for plutonium compounds, oxidation, and plutonium reduction.

C. S. Garner supervised the group until February 1946 when K. M. Harmon was appointed group leader. Harmon remained until June 1946 when he was succeeded by J. F. Lemons as acting group leader.

During August 1945, the group began to devote more time and personnel to finding a satisfactory chemical method for separating lanthanum from barium for the RaLa program. This action seemed justifiable from the number of promising leads uncovered in exploratory work.

Investigation of RaLa deposition on insoluble fluorides through the end of 1946 was based on a method studied by CMR-4 which involved CaF_2 as the insoluble fluoride. This work was a search for other relatively insoluble compounds having the proper solubility relation to LaF_3 to give more complete recovery of the lanthanum compound in the new form, faster conversion, and satisfactory filtration.

332

Initiator Chemistry

Group CMR-3 initiator production (under D. S. Martin until August 1946 when D. I. Vier became Group Leader) had been planned for DP East Site. However, construction and installation of the highly technical equipment took longer than anticipated, and it was not until September 1945 that production started in the new laboratory. The process was always highly difficult and involved using a material as hazardous as plutonium.

Radiochemistry

Group CMR-4 under R. W. Spence, continued its work on radio-lanthanum (RaLa) development and operations, Water Boiler chemistry, and tritium experiments (Chap. 17).

No change was made in the RaLa method (Chap. 17), which had proved satisfactory up to 2000 Ci. Operation refinements (including a new source container) and introduction of a new flow sheet, specifying and simplifying operations, allowed personnel to handle up to 3000 Ci, but the dosage was too high for continued operation at these levels. This redesigning of operations also speeded up extraction of the active material.

Work on the uranyl salt solution, "soup," for the Water Boiler continued under this group and involved routine analysis and purification of the gases evolved (Chap. 24).

As early as May 1946, the group prepared a foil of tritium in the form of water THO (half the hydrogen as the tritium isotope) absorbed on the surface of freshly prepared aluminum oxide, Al_2O_3. Preparations were under way in the summer of 1946 for further experiments to determine the half-life of tritium and to investigate the magnetic moment of the triton.

METALLURGICAL DEVELOPMENTS

Physical Metallurgy

CMR-5 was supervised by G. L. Kehl until February 1946 when R. D. Baker became Acting Group Leader in addition to being Group Leader

333

of CMR-8. Baker remained in that status until August 1, 1946, when F. M. Walters, Jr., assumed leadership in addition to his duties as Associate Division Leader.

CMR-5 took on the problem of developing fabrication methods for fast reactor rods early in March 1946 (Chap. 24). Extrusion methods proved most suitable, but although correct density was obtained, for some obscure reason the extruded bars were slightly tapered. The ultimate solution lay in making the rods 0.02 in. oversize and machining them to specifications. They could be machined to 0.0005 in. in diameter over a 6-in. length. Plans were made in June to transfer manufacture of these rods to DP Site to eliminate as much serious contamination as possible. By July, five extrusions had been made, four of which met specifications and were sent to DP for machining and coating.

Metal Fabrication

The Metal Fabrication Group, CM-7, was supervised by J. M. Taub, until December 1945 when it became CMR-6. There were no changes in the group functions or leadership. The group did extremely varied work for CMR Division and the other technical divisions. Its routine services included refractories, general foundry work with uranium and special alloys, plastic services, electroplating, and general powder metallurgy.

CMR-6 also conducted research on problems peculiar to their work and completed a new process for fabricating uranium spheres in July 1946. This process involved new vacuum casting furnaces and machining jigs and made it possible to make two, or possibly three, castings per day instead of one.

Production of special pieces for the fast reactor assembly (Chap. 24) was completed in June 1946. Development on the generating tower for the Van de Graaff construction included work on the Mykroy rings and shellac adhesives. This Van de Graaff fabrication, started in April, was still under way at the end of 1946.

Another interesting job was recovery of normal uranium metal from the shop turnings. Vacuum casting charges of 100% down to 50% briquetted turnings with the rest good uranium metal yielded only 50% metal recovery. In August 1946, a different method consisting of melting in air under a barium chloride flux and then bottom-pouring into a graphite mold was first used. This yielded recoveries up to 65% when one-third of the charge was virgin metal. When Ames biscuit metal was used as the virgin metal, the metal recovery increased to 77% of the turnings, with 100% recovery of the virgin metal assumed. Remelting of the uranium turnings (approximately 7000 pounds were on hand) began in October. With the equipment available, one man could process 60 lb of turnings per day. Larger briquetting dies were in process, which would at least triple this output.

Metal Production

The methods for recovery and purification of ^{235}U at the end of 1945 were not well worked out and were inefficient and unsafe.

Group CMR-8, supervised by R. D. Baker, carried the production load as well as conducting research in the field. In fact, CMR-8 continued to produce all the ^{235}U metal for the Manhattan Engineering District. However, it was impossible for the group to conduct intensified research until it had absorbed the personnel from CMR-10 in February 1946. After that, it began to develop processes for recovery of ^{235}U from all residues originating at the project.

A more efficient, less hazardous hydrofluorination process for converting purified oxide to the tetrafluoride made it possible to start putting large quantities of ^{235}U back into circulation.

Late in 1946, all the installations the group used in D and M Buildings were overhauled to reduce the contamination danger. Dryboxes and equipment for enclosing the reduction operations on ^{235}U were installed.

Metal Physics

The Metal Physics group (CMR-9, E. F. Hammel, Group Leader), established the following program of experiments early in 1946.

- Specific heat of plutonium from room temperature to the melting point,
- Thermal conductivity of plutonium at room temperature
- Self-diffusion studies on uranium

Part of D Building was set aside for these investigations, and furnace and control apparatus, a constant-temperature bath, a vacuum system, and auxiliary parts were designed Construction and installation of this equipment occupied most of the group's time for the rest of 1946.

A preliminary value for the thermal conductivity of a 3% gallium alloy delta-phase plutonium was obtained in July 1946. This was investigated

335

further during the last quarter of 1946. These tests were conducted at six different temperatures in the 0 to 60°C range. The thermal conductivity of plutonium was found to be 0.0195 ± 0.0005 g-cal/cm²/s for a temperature gradient of 1°C/cm per centimeter in the temperature range mentioned.

PLUTONIUM PRODUCTION

Plutonium production was charged to Group CMR-11. J. E. Burke was Group Leader. He remained in charge until March 13, 1946, when S. J. Cromer took his place. In November 1946, Cromer left and F. K. Pittman became Group Leader.

DP Site, the new production area, was divided into the East Area for polonium processing and the West Area for plutonium processing and production of bomb cores. (The details of design and buildings are discussed in Chap. 17.) Construction was largely completed by the middle of August 1945, but it was a month later before all the hoods and technical equipment were installed and operations began.

Conspicuous among Laboratory worries about plutonium production was the process used for purifying the plutonium nitrate slurries received from Hanford and converting them into metal. The established processing technique when the war ended involved an ether extraction of plutonium nitrate in glass columns. This was the process used when DP Site became the center of operations. The fragility of the glass columns, the explosive nature of the ether vapor, and the toxic properties of the plutonium combined to make DP Site a potentially extremely hazardous installation.

As previously mentioned, a new group, CMR-13, was formed in March 1946 to help develop a new plutonium purification procedure. This group worked harmoniously with CMR-11, the Plutonium Production group, on this problem.

For some time, research on alternate processes of purification and reduction had been under way. In June 1946, a process involving a simple oxalate precipitation, with the consequent entire elimination of the ether stage, was developed and incorporated in the production cycle. The new process not only increased the safety of the operation, but the resulting product equaled the quality of that produced earlier. As a result of this and other engineering and process developments, the backlog of

336

plutonium material from Hanford was consumed and converted to metal faster than new material was received. Consequently, by August 1, 1946, this backlog was entirely eliminated. In fact, by August, DP Site could handle at least twice Hanford's maximum production.

In August 1946, DP Site had to take on the isolation and purification of metallic gallium. The supply of this material was becoming increasingly critical and, to safeguard it, Los Alamos secured a large quantity of raw materials and Group CMR-11 began to develop methods for extracting the metal. It was planned to turn this process over to a commercial firm as soon as a successful method was developed.

In the fall of 1946, a new engineering and development section was formed in CMR-5 under Frank Pittman. Its main tasks were to improve existing operations, especially by redesigning plant equipment, and to make working conditions less hazardous. Additional research problems included new methods for recovery of reduction and casting residues, supernatant solutions from the oxalate precipitations, and metal scrap from metal fabrication operations.

HEALTH INSTRUMENTS AND INDOCTRINATION

The very important "watchdog" functions of Group CMR-12 included monitoring and decontamination in the Technical Area and DP Site (both West and East Areas), care and use of counters and meters for detecting radioactivity, and laundry of contaminated protective clothes and respirators.

W. H. Hinch was the Group Leader until he left the project on March 26, 1946. Then J. F. Tribby accepted the position.

Throughout 1946, the HI Group (CMR-12) increased its monitoring in the various division areas. Besides this "police" work, much effort was spent in educating personnel about the importance of the health safety rules and regultions. It was difficult to instill respect for some of the procedures in both the scientific and production personnel. This indoctrination was in collaboration with the Health Group, who had experienced this same lack of interest from employees (Chap. 22). This group also started investigations to develop types of radiation detection equipment that were more stable, rugged, and sensitive than those previously used.

337

CHAPTER 27

Explosives Division

ORGANIZATION OF X DIVISION

A glance at the early arrangement of X Division (Chap. 16) reveals a complex organization of many groups and subgroups. This complexity, as explained in Part I, arose from the division's rapid growth and the number of functions it absorbed from other divisions (primarily some of the Gadget Division and Ordnance Division activities).

The trend in X Division and other technical divisions was toward a simplified structure because of insufficient personnel and discontinuance of certain wartime programs. Groups had to be combined to use the diminishing staff to the best advantage and to concentrate on the most important peacetime research problems.

G. B. Kistiakowsky returned to Harvard University in October 1945, leaving Max F. Roy, as Division Leader, to cope with reorganization. Reorganization was completed in January 1946. The work of subgroups X-1B, Terminal Observations, and X-1C, Flash Photography, was transferred to the newly formed M Division. The personnel and functions of Group X-2, Engineering; Group X-5, Detonating Circuits; and Group X-6, Assembly and Assembly Tests, were transferred to Z Division. Subgroup X-1D, Rotating Prism Camera, became Group X-8 under A. W. Campbell, and subgroup X-1E, Charge Inspection, became Group X-1, Radiographic Research, under G. H. Tenney. Subgroups X-3A and X-3B were combined as Group X-2, Explosives Research, with E. R. Van Artsdalen as group leader. X-3C, X-3D, and X-3E were combined as Group X-3, Explosives Production, with L. E. Hightower as group leader. The functions of Group X-4 were changed, from mold design, engineering services, and consulting to a general investigation of materials suitable as slow explosives, with J. W. Stout as group leader. These changes left the following organization.

338

X-1	Radiographic Research	G. H. Tenney
X-2	Explosives Research	E. R. Van Artsdalen
X-3	Explosives Production	L. E. Hightower
X-4	Slow Explosives	J. W. Stout
X-7	Detonators	K. Greisen
X-8	Detonation and Shock Phenomena	A. W. Campbell

Several other changes occurred during the year. With the loss of senior personnel in Group X-4, its functions were transferred to Group X-2 on June 15, 1946. When Van Artsdalen left the project, M. L. Brooks became group leader of Group X-2 on September 16, 1946. L. B. Seely replaced K. Greisen in Group X-7 on June 15, 1946. A new group, X-6 Detonation Physics, was activated on November 1, 1946, under J. C. Clark, to study detonation and shock phenomena with flash x-ray techniques.

Radiography Group

Group X-1, under G. H. Tenney, continued its investigation of radiographic methods for inspecting explosive charges. It also developed radiographic inspection techniques for nonexplosive objects.

Research was increased further by special work on the radiographic possibilities of various radioactive sources available at Los Alamos. As a result of an April 9, 1946, meeting, a 14-Ci RaLa source was obtained for fundamental radiographic experiments on steel. The first tests with this source were performed on June 20. A lead cylinder, equipped with a conical lead cover and a special lead shutter functioning by remote control, had been constructed to hold the source. This preliminary experiment showed that RaLa, as one of the obtainable isotopes, could be used for industrial radiography. Further investigation on this phase of radiography was suspended until more personnel could be obtained.

Research was curtailed by a fire on November 19, 1946, which destroyed one x-ray room and one darkroom at T-Site.

The Explosives Research Group

The work of this group under Van Artsdalen and later under Brooks included studies on slow explosives, thermal properties of cast explosives, development of faster, more powerful explosives, measurements of the

339

physical properties of cast HE, and casting of special explosive charges. As previously stated, Group X-4, Slow Explosives, was absorbed by the Explosive Research Group in June 1946.

Explosives Production

Shortly after the war, Group X-3 production of full-scale lenses ceased except for experimental purposes and process development. This cessation resulted from lack of personnel and the unsafe character of the buildings in which this work had been done. The last full-scale charge (although not the last full-scale casting) was made in Building S-25 in October 1945. By this time, it was possible to make full-scale lens castings on a production line basis.

Group X-3 was led by Major J. O. Ackerman, until November 15, 1945.Then L. E. Hightower became group leader, and Ackerman acted as an advisor.

Production was stopped on December 19, 1945, because the water shortage had become so acute. Before this official closing, the group had been severely handicapped by low water and low steam pressure on several occasions.

The main problem confronting this group, after the water situation had cleared, was the hazardous condition of its production lines (Chap. 1). The casting line for experimental lens production, S-31, at last became so deteriorated that it was closed early in 1946 for renovation.

But in April 1946, at least two months before reconstruction was complete, S-25, the casting building for the full-scale lens, was closed because it was no longer safe, and all equipment was moved back into S-31. Building S-25 was closed for the rest of the year, and all production was continued in S-31, even though that line had not been finished.

By the fall of 1946, the Casting Plant at the Naval Ordnance Testing Station, Inyokern, California, had come into operation and relieved the group of the necessity of full-scale charge production. The Los Alamos group then concentrated on preparing scale castings for experimental work on process technology.

Detonation Physics

High-speed flash x-ray photography, which proved so useful during the war, was revived for study of detonation waves and shock waves in

340

substances arising out of high-explosive detonation. This work was the responsibility of Group X-6, organized in November 1946, with J. C. Clark as leader.

Detonator Production

Detonator production, Group X-7, under K. Greisen until June 15, 1946, when L. B. Seely, Jr., became group leader, concentrated on improved quality. Constant experimentation showed that by stabilizing the method of PETN precipitation and with proper attention to pressing, bridgewire soldering, and geometrical tolerances of the components, it was possible to produce better detonators.

In April 1946, standard detonator loading was moved from South Mesa Site to Two-Mile Mesa and only experimental detonator loading continued at South Mesa. This new location gave considerably more working space to the group.

Procurement difficulties arose in April and May when the Detroit Centerline plant was reorganized and operations temporarily ceased. It was impossible to secure detonator parts, but production was maintained by using the entire stock of parts on hand and salvaging others from rejected lots.

DETONATION AND SHOCK PHENOMENA

A. W. Campbell's Group X-8 continued its studies of shock wave and detonation phenomena. Experimental firing of full-scale lenses, experimental charges, and plane wave shots, and initiation of Composition B were all done at Anchor Ranch Site.

Among the most interesting phases of X-8 work was testing of special slow components. These studies were pursued vigorously from early 1946 until the end of the year, but the results were not conclusive.

The problem of increasing the intensity of the blast luminosity of the surface was solved materially in October 1946. Adding calcium peroxide on the surface of an explosive proved very intensifying. The only flaw was the need to improve cohesion between the calcium peroxide and the explosive. At the end of the year, experiments were still in progress to locate a better substance.

Experiments were started late in 1946 to test how cavities in explosives affected the emergent detonation wave.

341

CHAPTER 28

Ordnance Engineering (Z) Division

ORGANIZATION OF Z DIVISION

Although Z Division had been created in July 1945 (Chap. 9), it remained fluid until late September 1945 when a somewhat formal organization existed, with the following groups and subgroups.

Z-1	Experimental Systems	Commander N. E. Bradbury
Z-1A	Airborne Testing	Dale Corson
Z-1B	Informers	J. B. Weisner
Z-1C	Coordination with Using Services	Glenn Fowler
Z-2	Assembly Factory	Colonel L. E. Seeman
Z-2A	Procurement, Storage, and Shipment	Colonel R. W. Lockridge
Z-2B	Production Schedules, and Manuals	R. S. Warner, Jr.
Z-3	Firing Circuits	L. Fussell
Z-4	Mechanical Engineering	R. W. Henderson
Z-5	Electronic Engineering	R. B. Brode
Z-6	Mechanical Engineering for Production and Sandia	---

J. R. Zacharias remained division leader until he returned to MIT on October 17, 1945, then Roger S. Warner headed the division.

During the next six months, two administrative problems became evident. Operation of the Ordnance Engineering Division was becoming increasingly difficult because of the division of its activities between Sandia and Los Alamos. Furthermore, because of lack of housing at Los Alamos, the Laboratory was searching for almost any means of relief. Consequently, it was decided to move the entire division to the Sandia Laboratory as rapidly as possible. In March 1946, all groups except Z-4 were reorganized and a transfer to Sandia began. The housing situation at Sandia was little better than at Los Alamos and tended to retard the transfer, which was not completed until July. Group Z-4, Engineering, under R. W. Henderson, remained at Los Alamos until February 1947.

Z DIVISION PROGRAM

The Z Division program included testing, design, development, stockpiling, and bomb assembly.

Testing Program

Two weapons existed in August 1945: the "Little Boy" ^{235}U gun assembly and the "Fat Man" plutonium implosion assembly. The Little Boy program was discontinued, however, because the weapon had low efficiency relative to the amount of active material involved.

The Fat Man presented other difficulties. Its absolute degree of reliability was so unknown that the safety factor in each component was unknown, and all that had been proved in tests was that there had been no failure. The next drop, however, could never be predicted with confidence. Also, the weapon's mechanical, electrical, and nuclear complexity required men with the highest degree of training, responsibility, and experience for field experiments.

To get reliable statistical data on the performance of each component required devising some system whereby actual conditions experienced by the bomb in flight could be duplicated in the laboratory. Telemetering of flight information from the falling bomb was undertaken. Roll, pitch, yaw, vibration frequency, temperature, and pressure were recorded.

It was believed that once the conditions experienced by a bomb in flight were known, they could be duplicated in a test laboratory and made as severe as desired. The behavior of each weapon component could then be tested and its point of failure ascertained. The weak points would then be strengthened and a safety factor for the weapon, as a whole, would be established.

Designing Program

After the war, the Laboratory, without concrete guides from higher authority, faced the necessity of holding together its highly experienced group of design and development technicians. The Fat Man, as it was used at Nagasaki, could hardly be called anything more than a scientific

343

gadget; it was certainly not a weapon. Its assembly and use demanded the most experienced technical personnel in large numbers, which is certainly incompatible with modern warfare. It was decided that the Ordnance Engineering Division would re-engineer the weapon to make production easier, simplify the assembly at an advanced base, and minimize the number of highly trained assembly personnel required. With this as the primary objective, it was decided to do whatever was possible to improve the ballistics of the bomb, subject to the dimensional restrictions imposed by the dimensions of the B-29 bomb bay.

Development

Several new fusing systems to increase the reliability of detonation at the proper elevation were considered. The radar fusing system on the Nagasaki Fat Man was an adaptation of one of the first versions of airborne tail warning radars and was very crude.

When the bomb design was frozen, it was impossible to make further alterations in this radar. Instead, work concentrated on development of entirely new fusing systems, taking full advantage of all wartime developments. Two systems were under consideration, a nonradiating device of primarily academic interest, and a specially designed radar tailored to fit the actual conditions imposed upon a free-falling bomb. Either system, when fully developed, was to be considered for stockpiling, but the new radar system might be effective for several years and far superior.

Stockpile Program

One of Z Division's responsibilities was production engineering and procurement of all bomb components and material required for the national stockpile. This included testing all components, processing them through various tropicalization procedures, packaging, storage (in facilities controlled by others), and continuing surveillance of all material to determine packaging effectiveness and to guarantee against deterioration. The only bomb component not under the division's stockpiling jurisdiction was the pit assembly.

Bomb Assembly

Z Division's packaging operation included assembly of the high-explosive charge around the pit. A production line assembled all

344

incoming components into the proper units for long-term storage. The division also assembled bombs for use in proof-testing of individual components both in the Laboratory and in free flight.

In August 1946, a Mechanical Test Laboratory was established at Sandia in which a complete set of testing devices was installed to obtain statistical data on the performance of individual components. This work included vibration testing under all temperatures and pressures to which the components might be subjected in a tactical operation. A large altitude chamber could test a complete bomb assembly as far as pressure, temperature, and humidity were concerned, but lacked integral shaking facilities. The simulated altitude could be varied from 40 000 ft to sea level at 2000 ft/s, with temperature, pressure, and humidity independently controlled.

As early as January 1946, a great number of division personnel were enlisted in the preparations for Operation Crossroads (Chap. 21). The division leader, Roger Warner, went overseas late in March, leaving Dale R. Corson as acting division leader. By April, many of the Ordnance Engineer groups were entirely taken over by B Division. Most of the Informer Group staff, for example, were working on the Bikini tests, and the vibration study was suspended until after July 1946. This same situation was true in the Fusing and Firing Group. Owing to the interruption by Operation Crossroads, the telemetering program was very slow in starting, and only toward the end of 1946 did the telemetering devices begin to repeat what was fed into them from the pickup devices, rather than report their own "shake and shiver." The Assembly Group's being involved in overseas duty left scarcely any men trained in bomb assembly and testing. The gravity of this predicament led to a recommendation that assembly be turned over to a purely military organization with backgrounds in electronics, mechanical engineering, and high explosives. This group would be permanently assigned to Sandia and would engage in the laboratory production program.

In July 1946, a U. S. Army Special Battalion was formed to take over the surveillance, stockpiling, field tests, and assembly work, as well as field work on development of new models. It was planned to divorce the civilian organization from the Los Alamos Laboratory setup, and make it a permanent Civil Service adjunct to the Special Battalion as fast as possible.

This operational philosophy began to swing in the opposite direction during the first part of 1946, and it was accelerated with the arrival of

345

Colonel H. C. Gee as Area Manager in the fall of 1946. Instead of military operation of the development program with Civil Service employees in a supporting role, the trend gradually swung toward a completely civilian operation under the University of California contract. The reasons are many, but it was considered that temperamentally, as well as from a standpoint of primary interest, it was not feasible to have a military organization in a responsible developmental position.

Dale Corson left the division in July 1946, while Roger Warner was still overseas on Operation Crossroads, and Lieutenant Colonel E. E. Wilhoyt became acting Division Leader.

ORGANIZATION AS OF DECEMBER 1946

At the end of 1946, Z Division contained the following groups.

Division Leader		Roger Warner
Alternate Division Leader		Lieutenant Colonel E. E. Wilhoyt
Z-1	Field Test	Glenn Fowler
Z-2	Mechanical Engineering	R. A. Bice
Z-3	Assembly Training	Arthur Machen
Z-4	Engineering (Los Alamos)	R. W. Henderson
Z-5	Firing and Fusing	O. L. Wright
Z-6	Mechanical Laboratory	Alan Ayers
Z-7	Production	J. L. Rowe
Z-8	Informers	William Caldes
Z-9	Stock Piling	Wilbur Schaffer
Z-10	Supply	Henry Moeding
Z-11	Little Boy	Harlow Russ

CHAPTER 29

Documentary Division

INTRODUCTION

In development of the Laboratory from its inception in 1943, a substantial number of technical staff groups were formed to solve special problems. All these groups reported directly to the Project Director, J. Robert Oppenheimer, and later to N. E. Bradbury, Laboratory Director. Some of these groups were dissolved when their functions were transferred to other agencies or with the disappearance of the problems involved.

The technical staff organization functioned effectively because of full acceptance of responsibility by very competent group leaders with little or no supervision from the Director (Chap. 22). Many of these group leaders had outstanding professional backgrounds, which made them exceptionally valuable in other fields of endeavor. For example, Ph.D.'s and graduate students who had majored in theoretical physics, mathematics, and philosophy were engaged in technical editorial work, cataloguing, report writing, declassification procedures, and historical records.

Starting in the fall of 1945, many of these group leaders left for other activities. The functions of some groups were interrupted or left to junior members who needed direction. The relaxation of security, which had compartmentalized most activities, and the consequent greater exchange of information permitted establishment of central uniform Manhattan Project procedures for the handling of much technical information. This required alteration of the Laboratory's informal wartime practices. Consequently, the Director had to effect many new procedures while losing most of his unusually well-trained technical staff for handling these matters. The many other responsibilities of the interim period discussed in Chapter 21 kept the Director from giving these staff functions much time, so the groups tended to lose their effectiveness.

347

FORMATION OF D DIVISION

When Major Ralph Carlisle Smith, the local Patent Advisor, returned from his security and technical advisory assignment to the Operation Crossroads press ship, *U.S.S. Appalachian,* the Director asked him to consolidate the technical staff groups, other than the health group, into a division, and ultimately to become Assistant Director. On August 21, 1946, the Director announced the formation of the Documentary (D) Division to be responsible for Technical Series editorial work, the Document Room, the Technical Library, editorial revision, review and control of reports; information dissemination; declassification; history; and various other technical services, with Smith as division leader and Herbert I. Miller as alternate. Smith retained the responsibility for patent control, delegated to him by the OSRD Patent Advisor, Captain R. A. Lavender (U.S. Navy Ret.). Later, when E. J. Demson left the project in 1946, Smith also accepted the responsibilities of an Assistant Director, particularly Demson's legal duties. This arrangement removed a considerable load from the Director, and reorganization of the staff groups so as to use the available professional personnel in several phases of work improved the services of the staff groups with fewer employees. The responsibilities were generally divided as follows: Patents and Legal, Library and Document Room, History and Technical Series, Report Editing and Review, Classified Information Dissemination, Declassification, Drafting (Patents, Reports and Miscellaneous), and Technical Illustrations and Art Work.

D-Division organization during 1946 was as follows.

D-1	Legal	R. C. Smith
D-2	Library	G. F. Campbell
D-3	Review and History	R. R. Davis
D-4	Editorial	J. F. Mullaney
D-5	Design and Drafting	L. F. Jacot
	Records and Administration	A. M. Frazier

Most of D-Division's functions are covered in the discussion of technical staff groups in Chapter 3. A few phases, however, merit additional comment.

The Technical Series

In conformity with other sections of the Manhattan Project, a program was initiated to record, in accessible and edited form, the technical knowledge and gains of the Laboratory. In priciple, it was proposed to prepare a "Handbuch der Los Alamos" in analogy with the famous Handbuch der Physik. Titles for seventeen volumes were established in August 1945, as well as volume numbers, and, in some cases, chapters, sections; and editors. Difficulty arose in establishing a title for the overall work. The original name "Handbuch der Los Alamos" was misleading in its English translation, so the title Los Alamos Encyclopedia was substituted. But, because it was decided that "encyclopedia" implied an alphabetical arrangement, that too was discarded and "Los Alamos Technical Series" was finally chosen (Appendix F).

Hans Bethe and David Inglis were originally responsible for this compilation, with the following staff of volume editors.

Volume No.	Title	Editor
0	"Relation Between the Various Acitivities of the Laboratory"	S. K. Allison
1	"Experimental Techniques"	D. K. Froman
2	"Numerical Methods"	E. C. Nelson
3	"Nuclear Physics"	R. R. Wilson
4	"Neutron Diffusion Theory"	G. Placzek
5	"Critical Assemblies"	0. R. Frisch
6	"Efficiency"	V. F. Weisskopf
7	"Blast Wave"	H. A. Bethe
8	"Chemistry of Uranium and Plutonium"	J. Kennedy
10	"Metallurgy"	C. S. Smith
11	"Explosives"	G. B. Kistiakowsky
12	"Implosion"	R. F. Bacher
13	"Theory of Implosion"	R. E. Peierls
21	"The Gun"	F. Birch
22	"Fuzes"	R. R. Brode
23	"Engineering and Delivery"	N. F. Ramsey
24	"Trinity"	K. T. Bainbridge

Only Volumes 1 and 2 were considered completely declassifiable under existing standards. However, much of the information in some of the others would eventually be declassifiable and, except for the weapon data, the rest was to be distributable throughout the Manhattan Project for its general benefit.

Shortly after initiation of the program, David Hawkins and Robert R. Davis were assigned the responsibilities of the Technical Series because of the imminent departure from Los Alamos of both Bethe and Inglis. When Hawkins left the project in the late summer of 1946, Robert R. Davis took over the detail as a Group Leader in D Division.

The Technical Series compilation proceeded slowly from its inception. Exceptional delays resulted because many individuals were reluctant or unable, after their departure, to continue obligations taken on while at Los Alamos. A more understandable difficulty was experienced by active project personnel, who were faced with conducting an active technical program while writing about one accomplished in the past. By January 1947, Volumes 0 and 22 were completed and issued, and two-thirds of Volume 1 had been issued.

Design and Drafting Group

The Technical Series and Technical History Groups acquired a drafting section to illustrate their volumes. There was also a design and drafting section to aid the Patent Group. The Report Editorial Group collaborated with the Shop Group Drafting Section, A-3, and the Post Historian used the Post Operation drafting staff in the Army Civil Service organization. Furthermore, the Ordnance Engineering (Z) Division had a Technical Illustration Group to prepare exploded views of the weapons and their components, to do general art work, and to make instructive illustrations for the manuals prepared by Z Division on weapon assembly and handling, and by X Division on high-explosive and detonator production techniques. The Technical Series, History, Patent, and Technical Illustration Drafting Sections were combined into a single Design and Drafting Group to avoid duplication of effort and reduce overall staff group personnel requirements. In addition, the Documentary Division groups were no longer required to call on outside drafting agencies, thus relieving their workload.

It was found that half the drafting staff could carry the entire load and, in addition, the year backlog of work in the Report Editorial Group was completely wiped out, not only illustrations, curves, diagrams, and the like for the reports, but also the detailed and tedious printing of involved mathematical formulas. The Technical Illustration and Art Staff functioned substantially independently of the rest of the group, but by limiting its responsibility, it could concentrate on the primary assignment of preparing exploded views, manual illustrations, and art work so that considerably greater production resulted.

Declassification Program

Shortly after the war, many individuals requested permission to publish papers on phases of Laboratory research and development which they did not consider classified. The procedure for handling these items was not clear-cut and was generally unsatisfactory. A few items were released through the local Security Office by its Washington headquarters, but the informal and uncertain treatment left much to be desired.

After the Manhattan Engineer District adopted the Tolman Committee recommendations on a program of declassification, the Laboratory established in June 1946 a special scientific staff under the direction of Frederic de Hoffmann, on loan from Harvard College, to review all the Los Alamos formal reports to see which might appropriately be submitted for declassification. Many reports required careful rewriting to remove classified information or overcome indications of classified applications.

A procedure was established whereby the Declassification Group reviewed a report or a rewritten version thereof to determine whether it should be submitted for declassification. When a report was approved for processing, it was routed (1) to the technical series editor to be certain it was adequately covered in that compilation, (2) to a responsible reviewer, a senior member of the Laboratory scientific staff, to be approved for declassification according to a Guide prepared on the basis of the Tolman Committee recommendations, (3) to the patent advisor to be assured that the Government's interest was protected from a patent standpoint, (4) to the local Army Security and Intelligence Officer as a check against unnecessary revelation of physical security safeguards, (5) to the group leader of the Declassification Group to be certain that no releases were made on associated project work without permission from the project, and to send abstracts to the Manhattan Project Editorial

Advisory Board for approval of publication, and (6) finally to the Laboratory Director for general overall review and approval for submission to Oak Ridge declassification headquarters for declassification.

Although the foregoing routing seems involved, once the report was put in shape for submission for declassification, the processing could be accomplished in a day, except for the time required for detailed review by the responsible reviewer, and the patent advisor. Allowing ten days for review of a document by the Manhattan Project Editorial Advisory Board, declassification took about two weeks if the subject matter was clearly releasable. Of course, there were questionable cases, and some items were refused declassification.

About 320 documents had been routed for declassification by December 31, 1946. Approximately 700 more were considered by the declassification group but never assigned numbers for routing because they were not declassifiable. Of those processed, about 250 made the entire round and were approved for declassification in Oak Ridge before the end of 1946. It appears that about 50 of these documents were approved for publication or published in recognized scientific journals during the same period. The Laboratory was proud of its contribution to the country's scientific literature.

An incidental service established by the declassification group, with the cooperation of the Library and Document Room, was the loan of declassified documents to former Laboratory staff members engaged in research at other institutions. These loaned documents were not considered publications, only private communications. They were loaned, not only to aid research in the nation, but also to advise the former staff members on the extent of declassification and the limits of information that might be disclosed to others who did not have access to classified material. In the latter respect, it was thought to be a valuable security measure.

CHAPTER 30

Conclusion

Because national legislation developed more deliberately than anticipated, the Los Alamos Laboratory operated through 1946 on Bradbury's general interim philosophy expressed in October 1945, which was based on the local conception of the nation's present and future need for such a laboratory (Appendix A).

It went forward on a research program in all the technical fields bearing on development of the weapon, including nuclear physics, chemistry, high explosives, equations of state, radiation, hydrodynamics, and phenomena of solids. Since most earlier developments had progressed on an almost entirely empirical basis, attempts were made during these sixteen months to increase the understanding of the processes involved.

Progress was made, but the Laboratory had no clear-cut picture of its future in the field of atomic energy. This still was to be decided by the new commission. The mission of the Manhattan Engineer District had been completed. The Atomic Energy Commission would, from January 1947, direct the course of Los Alamos and the other projects.

To help the commission in this task during its first visit to Los Alamos in November 1946, Dr. Bradbury wrote a brief account of the Laboratory's history, accomplishments, and problems, and he offered his suggestions (Appendix G). The following paragraphs from that account reflect that once again the Los Alamos Laboratory faced a critical period.

"Your Commission now faces the problem of determining the character and future directives of Los Alamos. Unfortunately, the local project is so small that the problems of the community bear upon the character of work done by the Technical Area, and reciprocally, the existence of the Technical Laboratory determines the existence of the community. While these problems can be discussed separately, their simultaneous successful solution is required for the success of either.

"The Los Alamos Laboratory does not presume to indicate to the

353

Commission what the policy of that body should be with respect to the national need for atomic weapon development. Nor should the Laboratory as such express its views on the relationship of such a national program to the international scene. The discussion which follows is based upon the assumption that the United States will require, for an unknown time to come, a program in atomic weapon development and research. Such a program should be directed not only at maintaining an immediate superiority for the United States in this field, but towards maintaining general scientific progress and a concern for basic and long-range developments which will make for strength in the future. It is also assumed that the government of the United States must know what weapons might be arrayed against it for the proper formulation of its own national and international policies. The ensuing discussion is based in addition upon an assumption, which the Laboratory can only suggest, that the Commission shares with the established armed forces of the United States a responsibility for the security and defense of the country; that the atomic weapon plays a fundamental role in any security program set up at this time; and that, therefore, the Commission and the Army and Navy are jointly concerned with this problem.

"It has been noted that, up to the present time, the Los Alamos Laboratory has been responsible for the atomic weapon in its entirety. The atomic bomb has been employed by the armed forces exactly as received from Los Alamos and assembled with only Los Alamos personnel. There has remained, ever since the close of the war, concern as to the engineering reliability of the weapon as well as a conviction that engineering improvements were not only possible but desirable. The skepticism of the armed forces with respect to the ballistic determinations of Los Alamos personnel has already been apparent, and it may be anticipated that this feeling will grow to include the fusing and firing mechanisms and the complexity of weapon assembly. It is further noted that a demand is already apparent for weapons of somewhat different engineering properties—e.g., a weapon which will penetrate the surface of water and detonate at a predetermined depth. Other requests from the armed forces including the guided missile investigators may be expected to appear shortly.

"It is the belief of the senior technical personnel at Los Alamos that this Laboratory should not attempt to carry out these purely ordnance engineering aspects of atomic weapon development. Conversely, it is strongly suggested that these problems should be handled using the

Sandia Laboratory, the existing ordnance facilities of the Army and Navy, as well as additional laboratories that may have to be set up.

"It is suggested to your Commission that the Los Alamos Laboratory may be most effective if its concern is limited to the nuclear components of atomic weapons including, naturally, the technique of supercritical assembly of active material. The Laboratory would then be expected to carry out research on both long-range and short-range modifications in the nuclear structure of atomic weapons, but would not be expected to present to ordnance engineering laboratories more than a functional design for a weapon with the exception of those parts intimately concerned with the nuclear reaction.

"Such a division of responsibility will clearly call for the most active liaison between this Laboratory and such other laboratories as are carrying out the engineering development. While such liaison will present problems, they are not believed to be insurmountable. To maintain the present philosophy and localize Los Alamos responsibility for complete weapon development will not only result in a practical strangulation of effort devoted to long-range research, but will curtail the responsibility of the armed forces in a problem in which they are presumably able and anxious to participate.

"It is further suggested that Los Alamos retain the responsibility for testing the nuclear reactions for new atomic weapons, but that such tests as have a purely military significance be carried out by the armed forces. The distinction which is intended is that of separating a test of the "Alamogordo" type from a test of the "Crossroads" type. In view of the limited facilities of this Laboratory, however, the most active assistance of the armed forces would be required in subsequent "Almogordo" - type tests, but the directive responsibility would come from this Laboratory.

"Whether or not Los Alamos should be continued over a long period of time is doubtless a problem which will be considered by the Commission. This question has naturally received consideration here, and having received a tentative affirmative answer, has resulted in extensive programs of permanent construction. Many, but not all, of the activities proposed for this Laboratory should not be conducted near populated areas. The isolation of the site represents certain community problems which is largely if not entirely balanced for personnel now here by the attractions of the climate and of the present mountainous location. The isolation of the technical community is more easily handled by a policy of encouraging attendance at national and regional scientific meetings, both of regular scientific societies and within the Manhattan District. The absence of railroad connections has contributed to a somewhat higher cost of transportation of materials to the project. Not a negligible factor involved in a proposed change of location is the fact that a large number of technical personnel have remained with the project because they and their families enjoy this location more than urban communities. It is hoped, should a new location be considered, that its advantages will be conspicuous.

"Should the international situation develop to the point at which the United States may cease to have any concern for further weapon development or production, the Los Alamos Laboratory program would require careful reconsideration. Since, presumably, this is not a point at issue at the present time, it need not be considered here except to state that the operations involving plutonium, the basic chemistry and physics, the fast reactor, the large Van de Graaff accelerator, studies of materials at high temperatures, pressures, and radiation densities are all activities which will undoubtedly play a role in the peaceful applications of atomic energy no less important than the role which they play in a program whose objective is weapon research."

APPENDIX A

Bradbury's Philosophy

Notes on talk given by Commander N. E. Bradbury at Coordinating Council meeting, October 1, 1945.

I. What should be the philosophy under which we operate the project during this interim period?
No one can doubt that Government-supported research in atomic energy problems will continue.

This project will be taken over by a commission created by legislation.

This at once suggests our first difficulties. The first hurdle is legislation: this may be such as to make it impossible for individuals of the high qualifications required to work under any commission.

Given good legislation, the commission itself may be poorly selected—under these circumstances, again, the proper people will not work on an atomic energy project.

The legislation and the commission may come too late—the longer this is delayed, with its corresponding uncertainty, the fewer good people will remain.

Particularly, security regulations may be set up so as to make it impossible for people to work.

Consulting may be done in certain engineering matters, and the consultant's mind compartmentalized. This is very much more difficult in the fundamental fields. Many people feel that they would prefer not to know secret things if this requires going out and not being able to make use of them in a University.

The direction in which a University or an industrial firm will go in the next few years is predictable. The direction in which this business will go is not.

All of the above things make it necessary to be explicit about the philosophy which one wants for the project.

Such a philosophy has three parts.

1. We should set up a project to study the use of nuclear energy on an operating basis which is, as nearly as possible, operating in what *we* consider to be an ideal way, in which the emphasis is as *we* consider it should be, together with even the derivatives of this emphasis. In other words, we should aim to turn over to the commission the best possible project that we know how to make. The commission may have other ideas—our ideas may not be their ideas. But in any event, we will have to set up a project which to us seemed a good project for a peacetime, interim, immediately postwar period. If we do not do this, we cannot complain that the project of the future was set up wrongly.

357

2. The project cannot neglect the stockpiling or the development of atomic *weapons* in this interim period. Strongly as we suspect that these weapons will never be used, much as we dislike the implications contained in this procedure, we have an obligation to the nation never to permit it to be in the position of saying it has something which it has not got. The world now knows we have a weapon. How many or how good, it does not know. To weaken the nation's bargaining power in the next few months during the administration's attempt to bring about international cooperation would be suicidal. One hopes that weapon emphasis will decrease with time. We are not a warring nation—the mere possession of weapons does not bring about war. Will the administration attempt to bring about international cooperation in these matters? Who knows—if it is not, we are doomed anyway, but our doom may be delayed a few months or years by having bigger and better weapons. I think we must be hard in these matters. To bring peace by threatening war is possible; to bring peace by requesting and promising cooperation seems more dignified. But the request and the promise, and surely the threat, are both fortified by weaponeering *now*; and the results of weaponeering may be that it may never have to be done again.

3. The project will decrease in size as it goes from a wartime basis to a peacetime basis.

These are, therefore, the three things on which I believe that the project's modus operandi for the next six months must be built.

We will set up the most nearly ideal project we can.

We will not discontinue weapon research until it is clearly indicated that this can be done.

We will decrease the project in size so that it can be accommodated on the mesa on a civilian basis.

II. How does one go about setting up an ideal project to study the problem of the use of nuclear energy?

These problems of the atomic nucleus are extremely difficult; the best men are required to solve them; how does one get the best men?

A good man will not work unless there is intellectual stimulation in the work which he is doing. Therefore, we will set up in all divisions programs of fundamental research which are related, but may only be distantly related, to the problems of nuclear energy and the manner in which it may be released. In this respect, we will follow the policies of good industrial laboratories in which a man may set up his own field of research, but does not have to show either a profit or even a close connection with the business of his employer.

I shall accordingly request division leaders to present programs of research which are intellectually of interest and upon which good personnel may be persuaded to work. The extent to which these programs will be set up will depend upon the scale of the laboratory and to this question I will return. It must be noted that our borrowed tools of research must be replaced. This means the cyclotron, Van de Graaff, etc., ...

An immediate revision of our salary procedures is essential. Heretofore, personnel have been hired on a no loss-no gain, patriotic duty basis. This is all very well in war time. It is not applicable now. The project is just another employer and it must compete for its personnel with other employers who can offer quarters close to civilization or in it, a predictable future, work which may be published, and better, or at least different, living conditions.

How do we meet this competition?

First, we can state our moral assurance that atomic energy work will continue in some form. People now associated with it will presumably be the key personnel of future developments. (BUT—it may be in a form that is repulsive to all of us. This is the chance we take and must pay our personnel to gamble on.)

We have a salary scale which has never been fully applied, but which appears to me to have the possibility of meeting our immediate requirements if it is applied. In other words, I propose to adjust salaries of personnel who may be pursuaded to remain with the project in accordance with their responsibilities, and positions with respect to the project.

This I propose to do in advance of a threat from them to leave to take other jobs. However, on an emergency basis it may be necessary to meet offers from other institutions or industry on a competitive basis.

359

It will take time—at least a month or six weeks—to clean our financial house. Where necessary, we may offer contracts extending to 1 July 1946, although the University of California may not be in business that long. Nevertheless, the contract requires that the agency taking over assume the unexpired obligations of the University, and the General has told me that he will guarantee such a procedure.

I wish to digress a moment at this point while discussing salaries. We are in urgent need of a Personnel Director. Mr. Clausen has indicated to me that he has strong personal reasons for leaving in about a month. The problems of hiring personnel, terminating the employment of individuals no longer essential to the project and uninterested in taking jobs essential to the interim program of the project, as well as the placement services of the project—all these combine to place an extremely heavy burden upon the man accepting this responsibility. It should also be mentioned that the problem of hiring SED's on a civilian basis will shortly become urgent. Such men form one of our most obvious labor markets and as such, have the unusual advantage that we have had definite information about each man before we hire him.

A definite procedure is now being set up whereby all project personnel desirous of obtaining jobs on the outside are brought in touch with employers. No effort will be made to discourage this. In fact, the opposite will be the case. If the project cannot meet the offer made by an outside concern for any one of many reasons, then this man will be permitted to go. All the project will request is that he and we understand how his work is to be taken over if it is to continue, and when he may leave.

The project may make a counter offer—it will only make one; ultimately, it will endeavor to make the offer before it is a *counter* offer. The project's offer should not be used as a lever in forcing up offers from the outside, and we will so request personnel interviewers who are informing us by copy of all negotiations with our personnel.

I have dwelt at some length upon salary procedures, but this is not because I believe that money can be made to answer all

360

arguments. In many instances, the project would be unable to offer any salary at all that could persuade a man to stay. This is a feeling with which I am personally in the most hearty sympathy.

The argument of duty or patriotism can no longer be used. For myself, I feel that the bear which we have caught by the tail is so formidable that there is a strong obligation upon us to find out how to let go or hang on. For everyone to pack up and leave would appear to me to leave the more difficult problems of the future not only unsolved, but with no prospect of solution. This however, I will never use as an argument—if an individual derives some satisfaction from this feeling, very well—but it is not a duty and will not be approached as such.

In one respect, the members of the council have somewhat more responsibility to the nation than do the remainder of the staff. As key personnel, I must urge that if you concur in my belief that we must leave an operating project for the commission, it is then imperative that you consider the tasks of your groups, and that you advise me as to whether they should be continued or

discontinued in the light of the philosophy that I am expounding. If they are to be continued, then you do not leave until you see a reasonably acceptable way for them to continue.
To sum up so far—I have said that our philosophy is to:
leave the best possible project for our success;

continue weaponeering until it is clear that we can taper off; and

decrease the size of the project consistent with the housing facilities on the mesa.

To build the best possible project we must have
Good men—this means reasonable salary scale, reasonable employment practices, and a program of intellectually stimulating research but not directed toward weapons necessarily.

A group of good physicists, chemists, explosives experts, metallurgists, and engineers is not enough. The project must have

361

a sound overall program if it is to be the ideal project for our successors. Accordingly, I come to—

What shall our general project program be, as far as atomic bombs and atomic weapons are concerned?

...We will develop internal modifications, possibly in the method of fusing, almost certainly in the method of detonating. ...We will set up a more careful program of gadget testing so that we will know the degree of reliability of each component. We will set up surveillance tests which at least must have the *possibility* of extending over a considerable time. We will set up Sandia Field as a field test site. It may not last there for more than a year, but we will learn *how* the ideal field test site for weapons should be set up, and it can either stay there, be moved, or become, let us hope, unnecessary in the course of time.

We will initiate the engineering of a new weapon whose aims should be—although, again, we hope it will never need use—increased reliability, ease of assembly, safety, and performance; in short, a better weapon. Much as we dislike them, we cannot stop their construction now. Possibly in six months, possibly in a year—maybe in a few years, weaponeering will stop, but our present lead is our chief weapon in procuring a peace—we must not lose it until that peace and that cooperation is established. In all this, we will invite the cooperation of the established military services, at all levels and wherever they can contribute.

We will purpose subsequent Trinities. The TR bomb was a bomb and not a weapon, if you will permit the distinction. We are entitled to do this from two premises:
The use of nuclear energy may be so catastrophic for the world that we should know every extent of its pathology. How bad *can* this bomb (if it were made a weapon) be? I shall return to this premise again in connection with the Super. One studies cancer—one does not expect or want to contract it—but the whole impact of cancer on the race is such that we must know its unhappy extent. So is it with nuclear energy released in this form. It can be a terrible thing; we cannot hide our head in the sand; we must know how terrible it is.

362

The occasional demonstration of an atomic bomb—not weapon—may have a salutory psychological effect on the world, quite apart from our scientific and technical interest in it. Properly witnessed, properly publicized, further TR's may convince people more than any manifesto that nuclear energy is safe only in the hands of a wholly cooperating world. It also may be pointed out that I believe that further TR's may be a goal which will provide some intellectual stimulus for people working here. Answers can be found; work is not stopped short of completion; and lacking the weapon aspect directly, another TR might even be FUN.

We will propose that the fundamental experiments leading to the answer to the question "Is or is not a Super feasible?" be undertaken. These experiments are of interest in themselves in many cases; but even more, we cannot avoid the responsibility of knowing the facts, no matter how terrifying. The word "feasible" is a weasel word—it covers everything from laboratory experiments up to the possibility of actual building, for only by building something do you actually finally determine *feasibility*. This does not mean we will build a Super. It couldn't happen in our time in any event. But someday, someone must know the answer: Is it feasible? We have now contended that our ideal project will have good men obtained by a good fundamental program and good employment practices; and that it will have a weapon program; that it will have a TR bomb program; and that it will have a TR bomb program; and that it will have a feasibility of Super research program. Now I claim that it must also begin to worry about a program of research leading to the *peace time application of nuclear energy*. I am well aware that this has been worried out and carefully considered elsewhere. We must also do it here. For this program alone will receive the united support of all people everywhere. For the present I do not see how to fit such an effort into divisional, group, or sectional lines. Specific suggestions are needed as to how to go about this within the general frame of our present organization. At this moment I am too uninformed about the situation to do more than generalize.

III. I now propose to discuss the question of how the project will decrease in size during the next three to six months.

Some people are leaving now; others will continue to drift away; this will go faster and faster as long as policies are either unformed or unimplemented.

The post will probably continue as a military organization and probably with adequate personnel for at least as long as it takes to set up the commission.

The two-year service rule will begin to take our SED's in large numbers about Christmas.

Meanwhile, we will lose them more slowly by point discharge.

SED replacements will probably be more or less available to some extent. They will, however, be untrained and generally less useful.

I therefore conclude that the project will be—insofar as the technical area is concerned—on a largely civilian basis by next March. There will be SED's but they will be relatively fewer; they will live in barracks, and they will not present a housing problem.

These civilians will come from three sources: People now here; new people hired; and SED's hired as civilians after discharge. All of these poeple who are married will sooner or later demand quarters as the price of staying. It will accordingly be necessary to revise our housing policy in the following way:

> All people essential to operating a project must be housed in a way that will keep them here.

> This means, in addition to obvious personnel, that machinists, truck drivers, lower grade technicians will get quarters—unless key personnel wish to say that they can get along without them!

> We have approximately 488 family quarters. Of families living in such quarters, possibly 25% of the wives may work. Thus, we can house in family quarters about 600 technical and post civilian employees. *Possibly* we may have a one-to-one ratio of unmarried to married personnel. This means another 500 or so civilians in

dormitories. There will thus be a total of about 1100 technical and post civilian employees. We now have about 3000 in technical activities.

I therefore suggest that in about three to six months we must be prepared to adjust our scale of technical activity to about 1/3 of its present magnitude. I have not included SED's in this figure, for I believe that the necessity of giving many more lower employee classifications housing will balance the extra assistance we will get from SED's living in barracks. Accordingly, I will have to ask that all estimates of future activity be based on about this 2/3 decrease in rate of working. Each activity will be asked how many it needs in all classifications to go on working at this rate—then these men must be housed. The day is rapidly going when the good machinists will live in dormitories away from their families. A similar statement may be made for S site—true, they can hire people, but only with adequate living conditions.

It is curious that the activity of the mesa should be dictated by its housing, but I see no other alternative. I am sure the General will build no more quarters, as this would further commit the mesa to permanency. This I doubt if he will do.

What sort of personnel policies shall we have to bring all this about?

Fair treatment of personnel leaving. This has been widely stated and agreed to. Hiring policies, 30 day, 90 day, and contractual termination policies, as well as dependents and household effects to be carried on as in the past.

The matter of a 40-hour work week. This is under discussion at the moment—when do we get to it, I don't know. Many questions of policy are involved—all these are coupled with our failure to jump before or when the Civil Service jumped. Time is necessary to do these things. In order to make quarters available for long-termers, short-termers may be requested to terminate their connection with the project if they are unavailable for project jobs which now need doing. In all cases this will be in accordance with their employment agreement. However, the project will be cut in

365

size, and quarters must be available for people coming in for the longer term—by long term, I mean till 30 June 1946.

How about personnel who lack degrees?

Urge them to leave to get them, but stagger their leaving and make some plans to get the good ones back.

IV. Now for some rather specific questions which do not easily fit into project policies:

The Handbook must be prepared. However, I doubt if it is desirable for a person to write eight hours per day. In other words, having to write up work should not be an excuse not to take another job. It may be a part-time job and should be so considered.

The University must continue for at least this quarter, but in free time.

What about the general, organization of the project? In general, I think the divisional organizations will stay about the same. However, R and F might be combined; Z, G, X, and T might stay as they are, and C and M might be split.

Administration and Services are badly needed—probably split into three parts: Personnel and general administration, procurement, and technical services.

What about the project of the future? We cannot say where it will be located. Economic considerations seem to indicate that to locate the project here on a permanent basis might tie it down to expensive maintenance (living, salaries) forever. However, this is not our question. Certainly, it is difficult to see how it can start to move inside of eight or nine months and at least six or nine months for the moving which would take place gradually. DP Site will probably stay here until it is too contaminated to use. It could not be moved after it was started in operation. Thus the project will be largely here for at least another year.

I would like to see the project set up as the best type of industrial laboratory with much more emphasis possibly in fundamental things and with academic exchange thrown in for good measure.

What about SED's? I have indicated that I expect to lose them in large numbers about the first of the year. Otherwise, we can make no special effort to have them considered as different from any other SED's in the country, and particularly we cannot attempt to get them treatment which differs from that prescribed by Army regulations and discharge procedure. We all know what we would like to see done; and we will see that our SED's get the best possible consideration under the law; we will not attempt to have them treated above the law—we hope it will be a good one which will get them back to school as soon as possible. We cannot put them on ERC for practical reasons—we can't house them, and we can't let them go if we don't.

What about civilians with deferments? We must not set ourselves above the Selective Service Law. If we can certify that a man is needed to carry out the program outlined above, that he is actually essential to this program, we will continue to obtain deferments. Otherwise, we must release him. This means we must know our program. It is realized that this is unfair to the man—but war is by nature unfair. Some people get killed and some get rich quickly. Some people will experience this unfairness a little late, but no later than the boy who becomes 18 this month or next. Should not war be distributed over as many generations as possible to lessen the burden on any one?

How about security regulations? This is now set by the President. Liberal interpretations are coming as fast as possible. We can't close the box after the secret is out. Attempt to attain consistency. Fundamental problem has to WAIT.

The project of the future I would like to see has, with lifted security regulations, the possibility of exchange activities with academic institutions. People to come to the project for a year, and project personnel to go to academic institutions for a year. Maybe on a similar arrangement on a part-time basis. Certainly on a consulting

367

basis. All of this involves some lifting of security in fundamental fields plus even more fundamental problems of organization.

APPENDIX B

Groves-Bradbury Letter

WAR DEPARTMENT
P.O. Box 2610
WASHINGTON, D. C.

4 January 1946

Dr. N. E. Bradbury
P.O. Box 1663
Santa Fe, New Mexico

Dear Dr. Bradbury:

It was my belief that the making of long-range plans with respect to the future of atomic energy should be delayed until after the passage of legislation so as to avoid serious commitments which might hinder the actions of whatever commission or other body should be established to take charge of the work. Unfortunately, no legislation has been passed, and certain forces are at work the effect of which has been to delay any legislative program.

It has therefore become necessary for me to make definite plans, despite the fact that this will commit to some extent at least any future control body. Our wartime effort was to end the war. Everything was sacrificed to that objective. We counted on suitable legislation being passed promptly at the end of the war. We should not count on atomic bomb development being stopped in the forseeable future.

The Los Alamos site must remain active for a considered period. Taking into consideration the type of work which must be done here, there has been found no site that combines as many desirable facilities for our work as Los Alamos. If one should be found, it would require at least six months to plan, twelve months to build, and six months to complete the move from Los Alamos. The only conclusion, therefore, is to stay at Los Alamos for at least the next few years, and to improve the existing facilities to such a degree as is necessary.

The major factors requiring improvement are the utilities, housing, and community facilities, particularly recreational facilities for single persons. This transition from war to peacetime community conditions will start immediately. To do this intelligently, however, requires planning, and this planning has already started.

With the current interest in the water situation, I wish to state my exact expectations with respect to this. First, all possible steps will be taken to maintain the existing system at maximum efficiency. This will include the trucking of sufficient additional water for as long as is necessary to supply continuous water service to all housing and to operate S Site and DP Site. Second, careful studies will be made with a view to securing a year-round supply based on 100 gallons per person per day, which is considered adequate for a community with our industrial needs. Third, construction will be initiated promptly as soon as the plans have proceeded to the point where initiation of construction is feasible.

With respect to power and highway communications, it appears that the expected loads can be properly accommodated. If not, necessary steps will be taken to improve these facilities.

With respect to housing, we are assuming that DP Site will operate on a relatively permanent basis, and studies have been initiated with respect to layout and design of the needed family housing. It must be realized that there are certain legal restrictions that set a maximum cost of $7500 per unit. This means that the most careful designs must be made in order that satisfactory permanent accommodations will be achieved.

With respect to community facilities, in addition to the recreation for single individuals already mentioned, there should be a wide range of consumer goods establishments and the stimulation of concessionnaires in this line is necessary.

Sincerely yours,

L. R. Groves
Major General, USA

LRG/b

Catalogue of Los Alamos University Courses and Student Enrollment
September 17, 1945

TO: ALL TECHNICAL PERSONNEL
SUBJECT: CATALOGUE OF COURSES

REGISTRATION

PLEASE NOTE CHANGED HOURS AND LOCATIONS. Registration will be held from Tuesday, September 18 to Friday, September 21 inclusive. The hours will be from 8:30 to 11:30 a.m. and from 2:30 to 5:30 p.m. in Room E-210. There will also be registration facilities in the

High School from 7:00 to 9:00 p.m. in order to make registration possible for persons not employed in the Tech Area.

COURSES

UNDERGRADUATE — JUNIOR-SENIOR LEVEL

Chemistry

11. **Elementary Organic Chemistry.**
Lecturer: M. F. Roy
Hours: Section I. Tues. & Fri. 10:30-11:45 a.m. in Sigma 47
Section II. Mon. 7:15-8:30 p.m. in gamma 49, Thurs. 8:45-10:00 p.m. in gamma 49
Prerequisite: Elementary Chemistry
Textbook: None
Description of course: Study of the major general classes of organic compounds, their properties, reactions, and uses.

12. **Elementary Physical Chemistry.**
Lecturer: I. B. Johns
Hours: Wed. & Fri. 4:15-5:30 p.m. in Gamma 49
Prerequisites: Elementary Chemistry and Calculus; Elementary Physics desirable
Textbook: "Outlines of Theoretical Chemistry," Getman-Daniels (Required. Price $3.50)
Description of course: This course will give the student a working knowledge of the fundamental principles of physical chemistry, including the study of gases, liquids, and solids, the principles of thermodynamics, the theory of solutions, thermochemistry and its applications, the treatment of equilibria—both homogeneous and heterogeneous, chemical kinetics, and electrolytic theory.

13. **Advanced Physical Chemistry.**
Unfortunately, it has been impossible to secure a lecturer for this course. It will, therefore, not be given.

Metallurgy

21. **Physical Metallurgy.**
Lecturer: George L. Kehl
Hours: 10:30-11:45 a.m. Mon. & Wed. in Sigma 47

371

Prerequisites: Elemtary Chemistry and one semester of Elementary Physics.

Textbooks: (Recommended)

"The Alloying Elements in Steel," E. C. Bain

"Engineering Physical Metallurgy," R. H. Heyer

"Principles of Physical Metallurgy," F. L. Coonan

"Principles of Physical Metallurgy," G. E. Doan & B. Mahla

"The Science of Metals," Z. Jeffries & R. S. Archer

"Principles of Metallography," R. O. Homerberg & R. S. Williams

"Structure and Properties of Alloys," R. M. Brick & A. Phillips

Description of course: State of aggregation; origin of metallic structures; crystal structure; equilibrium diagrams of metallic systems and their interpretation; nonequilibrium conditions in metallic systems; plastic deformation and annealing; nonferrous metals and alloys; iron and steel; basic concepts of the heat treatment of steel

Physics

31. **Electricity and Magnetism**

Lecturer: R. Brode

Hours: Mon. & Wed. 10:30-11:45 a.m. in Gamma 49

Prerequisites: Sophomore Physics and Calculus

Textbook: Probably Page & Adams

Description of course: Detailed discussion of the properties of electrostatic and magnetostatic fields. Electric currents and their magnetic fields, alternating currents, inductance and capacitance, oscillating circuits, and electric waves.

32. **Modern Physics.**

Lecturers: B. Rossi and L. Parratt

Hours: Section I: Wed. & Fri. 9:00-10:15 a.m. Rm. B-223

Section II: Mon. 8:45-10:00 p.m. Rm. B-223

Thurs. 7:15-8:30 p.m. Rm. B-223

Prerequisites: Freshman and Sophomore Physics, and Calculus.

A course in Electricity & Magnetism is desirable.

Textbook: "Introduction to Modern Physics," Richmeyer and Kennard

Description of course: The experimental and theoretical development which leads to the present concept of the constitution of
matter. Beginning with the discovery of the electron, the course
will discuss various methods of determining Avogadro's number,
the structure of atoms, the atomic nucleus, and cosmic radiation.

33. Electronics

Lecturers: D. K. Froman and Elmore

Hours: Section I: Tues. & Fri. 10:30-11:45 a.m. Gamma 49
Section II: Mon. 7:15-8:30 p.m. Rm. B-223
Thurs. 8:45-10:00 p.m. Rm. B-223

Prerequisites: Differential & Integral Calculus, General College
Physics. A course in Electricity and Magnetism
(Physics) *or* a course in Alternating Currents (Engi-
neering)

Textbook: "Theory and Applications of Electron Tubes" (Recommended)

Description of course: *Electric Cirucits:* fundamental laws and their
application to complex circuits for D.C., sinusoidal A.C., and
transient currents. *Electron Tubes:* parts and their functions;
static and dynamic characteristics and their measurement; special
tubes; some basic circuits. *Electronic Circuits Design:* detailed
parts specification; applications to simple but complete electronics
circuits. *Basic Electronic Circuit Elements and Complete
Circuits:* emphasis on circuits for industrial control and scientific
measurements rather than on radio, television, and radar.

34. Micro-Waves

Unfortunately the offering of this course at the present time seems to
present insuperable difficulties generally connected with the con-
fidential character of some of the information. It is hoped to give this
course in the following semester if the courses are continued at that
time.

Mathematics

41. Differential Calculus

Lecturer: P. Whitman

Hours: Wed. & Fri. 8:30-10:15 a.m. Gamma 49

Prerequisites: Analytic Geometry; Trigonometry

Textbook: "Elements of the Differential & Integral Calculus"
Granville, Smith & Longley (Required)

373

Description of course: Differentiation of algebraic and transcendental functions; applications to slopes, maxima and minima rates, etc.; higher derivatives; differentials and applications to small errors, etc; integration of standard elementary forms and applications to simple areas, only if four-hour course.

42. Differential Equations

Lecturers: J. W. Calkin & D. A. Flanders

Hours: Section I: Tues. & Thurs. 9:00-10:15 a.m. Gamma 49
   Section II: Wed. & Fri. 7:45-9:00 p.m. Gamma 49

Prerequisites: One year of Calculus

Textbook: "Differential Equations," H. T. H. Piaggio (Required)

Description of course: Ordinary differential equations of the first order, linear equations, miscellaneous special equations, existence theorems, numerical methods of solution, solution in series, selected topics in partial differential equations.

GRADUATE

Chemistry

61. Thermodynamics

Lecturers: G. S. Kistiakowsky and E. R. Van Artsdalen

Hours: Section I: Wed. & Fri. 9:00-10:15 a.m. Sigma 47
   Section II: Mon 8:45-10:00 p.m. Gamma 49
       Thurs. 7:15-8:30 p.m. Gamma 49

Prerequisites: one year of Calculus; one year of college Physics and the elementary chemistry up to and including one year in Physical Chemistry.

Textbook: "Thermodynamics," Steiner (Recommended)

Description of course: This is a course in chemical thermodynamics and because of time limitations it will not deal with topics of largely engineering interest (heat flow, heat engines, etc.) or of interest exclusively to physicists (such as the theory of thermoelectricity, etc.). The meaning of the three "laws of thermodynamics" will be discussed and they will be applied to the calculations of homogeneous and heterogeneous chemical equilibria, vapor-solid equilibria, ideal and nonideal solutions, surface tension, etc. The interpretation of the "Third Law" in statistical terms will be briefly

374

discussed and the problem of the calculation of absolute entropies and of free energies of substances gone into in detail. The approach to all these problems will be of the type used by Gibbs (rather than the elementary approach as used by Lewis and Randall in their book, for instance) and hence, those taking the course are expected to be familiar with differential and integral calculus, including partial differentiation. Otherwise, the prerequisite is a course in elementary physical chemistry.

62. Radiochemistry
   Lecturer: J. W. Kennedy
   Hours: Wed. & Fri. 4:15-5:30 p.m. Rm. B-223
   Prerequisite: B.S. Degree in Chemistry or equivalent; or by special
      arrangement.
   Textbook: None. Some reference books recommended, including
      Davidson.
   Description of course: Natural radioelements, radioactive decay, nuclear transmutation, accelerating devices, radiations from radioactive substances, detection techniques, the study of radioisotopes, new elements, chemical behavior at very low concentrations, carriers, tracers, exchange reactions, biochemical studies and other applications.

63. Theoretical Organic Chemistry
   Lecturer: Mr. Lipkin
   Hours: Mon. 9:00-10:15 a.m. Rm. B-223
      Thurs. 10:30-11:45 a.m. Rm. B-223
   Prerequisite: One year Elementary Organic Chemistry and one year
      Elementary Physical Chemistry
   Textbook: "Theory of Organic Chemistry," Branch & Calvin
      (Recommended)
   Description of course: Electronic structure of organic compounds; effect of resonance on the properties of organic systems; the relationship between physical properties and the structure of organic compounds.

Physics

71. Theoretical Mechanics
   Lecturer: Mr. Keller
   Hours: Mon. 9:00-10:15 a.m. Sigma 47
      Thurs. 10:30-11:45 a.m. Sigma 47

Prerequisites: A.B. Degree in Physics, or equivalent amount of undergraduate Physics; Differential Equations.

Textbook: "Whitaker Analytical Dynamics"

Description of course: A course in the dynamics of particles, rigid bodies, elastic media, and fluids. Topics to be taken up will include vector analysis; particle dynamics; Lagrange's equations; Hamilton's equations; rigid body dynamics; vibrating systems; coupled systems and normal coordinates; dissipative systems; elastic media and hydrodynamics.

72. **ElectroMagnetic Theory**
Lecturer: H. A. Bethe
Hours: Tues. & Fri. 10:30-11:45 a.m. Rm. B-223
Prerequisite: Calculus, Differential Equations, and an undergraduate course in Electricity and Magnetism, or its equivalent.

Textbook: Abraham & Becker (Recommended)

Description of course: The course will start by setting down and explaining Maxwell's equations. Various phenomena will be derived from these equations; a relatively short time will be devoted to electro-statistics, an extensive treatment will be given of stationary currents and their magnetic fields and of high-frequency electromagnetic waves. Electromagnetic cavity resonators and wave guides will be discussed. Relativity electrodynamics will conclude the course.

73. **Statistical Mechanics**
Lecturer: L. I. Schiff
Hours: Mon. 9:00-10:15 a.m. Gamma 49
     Thurs. 10:30-11:45 a.m. Gamma 49
Prerequisite: Theoretical Mechanics and Modern Physics; Quantum Mechanics desirable.

Textbook: "Statistical Mechanics," - Tolman (Recommended)

Description of course: First part, General Theory (8 to 10 weeks). Introduction; classical statistical mechanics; detailed balance and the H-theorem; quantum statistical mechanics. Second part, Application (6 to 8 weeks). (It will probably be possible to discuss briefly three or four of the topics listed below; these will be selected in consultation between students and instructor.) Free electron theory of metals; specific heats; electromagnetic radiation; fluctuations; imperfect gases; atomic nuclei; cooperative phenomena; equilibria in gases; reaction rates in gases.

74. **Elementary Quantum Mechanics**

Lecturer: E. Teller

Hours: Mon. & Wed. 10:30-11:45 a.m. Rm. B-223

Prerequisite: Theoretical Mechanics; Electromagnetic Theory; Differential Equations

Textbook: None for the time being.

Description of course: A systematic description of the laws of quantum mechanics and their relation to classical physics. Specific topics to be discussed: correspondence principle, wave-particle dualism, uncertainty principle, Schrodinger—and matrix-formation of quantum mechanics, and electron spin.

75. **Nuclear Physics**

Lecturers: Manley and Weisskopf

Hours: Tues. & Thurs. 9:00-10:15 a.m. Rm. B-223(Section I)
Wed. & Fri. 7:45-9:00 p.m. Rm. B-223(Section II)

Prerequisites: One semester Quantum Mechanics; Modern Physics (Atomic Spectra, Structure Elementary Particles.)

Textbook: None

Description of course: 1. Elementary particles and properties; 2a. Systematics of Nuclear Structure; nuclear reactions; alpha decay; fission. 2b. Observational methods. 3. Deuteron system, p-n scattering. 4. Theory of beta and gamma decay. 5. Theory of nuclear reactions.

76. **Neutron Physics**

Lecturer: E. Fermi

Hours: Tues. & Thurs. 9:00-10:15 a.m. Sigma 47

Prerequisites: Differential equations; introduction to theoretical physics; a knowledge of the elements of Nuclear Physics; Introduction to Quantum Mechanics desirable.

Textbook: None

Description of course: Neutron sources (radioactive sources, accelerating machines, piles) (1). Neutron reactions (capture, scattering, etc.) (2). Neutron detection (fast detectors - radioactive detection, counters, fission counters, etc.) (3). Slow neutrons

377

(include diffusion theory, velocity selector) (6). Fission by neutrons, Chain reaction (2). Slow neutron piles (10). Fast neutron chain reactions (6).

77. Hydrodynamics
Lecturer: R. E. Peierls
Hours: Wed. & Fri. 4:15-5:30 p.m. Sigma 47
Prerequisites: Theoretical Mechanics; Differential Equations.
Textbook: LA-165
Description of course: Kinematics of continuous medium. Lagrange and Euler variables, equation of continuity; hydrostatics, stresses, definition of ideal fluid; Euler's equation; Bernoulli's theorem. Conservation laws. Vorticity; Thomson's theorem; Irrotational flow. Potential theory; method of images; complex variable; flow around sphere and cylinder; mapping; flow around a corner; airfoil theory; application to free surfaces; vortices. Viscosity; equation for viscous flow; Poiseuille formula, flow between plates; Stokes law; turbulence; laws of similarity; Reynolds number; examples of critical Reynolds numbers; resistance coefficient; boundary layers; heat transfer; theories of turbulence. Compressible fluids; sound waves; Sound waves in medium of varying properties. Supersonic flow; Mach angle; characteristics; short waves; Hugoniot conditions; Rayleigh-Taylor theory. Interaction of short waves.

STUDENT ENROLLMENT

October 1945 - June 1946

Course Name	Students for Credit*	Auditors*	Students Receiving Credit
Theoretical Mechanics	14	16	7
Electromagnetic Theory	24	36	5
Theoretical Organic Chemistry	14	9	10
Electricity and Magnetism	19	21	5
Modern Physics	32	36	13
Statistical Mechanics	5	12	5
Differential Equations	28	11	22
Elementary Physical Chemistry	6	7	5
Thermodynamics	26	28	6
Radiochemistry	18	27	13
Differential Calculus	13	15	5
Physical Metallurgy	5	12	4
Electronics	13	22	5
Quantum Mechanics	18	31	11
Neutron Physics	15	21	4
Hydrodynamics	10	28	3
Electrodynamics	18	35	10
Nuclear Physics	1	32	1
	279	399	134

*Approximate figure

379

APPENDIX D

University Affiliations Conference

CONFERENCE REPRESENTATIVES FROM LOS ALAMOS

Representative	Institution
Bonner, T. W.	Rice Institute
Brewster, Ray Q.	University of Kansas
Buchta, J. W.	University of Minnesota
Colby, M. Y.	University of Texas
Dempster, R. R.	Oregon State University
Dodson, Richard	California Institute of Technology
Gingrich, N. S.	University of Missouri
Glockler, George	University of Iowa
Gustavson, R. G.	University of Nebraska
Hughes, A. L.	Washington University of St. Louis, Missouri
Jacobs, James A.	University of Iowa
Kirkpatrick, Paul	Stanford University
Larsen, H. D.	University of New Mexico
Marvin, H. H.	University of Nebraska
Nielsen, Jens Rud	University of Oklahoma
Pietenpol, W. B.	University of Colorado
Regener, Victor	University of New Mexico
Smith, Sherman	University of New Mexico
Smythe, W. R.	California Institute of Technology
Stewart, M. A.	University of California
Suttle, John F.	University of New Mexico
Van Atta, C. M.	University of Southern California
Weniger, Willibald	Oregon State University
Worcester, P. G.	University of Colorado

Dr. N. E. Bradbury Director
Colonel L. E. Seeman Associate Director

Clark, J. C.	King, L. D. P.
Conard, D. B., Major	Manley, J. H.
Froman, D. K.	McKibben, J. L.
Graves, Alvin	Metz, Charles
Hall, David	Morrison, Philip
Hill, E. L.	Nicodemus, David
Hoyt, Frank	Reines, Frederick
Jette, E. R.	Richtmeyer, R. D.
Jorgenson, Theodore	Spence, R. W.
Kellogg, J. M. B.	Taschek, Richard
Kelly, Armand	Whipple, Dr. Harry

July 19, 1946

MORNING SCHEDULE

7:00—8:00 Fuller Lodge - Breakfast
8:30 Conference Room (B-223)
WELCOME ADDRESS - Dr. N. E. Bradbury
Los Alamos Facilities:
General Description Physics Division - Dr. John Manley
General Description Chemistry Division - Dr. E. R. Jette
10:00—12:00 Tour of Laboratory
12:45—1:30 Fuller Lodge - Lunch

AFTERNOON SCHEDULE

1:45—2:45 Conference Room (B-223)
UNIVERSITY COOPERATION - Dr. N. E. Bradbury
What We Propose - I: Dr. Frederic Reines
What We Propose - II: Dr. R. W. Spence
3:00—5:00 General Discussion, guided by Dr. N. E. Bradbury

381

8:00   Fuller Lodge - Dinner
        AFTER DINNER DISCUSSION
        (Dr. N. E. Bradbury presiding)
        The Water Boiler: Dr. L. D. P. King
        Isotopes and the Water Boiler: Dr. R. W. Spence
        Fast Reactor: Dr. Philip Morrison

OPENING REMARKS

N. E. Bradbury

As you are probably aware, the Laboratory has a very definite academic tradition, in spite of the fact that we are only about three years old. The entire staff of the Laboratory has been drawn almost without exception from the staffs of academic institutions and from their graduate students. This was true when the Laboratory was first set up and continues to be true. For many of the personnel here, the absence of academic contacts during the war years was a source of regret. The present possibility of establishing such contact with the universities of this region is, therefore, a particularly pleasant prospect for us. I hope that today, in the course of this conference, we can work out the problems and techniques whereby this cooperation may become a reality.

I would like to start the meeting this morning with a brief résumé of the reason Los Alamos exists and what we hope to accomplish both for the Laboratory and the universities by this proposed program. Los Alamos was set up to investigate the possibility of creating an atomic bomb. If theoretical and experimental physics showed this to be possible, we were to design and construct such a weapon. As you know, the weapon was shown to be feasible and it has been constructed. With the termination of the war, the immediate problem of constructing weapons could take a very much lower priority. The Laboratory still has, in the absence of international legislation, a definite weapon aspect. Despite this, the emphasis can be shifted somewhat and we can now turn our attention to problems which are more or less fundamental in establishing how and why nuclear energy can be converted into either weapons or power or used in other ways. While our interest in weapons has had to be maintained we, nevertheless, now believe it proper that we so attempt to broaden our understanding of the fundamental physics and chemistry

382

which is involved. For this reason we hope to enlarge our research program in basic nuclear physics and nuclear chemistry. From a certain standpoint of view, this is weaponeering. It is well known that the success of this laboratory as well as the success of all other technical laboratories during the course of the war was due in a large measure, if not entirely, to the extraordinary backlog of scientific information and scientists in the country. Without this backlog of techniques, personnel and information, it would have been impossible for this country to have carried out the developments so important to victory. We feel quite strongly, therefore, that a laboratory such as this, which at least for the present has a definite place in the scheme of national defense, must take part not only in a short-range program but must also be a part of those activities which have a definite long-range aspect. These long-range aspects to our thinking include the training of future scientists, the training of technical personnel, the broadening of our basis of understanding of physics and chemistry, and the conduct of fundamental research in these fields, not only for the immediate benefit of the Laboratory but also for the benefit of the scientific population of the country at large.

Now the express purposes which we have to foster in the course of this program of university cooperation are as follows. It has been stated, I believe, in the Smythe report, the bible of all security officers, that this Laboratory comprises one of the most excellently equipped physics laboratories in the country. I believe this statement is true or very nearly true. In common with other laboratories at this time, because of the necessary return of a large portion of our staff to their academic organizations, to their teaching, and to their graduate studies, the Laboratory now finds itself in a position in which its facilities are not receiving their widest use. We feel, and I think justifiably, that these facilities which can be devoted to fundamental research should be fully employed in this pursuit. In addition, during the course of the war we have developed here many techniques, both instrumental and theoretical which will ultimately be published and be available to the country at large. However, the process of publication is a slow one, it takes time, to prepare the necessary manuscripts, it takes time to accomplish the mechanics of publication. We believe that these techniques, which can be released at this time, should be as widely disseminated as possible and so get into general usage without delay. In general, scientists do know these things exist but they do not know how far they have progressed, they

383

don't know the developments that have actually taken place in laboratories such as this.

In addition, we are aware of the facts that university laboratories at the present time are crowded, and the staffs of universities are overworked. An enormous number of graduate students have returned to academic universities to complete their graduate work. We believe we can be of assistance in this matter by providing both space and direction for graduate research in physics and chemistry.

Then there is the matter of financing research. I think you are all aware that the present requirements for fairly rapid research in nuclear physics require expenditures of large sums of money for the necessary equipment. Many universities feel that it is impossible, unwise, or inexpedient to invest such sums at the present time. The equipment is available here, the funds are available here, and we think we may, therefore, be of assistance along such lines.

In general, we expect that the lines of research that we will carry out here will be in the broad fields of nuclear physics and chemistry. It is quite apparent to anyone who has read the Smythe report that the neutron is the key to the entire atomic energy program. That means, to take the case of physics, that we are obviously interested in all problems that involve neutrons, encompassing the entire periodic table. Since fission is also involved here, it goes without saying that we are interested in the basic problems of fission. We are interested, therefore, in the elements that appear at the upper end of the periodic table. We hope to be able to obtain here a much more complete understanding of the mechanics of fission than now exists. The fact that the elements at the very low end of the periodic table are known to possess the possibility of nuclear reactions similar to those which take place in the sun, i.e., thermonuclear reactions, indicates we should increase our knowledge of the behavior of these elements just in case there should be something there that we should learn. As far as chemistry is concerned, we are, of course, primarily concerned with the chemistry and radiochemistry of the elements that are of particular interest to this laboratory. These will naturally include the elements at the upper end of the periodic table. There are also metallurgical problems connected with the fabrication of these new elements.

We hope from this program of cooperation that we will, therefore, obtain personnel to work in our laboratory to carry out research problems that we may suggest, that may be suggested to us, or that we

384

feel lie in the province of interest to this Project. We may obtain from the universities assistance in the guidance of these students by having members of their staff here both to conduct their own research and to assist in the direction of these students. In this way, we expect to increase the amount of research done by this laboratory that is of interest to this laboratory and to thereby increase our understanding of the basic physics and chemistry problems pertinent to our problems. We hope the universities will acquire from these arrangements additional facilities for the conduct of research in their graduate development. We hope to establish contacts so that they will know what facilities we have here and what problems they may undertake here. We do not propose to go in the undergraduate student business. We propose to give an occasional course on a completely voluntary basis both by our own staff and such members of cooperating university staffs who come here. They will be courses that will be given along lines of our particular problems here; for example, neutron physics and nuclear physics. These courses would treat of certain theoretical and experimental fields in which we have done a lot of work here and in which we hope to do more. These courses will be more appropriate to the three-year graduate student and will simply add to his fund of information. We are not concerned as to whether they actually receive academic credit.

There will obviously be many problems in setting up such an arrangement: security problems, economic problems, theses problems, problems concerned with giving doctorate examinations. These are problems we hope to be able to discuss this afternoon so as to arrive at some solution. I am willing to guess at this time that our arrangements will have to be quite flexible; different universities will require different arrangements. I see no essential difficulty in this. The economic problems we will propose to solve by making such graduate students regular staff members of this laboratory employed by us. They would be doing a job we want done, and therefore, there is no reason why we should not employ them. This solves several problems and I think it will turn out to be quite attractive to students in question. This afternoon, we will discuss among other things the security and declassification problems. I have obviously left unsaid many things that you would probably like to have heard said about the laboratory. However, the program this morning is quite tightly scheduled and I would like, therefore, to conclude my remarks on the general purpose of this meeting.

DESCRIPTION OF PHYSICS DIVISION ACTIVITIES

John H. Manley

Dr. Bradbury has already told you that Los Alamos has had something to do with the atomic bomb and that, in particular, there has been a great deal of work done here on neutron physics and nuclear physics. In discussing the work of the P Division I believe that since you will be making a tour of the Laboratory later this morning, it will be more profitable to tell you about the facilities and equipment of the Division rather than to tell you about the work from the point of view of the physics problems at hand. I will occasionally, however, mention certain problems so you will be able to get an idea of the nature of the work. The organization of the Division is built up around certain pieces of equipment and this organization is given here, together with the names of the individuals in charge of the groups.

Electronics, Mr. Watts
Water Boiler, Dr. King
Cockcroft-Walton, Z Bldg., Dr. Jorgensen
Cockcroft-Walton, U Bldg., Dr. Bretscher
Fast Reactor, Dr. Morrison
Van de Graaff, 2-1/2 MeV, Dr. Taschek
Van de Graaff, 8 MeV, Dr. McKibben
Betatron, Dr. Ogle

The electronics group is charged with the responsibility of designing and building all sorts of electronic equipment. One of the chief problems in nuclear physics, as you know, is the measurement of transients and pulses. Consequently this organization produces many different types of pulse amplifiers, scaling units, and so on, for this purpose. Also, there are many problems connected with the measurement of short time intervals, even down to the order of fractions of microseconds. I would like to single out two pieces of equipment to indicate the type of advance that has been made during the course of the war: 1) the voltage analyzer, and 2) the time analyzer. The voltage analyzer will sort pulses into 10 channels according to their magnitude; that is, it will sort and record the number of pulses between 2 and 4 volts, 4 and 6 volts, and so on. Therefore, if it is desired to examine the energy distribution of fission fragments in an ionization chamber, 10 points can be taken simultaneously with this equipment, thus decreasing the time of taking data

386

essentially by a factor of 10. Similarly, the time analyzer records the number of pulses occurring in a given set of time intervals; for example, the number between 0 and 0.4 microsecond, the number between 0.4 and 0.8 microsecond, and so on. A practical application of this would be the measurement of the reverberation time of neutrons in a large block of material. A burst of neutrons is introduced into the block, and a counter, together with the time analyzer, will then show directly the exponential decay of the number of neutrons in the block. It is to be noted that the electronics group is not engaged in standard production work, since physicists and chemists are always requesting unique types of circuits. The electronics group is kept on their toes to meet demands of this nature.

The next item is the water boiler, which is a chain-reacting unit with an enriched material. Since Dr. King will tell you more about the actual design of this instrument this evening, I will confine my remarks to the statement that it is a piece of equipment that gives a neutron spectrum from high energy all the way down to thermal energies. The thermal flux is of the order of 10^7. Let me remark here that in the old days, a standard neutron source was formed of an appropriate mixture of radium and beryllium. If one takes a curie of radium and mixes it with beryllium, one gets a source that gives a flux of the order of 10^5 neutrons per second at a distance of one centimeter. One of the interesting things that has been done in connection with the water boiler is to filter the thermal neutrons through additional graphite. The graphite is used to slow down the neutrons and if they are filtered through still more graphite they will be very "cold," actually having a temperature of about 20 kelvins. There are interesting experiments that have been done in this region of what we call super-cooled neutrons.

It is possible with this reactor to get a total flux of the order of 10^{11}. There is a hole completely through the chain-reacting material so that if, for example, very short irradiations are desired at a flux of this order of magnitude, it is possible to shoot with an air gun whatever sample is desired through the reactor and catch it on the other side.

In connection with the activities of this water boiler group, there is at present, under construction a thin-lens beta-ray spectrograph. Not only is it desired to study the beta rays but also to investigate the energies of the gamma rays that are emitted in the straight neutron capture process.

The two Cockcroft-Walton machines listed are useful in looking at reactions at low accelerator voltages and also as neutron sources. The top voltage of the equipment in Z Building is about 200 kilovolts. This accelerator gives an atomic beam of the order of 20 microammeters. It has been used with the D-D reaction as a neutron source and also to investigate the properties of that particular reaction in that energy region. The other two accelerators have top energies of about 125 kilovolts and of about 50 kilovolts. One experiment I might just mention that would be of interest as a graduate thesis problem is to study the range energy relation for hydrogen at very low energies. There seems to be no good data in the literature on this low-energy region.

The next item is another reactor, the so-called fast reactor. Dr. Morrison will describe this particular neutron source in greater detail this evening. I may remark here that this will be the first reactor that makes use of the new element plutonium and that it will give a neutron flux of the order of 10^{13}; in other words, about of a factor of 100 larger than the water boiler. It is called a fast reactor not only because it operates quickly, but because its spectrum has a higher neutron energy than the water boiler spectrum. It will be quite a useful piece of equipment since a factor of 100 in intensity is worth having. Experiments on the properties of materials in high radiation densities will be possible. Bombardment by neutron and gamma rays causes structural changes, as many of you know.

To turn again to accelerating equipment, the Laboratory has one Van de Graaff in operation, which gives 2-1/2 million volts at currents as high as 60 microamperes. This extremely versatile instrument is provided with a precision voltage control good to 1.5 kilovolts. Thus it is possible to get very accurate results. It can be used in the investigation of all types of nuclear reactions within its energy range. It is also useful as a neutron source, particularly with the use of the lithium p-n reaction in which it gives monoergic neutrons from a low limit of 5 kilovolts up to the order of a thousand kilovolts. One has considerable flexibility in obtaining neutrons of precise energy in this range. Of course, using the D-D reaction the range can be extended still higher and other reactions will fill in between the lithium p-n and D-D reactions so that one has essentially a continuously variable neutron source up to the order of 4 or 5 million electron volts.

The second Van de Graaff, which is now under construction, is designed to give 8 million volts according to Dr. McKibben's conservative estimates. It also will have a precision voltage control and will extend the range of operation of the project's Van de Graaff machines.

The project's cyclotron some of you may know as the Harvard cyclotron. It was loaned to this project and arrangements are now under way for its purchase. It will give on the order of 50 microamperes of 10-million-volt deuterons. Its use during the war was primarily as a neutron source since our chief lack of knowledge was in the properties of various materials with respect to their reaction to neutrons. One thing which made it extremely useful was its modulation equipment, which permits bursts of neutrons from the cyclotron to be sorted according to their velocities, thus again providing an essentially monoergic neutron source. As one goes to higher energies the resolution of the equipment falls off, but up to about 10 electron volts the examination of the locations of resonances and other phenomena associated with slow neutron processes is possible.

The next item on this list is the betatron, which has a top energy of 22 million volts and is used primarily for investigating γ-n processes.

There are a few auxiliary things in the Physics Division, which do not appear on this list and which I would like to mention. For example, we have a remarkable collection of natural radioactive sources. The sources in this category are radium-beryllium, polonium-boron, radium-boron, and so on. We sometimes fail to realize with our wealth of other equipment that even 5 or 8 years ago any laboratory with these natural sources would be extremely well equipped. Also, we have a special laboratory for very low counting rates where things can be examined which give pulses of the order of one a month under suitable conditions. There is also a laboratory for the standardization of sources. It is primarily a graphite column and is an extremely valuable piece of equipment for calibrating other sources in terms of a standard or for calibrating detectors.

There is one other activity in which the Laboratory has been engaged but which is not at present the responsibility of any particular group, and that is the investigation of various types of ionization particle detectors. There is still a great deal of work to be done along these lines, in spite of the progress made during the war, and it should be a fruitful field of work for graduate students.

I think Dr. Richtmeyer of the Theoretical Division would like me to say a word about the physical equipment of that Division. In addition to ordinary calculating machines, it is equipped with a complete set of International Business Machines that are used in connection with complicated calculations.

ACTIVITIES OF CMR DIVISION

Eric R. Jette

The work in the Chemistry and Metallurgy Division is organized rather differently from that in the Physics Divisions. It is organized around material rather than around equipment. We have a very well-equipped laboratory here and we have or can get any standard manufactured equipment that exists on the market. In addition, we can make a great deal of equipment of a special nature that we need.

The Division is concerned with the handling of rather large quantities of radioactive or fissionable materials, except in one branch that I will mention later. We are generally not interested in any problems that can be handled only on the microgram or even on the milligram scale. The necessity for handling large quantities of such materials—for example, all the plutonium ever made has passed through Los Alamos—gives rise to many problems. The element polonium is one of the nastiest to handle, and that raises some very special problems concerning the protection of personnel. For that reason, we have had to develop a good bit of laboratory furniture to protect the workers from the contaminated dust. In view of this, practically all our experiments are done in closed systems. You will see some of these when you tour D Building. You will see boxes having glass fronts with arm holes fitted with heavy gloves in which the men reach in to work.

One of the many things we have to study is the chemistry of plutonium. There has been a great deal of work done at Clinton, Berkeley, and Chicago on the chemistry of plutonium. The difference between the work done at those places and here is largely one of scale. While they work with milligram quantities as a maximum, we work well up in the gram scale. This enables us to study many of the chemical reactions in a much more thoroughgoing fashion than is possible at the other laboratories.

So far, our main efforts have been devoted to studying those reactions of interest in the production of plutonium. We have also gone into the study of uranium reactions, but because the uranium is an old and well-known element and working with it does not involve the health hazards that plutonium does, we haven't done much work on its chemistry. There is still very much to be done in the study of the chemical reactions and the preparation of plutonium compounds. The physical chemistry of the

390

reactions is particularly important. The reactions are complicated by the existence of several oxidation states, and the conversion from one state to another is very frequently complicated by rate phenomena. The ordinary kind of data one needs for production purposes are also basic chemistry, such as oxidation reduction potentials and solubilities of other materials and the plutonium salts. These solutions contain large quantities of other materials and have been worked up to the point where we can handle our production problems, but we still don't understand the reactions very well. It is one of the objectives of the Division to get our knowledge to the point where we understand what we are doing.

Another element worked with here is polonium, and there are very many chemical problems involved with that. There again, we handle larger amounts of the material than has been possible in the past. Although gram-wise they are still trivial, energy-wise they are somewhat fearsome. Again, special techniques have had to be developed, and we have gotten to the point where we do what we are supposed to do without knowing too well why those operations work. Polonium chemistry is almost a wide-open field and it is one of the things we hope to go after rather actively.

The third field is one of radiochemistry. Here again, the Laboratory is able to secure radioactive materials—it has access to many isotopes and radioactive material that ordinary laboratories don't get very easily, and we have the use of some of the physics equipment such as the cyclotron and water boiler to make certain of these materials. This radiochemistry is essentially tracer work, and we are equipped with the necessary counters, etc. This is a field that has been pushed along rather sporadically; at times, we have needed some techniques and results and then the work has been pushed. But now the Laboratory plans are to get into these fields rather steadily. Radiochemistry as it is at the present time is a relatively new field. The old men in it in this country have been in the game only 5 or 6 years longer than the young men. There are very few places where training is given. We expect to be able to train men and give them experience in working on the tracer scale with fissionable materials. Now, there are a few other things we are very much concerned with; the physical properties of some of these elements such as plutonium and polonium, for example. We are now in the process of determining such characteristics of plutonium as heat capacities and thermal conductivities. Electrical conductivities and magnetic properties will be studied later on. In general, we will have to produce here a complete table of quantitative values for the various properties. The same is being done insofar as is possible with polonium. One of the jobs we have recently completed was the determination of the crystal structure of metallic polonium.

In the metallurgical work, the physical and mechanical properties, as well as the structure of the alloys, will have to be determined. As I mentioned before, the equipment we have for this type of work is not particularly exciting in the same sense that the water boiler is exciting. We have spectroscopes, x-ray diffraction equipment, and electrical instruments, and all of the usual chemical equipment. We go in a great deal for high-vacuum work, and we have a great many high-vacuum installations with relatively high capacity.

One of the developments was in the field of analysis where methods were developed for determining carbon and oxygen in very small samples in spite of the fact that the contents of these impurities were running in a few parts per million. In work in analytical chemistry, our analytical methods on plutonium and uranium are not in too good a shape, and there are very many problems along that line.

While we have a good deal of metallurgical equipment, there is nothing unusual about that equipment. Any good metallurgical department has about the same sort of equipment. We don't undertake here systematic or long-range investigations on just any alloy system that one might think of. We try to get other universities and laboratories to handle most of the problems. The problems that require special handling and special techniques characteristic of the work on plutonium are the kinds we handle here. For example, if someone wanted work done on the gold-plutonium system, we would certainly want to do it here as we have the protective measures and techniques necessary to handle it. But on the other hand, work on gold-beryllium system we would probably try to get another institution to handle.

To indicate the types of problems in metallurgy in which we are interested and in which we are prepared to carry graduate students, there are metallurgical problems in physical and mechanical structure and in phase relations of plutonium alloys. We are also interested in the study of the thermodynamics of many of these plutonium reactions. These include not only reactions that take place in solution but also a number of high-temperature reactions involving reactions with refractories, the gas-solid type of reactions for making anhydrous compounds, and that sort of thing. Thermodynamics questions include a very wide range of problems, including such matters as electrical potentials, solubilities, activity coefficients, and the like. We are very much interested in rate phenomena because they are very definitely important in our production plant and, in addition, have scientific importance. The particular reactions we are first interested in are those that are involved in the production plant. That doesn't lessen their intrinsic interest. We had to pick a starting point and we picked that one.

In analytical chemistry, as I have said, there are quite a number of problems. We are able to handle radioactive materials here and we can work quite safely with them. The whole Laboratory has been built around the handling of that kind of material, and these, plus the availability of such materials, constitute the most unique features in the work of the chemistry and metallurgy division.

ACTIVITIES OF M DIVISION

Darol K. Froman

As Dr. Bradbury said, the name of this division is the Experimental Physics Division. It isn't implied by that name that other divisions such as the Physics Division do not do experimental work—in fact, a great deal is done in chemistry and physics, and we have no corner on the experimental research done at Los Alamos.

Incidentally, it may be that during the day you will hear some explosions, and don't be startled if you do. There is a certain amount of HE (High Explosive) in the neighborhood of this laboratory, and work that goes on with explosives is carried on a considerable distance from these buildings from the point of view of safety.

There have been quite a number of techniques developed or improved for the study of phenomena connected with high pressure and high temperature—namely, those things in connection with explosives. There is at Los Alamos an Explosives Division, which is represented at this meeting by Dr. J. C. Clark, Associate Division Leader. Dr. Max Roy, the Division Leader, is unable to attend. The Explosives Division and the so-called Experimental Division have quite a number of techniques worked out for the study of physical phenomena apart from the fields of nuclear physics. Since partly because of the lack of time you won't be visiting the laboratories of these Divisions where this work is going on, I would like to mention the sort of apparatus at our disposal. We have high time-resolution measuring equipment such as flash x-ray techniques, flash gamma-ray techniques for taking photographs in the microsecond region, flash photography in the same region, and electrical contact magnetic methods. In addition, we make use of piezo-electric crystals and so on for measuring the motions of material impelled at high velocity, the pressure involved therein, and such quantities. It is quite feasible to measure material of shock velocity of the order of 20 000 to 40 000 meters per second in distances of a few millimeters with a precision of 1%. This means that the timing equipment for measuring a displacement time curve, under fair conditions, is good to a hundredth of a microsecond, and under best conditions it can probably be pushed to a few thousandths of a microsecond.

Various interesting phenomena arise that are associated with these high velocities. If, for example, you excite a shock in most gases, they become luminescent and the spectra of shock-induced luminescence is an interesting subject about which a great deal is not known. I think this is one of the fields in which we might interest some graduate students. Again, the subject of making nuclear physics measurements on the very light elements has been mentioned. It is conceivable that the high velocities obtainable in this field of physics might produce motions of materials of such high temperature that we could produce thermonuclear reactions mechanically. If this is so, it should become a field of considerable interest to this laboratory. Since you, unfortunately, won't be seeing this equipment, I didn't want you to go away without having some idea of it. This unique field is a difficult one to investigate in most universities because of the geographical location.

Dr. Bradbury comments on the above statement:

I would like to add to Dr. Froman's statement that the field of research that has been discussed, namely those fields having to do with the propagation of shock in metals, the mechanisms of shock, mechanisms of detonation of HE, are fields that interest us but, of course, are not widely known and are not generally studied at an ordinary university. These are fields of research that pertain in considerable measure to classical physics and are intimately related to the equations of state of materials under very extraordinary conditions of pressure and temperature. These classical problems have been approached with quite modern techniques

394

and provide experimental fields representing another aspect of the Laboratory's work that I think should prove to be of great interest.

REMARKS CONCLUDING THE MORNING SESSION

N. E. Bradbury

We have endeavored to describe for you in a brief and rather sketchy fashion the activities of the Laboratory in that we believe graduate research could be done by graduate students. This had to be sketchy and incomplete in view of the fact that the present employees of the technical area in which this work is done number over a thousand. The presentation you have heard this morning has, of necessity, covered the general ground.

I would like to conclude the meeting this morning with a few observations as to the general philosophy of the Laboratory in connection with this proposed program. As has been described to you, there are two possible ways in which a given experiment might be undertaken here by a graduate student. One would be for us to suggest an experiment directly to the man, or to the university or to the department, and another would be for the university or the department to suggest a man and an experiment. In the latter case, we would get together and decide if it is an experiment that falls in the normal field of activities of this laboratory, as described this morning. I think you will have gathered from what has been said that in the fields of basic chemistry of the fissionable materials and in basic nuclear physics, our interests can be and will be quite broad. We don't claim in any way to have a unique monopoly of the bright ideas for experiments in these fields. We hope and expect that there will be suggestions made to us for jobs that we like, for which we have the equipment, and for which a man is suggested to carry them out. I wish the impression to be quite clear that we are not suggesting that the experiments would be dictated by this laboratory. Yet, for purely practical reasons that primarily involve

395

the expenditures of government funds, the experiments carried out here will have to be experiments that the Laboratory approves. The Laboratory will approve experiments that fall in the general fields of the work described here. I am trying to make the base of this pyramid of knowledge as broad as possible. I believe we can look with favor and interest on a wide variety of experiments that fall in the general outline of the work described.

I am sure, also, that many of you will be concerned as to the general philosophy of the publication of results as it exists at present. We will go into more detail this afternoon. I merely want to mention this morning that the Manhattan District has adopted the point of view presented to it in the so-called Tolman report. This is a declassification guide or report prepared by Drs. Richard Tolman, J. R. Oppenheimer, R. F. Bacher, and others, which recommended an order of release of the scientific research done by the various Manhattan District projects. The Manhattan District then secured the approval of the President of the United States to carry out the declassification. I think it is not generally known that the security imposed on the Manhattan District and for which it has been very frequently and widely criticized was not the choice of the Manhattan District but was imposed on it by presidential authority, and it was necessary to secure the presidential authority to release the material on work done by the Manhattan District.

This declassification program is now under way, and we are now releasing scientific articles at a considerable rate. When I left for Bikini several weeks ago, there were at Los Alamos something on the order of 100 articles or documents undergoing the declassification process. It is becoming a formidable task from the point of view of the work associated with it. The declassification guide at the present time allows the release of the details of instrumental techniques, a large variety of fundamental physics, and chemistry dealing with all the elements except those exclusively involved or having some particular interest in the construction of the weapon. The fundamental physics of some of these elements, such as plutonium, will probably be reserved for release at a somewhat later date. The actual techniques of how to make an atomic bomb will be, I presume, one of the last things that will be revealed.

I want to make it clear that the process of declassification and the release of scientific information is going on now, and I have every reason to believe it will continue at an accelerated rate. This will, of course, depend somewhat on the character of legislation, which is at present pending in Congress. The background work is already done.

The progress of physics is dependent upon a free interchange of basic nuclear physics data. Yet, even without this free interchange, I think some spread of knowledge is bound to occur by the process of diffusion if nothing else, and to attempt to stem this particular tide is quite futile. Since it will happen in one way or another, well it is that it is happening now in an orderly way; the things that are most obviously declassifiable are being released now, and the things about which there are questions will be released a little later on, and so on. The question of what would happen to a thesis or a piece of work done here by a graduate student we will discuss in more detail this afternoon. At the present time we must recognize that work in some fields would be releasable and publishable immediately, but that others might have a higher classification and not be released for some time. I wish you to understand that this problem is, in general, being handled in a reasonable and logical way.

Another word about the philosophy of the Laboratory insofar as it concerns the actual construction of an atomic bomb. The principle has always been to avoid what might be called compartmentalization. All members of the Laboratory staff (a man with a B.S. degree or better in science) have essentially complete access to all information. Even during the early career of the Project there was a very definite widespread knowledge among staff members of what was going on. I would like to continue this. It is, therefore, desirable that the graduate student doing a research problem in physics or chemistry be a member of the Laboratory's regular staff as an employee. This, of course, means that he must undergo the ordinary security investigation; when a graduate student comes here, he would simply be a member of the Laboratory along with everyone else. The only difference is that he would not be subject to diversion from his thesis activity. In other words, if an emergency arises where we would have to have some job done, the graduate student on this basis could not be taken off his thesis and put on some Laboratory work.

We will have to assume, then, that the graduate students are members of the Laboratory staff. They will not be deliberately lectured on how to make an atomic bomb, nor will their ordinary work lead them into contact with matters of bomb operation and design, nor will they be encouraged to find out these things. But there will be no outside restraint put upon their activities, and I think it inevitable they will find out in a general way what makes an atomic bomb tick. They will then, of course, be subject to the ordinary security restrictions about discussing this, and since they will be reasonable people and discreet, this will be no source for concern.

This afternoon, we will go into more detail on what we propose as a starting point for this program of cooperation. Undoubtedly, as I have said before, there will be things that we will not have thought of, there will be problems that you will point out, but we will make a certain general proposal as a starting point.

AFTERNOON SESSION

OPENING REMARKS

N. E. Bradbury

We will now hear from two members of the Los Alamos staff what we propose along the lines of university cooperation. I would like to emphasize that the mechanism of affiliation will have to be extremely flexible. It will have to take care of a wide number of different university policies, it will have to take care of individual requirements, and I see no reason why this cannot be worked out.

WHAT WE PROPOSE—I

Frederick Reines

The brief talks that Dr. Spence and I are going to give now are for the purpose of indicating the sort of thing we have considered and to provide a starting point for the discussions. After all, the real reason why we are here is to see what you think about all this and to learn from you what changes in our ideas are necessary in order for us to make this a practical arrangement.

The first item of interest is the scope of this plan. We are interested in supplementing and assisting with the training of graduate students. We are not interested, as Dr. Bradbury clearly pointed out, in competing with universities. This is not the idea at all. We would perhaps take students who had a B.S. degree plus two years of experience in graduate studies to do their research here. There might be courses in specialized topics—say nuclear physics, neutron physics, radiochemistry, and the chemistry of the heavy elements. We are not going into the business of educating in a formal way, and we recognize that it is entirely the university's obligation to take care of the matter of accrediting courses and accepting theses and so on. We do have staff members here with the proper qualifications to teach these various special courses, but we do not intend to carry on a full-fledged graduate teaching program.

The main purpose of the student's coming here would be to do his thesis research. This naturally brings up the problem of who suggests the research and then who guides it. It might be that the problem is initiated by a member of the Los Alamos staff. He might say that there is a research problem which would take about a year, for which he would like to find a willing graduate student. He then makes the suggestion to the university by whatever formal procedures are adopted. On the other hand, if people from universities are aware of the sort of fundamental research going on at Los Alamos, they might have some idea and suggest it and a student whom they think is suitable to work on it. There is only one catch—if you want to call it a catch—and that is, the problem has to be of some interest to the Laboratory. It is pretty hard to find things that aren't of interest to the Laboratory if you stay in these general fields. One can't tell in advance, but if it is in nuclear physics or radiochemistry we are probably interested, although the division of the Laboratory in whose province the problem lies would be called upon to decide in each case.

The next question is the status of the student at Los Alamos. Certainly the student would be recognized as such and, academically, would be under the jurisdiction of the university. He would, in actuality, have a sort of dual role, and this perhaps might make things a little difficult. As far as the university is concerned, they would recommend a specific man for such a fellowship to a committee that is set up here. Then, after the man has been cleared, it is decided whether or not he is one of the better students.

399

In starting, we would set an upper limit on the number of students we would accept, say perhaps 15. This number is not to be taken as fixed. It depends, first of all, on whether the student is in existence; it may be that there are no students to come to Los Alamos this fall. It depends on whether the students that would like to come are considered to be properly qualified by the university, whether they can get clearance, and whether there are problems on which a student cares to work. One has to settle all these things. One would hope to start out on a very modest scale so we can learn about the problems that we may expect to encounter in connection with such an arrangement.

So far as the university is concerned, we might call this man a Fellow. As far as Los Alamos is concerned, he would be an employee, a regular member of the Laboratory staff. He would receive the customary clearance and so would have access to all documents regardless of classification.

We are very anxious that the student be treated as a member of the staff but also that he not be simply taken off his thesis research as can other staff members to do other things of interest to the Laboratory but not concerning his thesis work. Actually, one should not draw too much of a hard and fast line in this direction because a student receives much of his training by working in groups with other people on other things. He might, as a result of his group activities, get his name on his own thesis and perhaps get his name, along with several others, on something else. It is certainly a good idea for him to work on a few machines in order to get experience. However, it is our intention that he would not be taken away from his research to do some job that wouldn't be of direct value in his graduate training. If he wishes to depart from strictly theses activities, it would have to be with mutual consent of the local advisor and the student. This is to make sure he is a student and not a member of the staff hired through the university to work at Los Alamos exclusively for our own use.

Now of course, it is desirable to make these things attractive so we can get fifteen or so good men from the affiliated universities. To get good men and relieve them of economic stress, we feel that they should be properly treated economically. The present salary range for a member of the Laboratory who has a B.S. plus two years is from $250 to $300 a month. In view of the fact that this man is getting special consideration, one might wonder about invoking the higher end of this scale. Perhaps one might set the lower limit at $200 a month. This is below the minimum scale for a B.S. plus two years, but just how much the student receives is a matter for discussion. One should probably make an allowance—say $50 per month, for the married student.

Then there is the question of another possible student classification. The idea is that there are people at Los Alamos who have a B.S. plus two years who are good men and who would like to go back to universities to finish the requirements for the Ph.D. This is certainly to be encouraged, but it does lose us trained men. Now if one of these men had a Ph.D., he might be interested in coming back to Los Alamos to work. From our point of view, it would be good to have those men come under the sort of plans we are discussing. As far as the university is concerned, they would be treated as any ordinary students who sought fellowships. As far as Los Alamos is concerned, they could get a leave of absence to go take their course and then come back to Los Alamos as employees to do their research. The university would first have to agree to take these men on as students. All these are suggestions, and how they are to be carried out is a matter for much discussion. Another interesting question is the status of faculty members of universities with respect to this plan—how, for example, they might guide their students. Perhaps the university will not send a man here full time, but they nevertheless will want its faculty to guide him personally. They might serve as consultants and be really responsible for the work but appoint a person here to guide the day-to-day activities of the student. There are classification and security problems that Dr. Spence will discuss in this connection. The supposition in this is always that what is being done is of interest to the Laboratory because it is the Laboratory's work being done and it is of interest to the university because they are training a graduate student who is going to get a degree from that university. Another possibility is that the member of the faculty goes on leave and comes to Los Alamos for a year or for a shorter period of time such as the summer. These are some of the possibilities as we see it. Of course, each appointment would be treated individually. Another possibility would be the exchange between Los Alamos and the affiliated university of men of similar caliber. I would like to close my remarks with another comment on the question of guidance. It might be that an affiliated university sends a student here but has no one to guide him. However, it may be that they are perfectly satisfied as to the qualifications of the members of the Los Alamos staff and are perfectly satisfied to have a member of the Los Alamos staff to guide the student. It might simplify matters if the Los Alamos staff member were appointed a member without pay of the particular university but residing at Los Alamos.

These are some of the general ideas we have on the subject of university affiliation. Dr. Spence will now discuss some of the problems associated with this plan.

WHAT WE PROPOSE—II

R. W. Spence

Dr. Reines has given a sketch of the proposed plan for university affiliation. I would like to discuss briefly some of the possible problems that will arise under this plan.

The first concerns classified material and its use by a student for thesis material. Under the suggested plan, the students will be cleared and we should expect no difficulty so far as the students are concerned. Faculty members of universities who are consultants or who come here on leave of absence or part-time during the summer will also be cleared. Difficulties will usually arise in connection with thesis material and examinations, and I will shortly propose ways of meeting these difficulties. The subject matter that is declassified is now in a state of flux; more and more material is being cleared. The items on the sheet that I passed out show material that has been declassified at the present time. We can expect a much wider range of subjects to be cleared in the future.

Thesis subjects may or may not be classified. Some will clearly not be classified and so no particular difficulty arises. Others will be clearly classified, while still others will be borderline cases, and at the beginning of a research, it will not be clear as to whether they will become classified or not. At any rate, it would be advantageous if at each university some plan could be worked out to handle classified theses. This becomes particularly true for the doctorate examination. Let us suppose that the thesis material is classified; the question comes up about the candidate's doctorate examination. Often such an examination consists of two parts; examination on thesis material itself and then an examination of the competence of the graduate student in the general field in which he has worked. It is quite possible that it can be arranged for a subcommittee to be formed of people who are cleared for the thesis material and that the examination on the thesis work be given by such a cleared subcommittee; the examination on the general field in which the student is working can be given by the regular examination board in the usual university fashion. The cleared subcommittee might be made up of faculty members who serve as consultants to the Los Alamos Laboratory or faculty members who have come to Los Alamos for part-time work or on leaves of absence, or Los Alamos staff members who have been approved by the university, or a combination of all such cleared people. I believe that problems concerning the doctorate examination can be worked out with each university, although the same plan may not work in each individual case.

The second question concerns the publication of classified work. Publication within the Manhattan Project is assured; the work can very well be written up as an ordinary project report. Publication outside the project will be withheld until the material can be declassified. It is possible that in some cases a declassified abstract can be written that will satisfy university requirements regarding abstracts of theses. There may be cases where no such abstract can be written but it is a possibility.

There are certain university requirements that, of course, are primarily the concern of the university only—the question of fees for the graduate students, residence credit, provision for graduate work in absentia. Obviously, for such a plan as we have outlined to work, each university must be willing to let the graduate student work in absentia for a year. There will be certain problems that each university will have, which have not been covered but which I hope will be brought out in the discussion.

Another point has been raised about individual versus group research. Generally speaking, the fields of nuclear physics and chemistry involve cooperative effort. That is to say, a person doesn't do all the work on a Van de Graaff, such as keeping it running and in order and do his research work at the same time. I think we will have to recognize in this field that a certain amount of cooperative or group research is inevitable, but we do not propose that the work a graduate student does be group research. The student will be primarily responsible for his own problem. If a student works on the Van de Graaff, for example, other members of the Laboratory also using this machine will cooperate with him in certain phases of the work, especially those concerned with the running of the instrument. He may lend a helping hand in some types of investigation, and in return, he probably will receive such cooperation, but I don't believe it has ever been proposed that the research would be of such

403

character that the graduate student would not himself be primarily responsible for his own thesis and the bulk of the work connected with it. He must still gain in his graduate work the usual amount of competence and independence in his thinking and in his work that would be required at any university.

Another problem in which you may be interested is how many students we can actually accommodate here. I can only speak in general numbers. A survey of the Laboratory, its facilities, and men qualified to handle graduate students would indicate that we could probably handle 15 to 25 all told. This is in physics and chemistry both. Roughly, the proportion is about equally divided between physics and chemistry, but slightly lower in chemistry. This is about the total amount we guess we could handle, although no definite figure can be put down.

I think you will see that most of the individual difficulties are those concerning either classified material and special ways to get around the concomitant problems, or difficulties of adequate supervision of the students to the satisfaction of the university staff. It is quite possible that a university staff member has a problem and a man in mind to work on it, and would like the work on the problem done at Los Alamos, but he himself does not have enough time to come here to supervise the research student. I believe that some satisfactory arrangement could be worked out so that periodic visits to Los Alamos could solve the problem. The direct supervision might be delegated to a Los Alamos staff member, and the faculty member who proposed the problem, or who is primarily interested in the problem at the university, could come here to check the progress of the student and discuss the problem just as if he were a regular consultant on this project. I think this arrangement could very well work out. I think it would be preferable from all points of view if the faculty member could be here on a year's leave of absence and guide the student in a direct fashion, but we are anticipating in the next few years that a shortage of scientific personnel will make this a very difficult thing; that it will not be possible for a university staff member to leave his university for a year to come here and work. During this interim time, we had hoped that some delegation of supervision could be worked out with qualified Los Alamos staff members.

As things now seem to us, the procedure might be something as follows: we would be recipients of a letter from a graduate dean or head of a department, saying that he had a man of certain qualifications. The man might have the subject on which he would like to work for his thesis,

404

or there might be more than one man of this nature. There might, in addition, be a staff member who would want to come along to supervise the research. At any rate, we would be made aware of an individual, or individuals, with or without specific problems, and with or without various amounts of direction from the university proposing the man.

The first thing we would then do would be to examine the qualifications of the man; see how he compared with others who might be presented, and see to what extent we were able to absorb this particular man, and the problem, if any, that was suggested. If the problem seems to fall in our province, and is one which we would like to have worked on, and the background and recommendation of the man seem to indicate he is the type we would like to have here, we would then at once proceed to have the man cleared. This is a province of security—G-2, if you wish. Clearance of an individual generally consists of looking into the background of the person concerned: where he was born, where his parents were born, what type of work he has been doing, his reputation—does he drink? is he discreet? who are his friends? with whom does he associate? is he a campus radical? is he a soap-box orator? etc. These are not necessarily against a man, but, in a general way, security wants to know what kind of a man this is. In other words, if he knows something, can he be trusted to keep it to himself? In only an almost infinitesimal percentage of cases do we run into security difficulties. Once in a while we find a man who, for some reason, cannot be cleared. Generally speaking, it turns out to be somebody who has announced in public that he is not in sympathy with the United States Government, or something of this sort. Since this is a government project, and the government is the employer, in the last analysis, it has to be satisfied with people it takes on. However, this refusal of clearance is something that occurs very rarely, and I don't anticipate any difficulty. It is something that takes a matter of from four to six weeks to carry through, depending on how many places he has lived, what he has been doing, and how far the investigation has to be carried through.

Let us assume the investigation shows the man is cleared. We then request the man to come here for an interview so that we can see him and he us. This, in fact, could be conducted at our expense, and would be called a pre-employment interview. Our project employment policy provides for us to pay this expense. The same thing would, of course, be true relative to any university faculty member we would sign as a

405

consultant, who might be responsible for the research that was to be carried out.

Let us assume that all interviews work out to everyone's satisfication and the man is employed and starts to work. Initially, he would be placed as a member of some active group; he would have a certain period of training, learning where things are located, techniques, etc., which would take from one to three months and would be characteristic of any new man starting in a new department. He would have to find out the mechanics of carrying on the work here. He would be required to learn the operation of the cyclotron and Van de Graaff. This would occupy only a part of his time, and his particular job would be his particular responsibility, although he would contribute and learn techniques of other jobs going on simultaneously in the Laboratory.

We do believe that a man should be able to accomplish a reasonable graduate thesis in about a year's time, probably from nine to fourteen months. But, if the problem assigned to a man should reasonably be accomplished in a year, and is not, then either it is too hard, or the man is not capable, and we ask him to terminate. We do expect to assign problems that could be accomplished in about that time, although we would be glad to have this particular point discussed.

Then, we would give an occasional course, but whether these courses were accredited would not be a matter of concern locally, although it might be of concern to the man. On this account, we would make an effort to see that courses were accredited. We would be primarily concerned with the work of the student here, in his rapid indoctrination and training in the particular work we do, and in adding to the general knowledge of students who might be working in other fields. This instruction would probably be done out of hours.

Salary arrangements would be made on a basis that would depend on the man's background, his family responsibilities, and things of that sort.

Having concluded his research to our satisfaction, and to the satisfaction of any faculty representatives or direction from his own university, the thesis will be prepared. What happens to the thesis will depend upon its classification; if it were declassifiable, there would be no reason why the thesis should not be returned to the university to be read by the university committee; if it were a classifiable thesis, one which could not be read by everyone, then there are various clearance possibilities which would have to be looked into for a solution. One possibility used in the past is for me to appoint a local committee containing, if possible, representatives of the university in question who happen to be on our current staff. This committee might be consultants from the university in question. This committee then reads the thesis and reports upon it to me. I inform the university of the general field and general title of the thesis, the names and academic affiliations of the committee reading, it, and their report upon it. This particular arrangement has been used and proven to be satisfactory for several universities at the present time; otherwise, it may be necessary to establish consultant members to this laboratory from the university in question. These would be staff members of this laboratory on a consultant basis, and thereby cleared to read material on this particular thesis. They would then report to their university what they considered to be the character of the work. The student's academic requirements, other than the thesis, we would have to leave entirely up to the university. These matters are the province of each individual university, except when we come to the question of what to do about examining the man on a classified thesis. Various methods have been suggested as to how it might be carried out. It may be that the reading of the thesis will be sufficient; it may be the qualified committee can be set up to read the thesis; and it may be that examination can be conducted so that the specific numbers to which the classified thesis would refer would not have to come into the examination.

I suspect, in this particular circumstance, we are possibly making a mountain out of a molehill, and that this may turn out to be an easier problem than we see here.

If the thesis is classified, then it becomes a Manhattan publication and a certain number of copies are printed locally and distributed in accordance with the rules of the Manhattan District; but until they are declassified, they could not leave the Manhattan District. If the material should, in the course of time, become declassifiable, we would declassify it and notify the man and his university, and if the man wanted it published he would put it into publication form, but this will take longer. I think the only thing we would request is that somewhere in his published material the man specify that the work was done in Los Alamos; but the man's present residence would be given as his residence. In other words, we also would like to get credit outside for doing some of these jobs.

I think that is about the situation, as we see it. Admittedly, there are

407

problems that we haven't made clear. If so, these are the questions we would like to have brought up now.

Dr. Stewart (California): I think probably the University of California has thought as much about this type of problem with which we are faced here as any other university in the country. I think, Dr. Bradbury, it might help in crystallizing the discussion if I could speak as a graduate dean. After all, as one of those unfortunate persons, I see certain difficulties that, at least, our faculty would not see unless they were pointed out.

I think every university will admit, immediately, and without any reservations, that the talents of the staff here and the type of facilities for research which can be done are not equalled anywhere. But the problem of doing graduate work away from a university, unfortunately, is not so simple as to say that with the distinguished staff, and with almost incredible facilities, all questions are answered.

The first problem that occurs to me, and I want to make it clear that I am not speaking against this proposition, is that any university awarding a degree is obviously awarding that degree for work done at the university in question, under the supervision of appropriate members of its faculty, in accordance to its own individual regulations and restrictions. Now, that is almost as formidable as it sounds. In other words, there must be very strict control on the part of the university of all graduate work. It isn't going to be simple for the university to say Los Alamos is the one place where it can be done. We are going to have for consideration comparable propositions from Naval Ordnance, for example, and other agencies. We are going to have to be very careful that our graduate work doesn't become, more or less, of correspondence course caliber, and we're going to have to be particularly careful to maintain the distinctions between an earned degree and an honorary degree. As a matter of fact, it is quite possible that some of the work that has been done and is and will be done, is work that justifies an honorary degree and a meaningful honorary degree. But it isn't necessarily so that, because it is work of value, it can be recognized in partial fulfillment of the requirements of an earned degree. I bring that point up first because I'm sure the graduate deans of your respective universities will think in much the same distorted fashion as I am thinking, because that's a disease of all deans. We become glorified policemen despite our good intentions.

Now, you see, that leads logically to a question already presented here;

408

namely, the matter of appointment of an individual who is going to personally supervise the work that is being done. I know of no university that will permit its students to work elsewhere in another university, e.g., and get credit in its university for work done there, regardless of the reputation of the university in which the work is being done. The degree represents the university that awarded it, and I am quite certain that my graduate council would not consent for a moment to recognize thesis work, regardless of the excellence of the man under whom it was done, unless it was done under one of our own faculty members. We have not had the problem to face in the past with reference to thesis work done at Los Alamos, because at least one of our men was here a good part of the time; but that's an important point.

Dr. Bradbury has suggested, and it is a suggestion that deserves thought, that a staff member of this laboratory might be appointed as a research professor. If that could be done, it would answer one of our problems, but, as most of us know, it is frequently difficult enough to get a new appointment for a man who is going to be a regular professor in residence, to say nothing of a research professor not in residence. I think, however, that it is really well worth full consideration.

The matter of residence was raised. I am inclined to think that possibly that point has been overemphasized by the men of the staff here. All of us have a certain minimum residence requirement that must be fulfilled. You have already stated that your idea is that a man will have two years of residence as a graduate student before he comes here. That's going to put him in pretty good shape if he is the type of man you want. He should have about finished his course work; he should have completed his departmental and qualifying examinations. I think you would probably want to make that a requirement for a man coming, and that is not an unreasonable requirement. At the University of California, we require a year of residence and candidacy, which is a common requirement; that is to say, a man is elevated to candidacy after passing his qualifying examination. The problem of satisfying the residence requirement is not a very serious problem. If a man could come here and do research leading to his dissertation in the period of one year, he would not be penalized at all by returning to his university for a year, during which time he might be analyzing the data he had collected here or doing other work. There may be difficulties, of course, with respect to security in that particular matter, but I do feel that, with our students in physics and chemistry, at least to the University of California, such students would have to return to the university for one year.

We have had other requests from students who have been doing work in Naval Ordnance. In several instances, with a considerable degree of reluctance, we have accepted, as thesis, classified material. Now, there were particular circumstances accompanying these special dispensations. In each instance there was a regular faculty member of the university at the place where the work had been done, and this faculty member was closely connected with, if not actually directing, the research in question. Each of these students proved to be exceptionally good, and furthermore, we felt rather sympathetic to them because of their contributions to the nation in time of war, and because their regular program of graduate studies had been interrupted by the war emergency. I mention that because possibly our council was a little more lenient than it will be in the future. We had no difficulty with the committees to examine the thesis because in each one of these instances upon which we acted favorably, we had an adequate number of faculty members who were privileged to examine thesis material.

Of course, we did meet with some objections because we rather sanctified, possibly, our regulation that the two copies of the thesis must be duly filed in the library of the university and that there must be the usual number of abstracts, which I'm sure no one reads but the student himself. The matter of classification of material, it seems to me, at least so far as the University of California is concerned, would present as a special problem the matter of filing the copies of the thesis in the library. I am quite convinced, however, that at least for several years to come, we shall have members of our faculty who are permitted to examine thesis material and to examine the students orally upon such matters.

One of the major questions in my mind is whether the graduate student would be glorified Diener, or whether he would be an investigator of the type we normally expect of students working for a doctorate. I believe it has been stated twice that a graduate student here would be expected to demonstrate the initiative, the independence of thought, and the independent analysis with appropriate direction and criticism that we would expect if his work was done in a university.

Another question is this: the cost consequent upon a faculty member who had been appointed as consultant coming here at frequent intervals to check up on his student?

It was my understanding there would be provisions to take care of that

410

faculty member's travel costs. There is, of course, connected with it another problem that the professor himself might present; namely, the amount of time required for the supervision of a single student. He might feel that he did not have time to devote to travel. That would be up to the individual professor, in most instances. I might say, in connection with this, that the leaves of absence of faculty members to come here for a semester, or year, is, I believe, going to be somewhat more difficult in the next few years than it has been in the past. Administrative officers of the universities are a little bit tired of receiving requests for a leave of absence, and they're showing greater reluctance to meet those requests. This reluctance is strengthened by the fact that the number of students is increasing alarmingly, and that is particularly true in departments of physics and chemistry. As a graduate dean, I'm almost convinced that everybody wants to do graduate work in chemistry and physics, and everybody wants to come to the University of California.

Now I have presented, I believe, the major points I had in mind, looking at this through the eyes of a dean who is ultimately responsible for the administration, so far as the university is concerned. It has great merit. Something that has not been mentioned today, but which is deserving of consideration by our administrative officers, is the need of prevention of a vacuum of young men qualified to continue the work of the Manhattan Project. I suspect, as an outsider, that you may be faced with a real problem there, and one which may be, with a little prejudice, presented to universities as a patriotic duty. That may be a shot in the locker, Dr. Bradbury. I'm trying to see this thing entirely as an administrator.

I would like to ask about library facilities. Dr. Bradbury mentioned a few moments ago that many would want to read. I should be inclined to say all should be expected to do a great deal of serious reading. May I ask about the library facilities?

Dr. Reines (Los Alamos): I spoke to the librarian because I recognized that this would be an important question, and the status of the library is this: the University of California had loaned many books to the library to be used during the war and, at the same time, many books were purchased for the use of the library. Now that the war is over, many have been recalled, but they are, at present, being replaced, and there is a buying spree on with respect to necessary books that would astonish a librarian if he weren't at Los Alamos. There is a very serious program now of building up the library so that it will be adequate for our research

411

here, as well as for any students that come. At the moment many of the shelves are empty because the books and journals are at the binders, but within two months we expect we will have an adequate library.

Dr. Bradbury: The library has been, and will be, maintained at a level which is comparable, technically, with the rest of the Laboratory. It had to be set up in the beginning as such, as was early recognized. It is purely technical; that is, we have nothing besides chemistry, mathematics, and physics books, with a few closely related sciences, some medicine, and things of this sort, primarily relating to these fields. This is a highly technical library.

In connection with the matter of either periodic consultant visits between actual members of your staff, you pointed out, Dr. Stewart, that it is questionable whether a man would want to spend his time entirely with a student. This I quite agree with; however, I think a man of such caliber to whom you would wish to give a task of this sort, would be a man whom we would like to have as consultant to us. He would come here as a part of the Laboratory—a consultant to other members of the Laboratory staff on other problems.

Dr. Stewart (California): Which in turn would make it more profitable for the individual professor, in view of the fact he would come here as consultant.

Dr. Bradbury: Quite so, the man would be appointed to do a consultant job, not to do a specific job, but as consultant to the project. He may do, and will do, many other things. The question of expenses can be handled on a consultant basis—this is a payment of travel, plus so much per day.

Dr. Stewart (California): I would like to add one thing—I know that several other gentlemen here have the same point of view. We have taken official action at the university with respect to the consideration of requests from Los Alamos as well as from Naval Ordnance. This is a definite policy, for it is the thing that has been in effect and will continue in effect. We feel that in such an arrangement as you propose we must avoid a bracket group—that each student must be considered individually by us, as well as by you. That is, we would not be willing to say that any student recommended by our department, and accepted by the project, would automatically, upon a fulfillment of requirements, be accorded a special dispensation. Not in the least. We would have to consider every case separately and as an individual case. However, I suspect that does not present a problem because we would be dealing

412

with a restricted number of students, say, a top of twenty-five. Obviously, there would be probably not more than two students at any one time from any single university. I think such a diversity of sources would be an advantage to the Manhattan District.

Dr. Bradbury: Of course, the particular facilities of the University of California, along these lines, means that you will have facilities on your campus, and facilities equally extensive to our own in many fields. I see no reason why your men should come here when the job can be done there. But there may occur occasions where we have facilities which would not be duplicated in the Radiation Laboratory of the University of California and for which there exists a desired use.

In this particular case, the more contact the university can have with this project, in the direction and understanding and full acceptance of the work that is done, the more easily will it get around such difficulties. As a result, we feel very strongly that the more of that sort of contact we establish, the better for our project program. To some extent it may also be advantageous for the universities to be equally aware of what is going on in a laboratory of this sort.

Dr. Buchta (Minnesota): I speak, not as a dean, but as one of the members of the council working with deans in the graduate school. Many questions have been answered, but I would like to point out, from our point of view, that we feel there are many difficulties. However, these are difficulties which might be called red tape. The fundamental approach should be pursuit of ways of advancing science and training students. Having that as our goal, many difficulties will be ironed out. We do have regulations concerning candidates—many would be met at once if a large amount of work were declassified. This would indicate we had gone a long way, to say the least.

Have you found a possibility whereby a man may be here less than a year?

Dr. Bradbury: We have given thought to that particular problem. I think there is a possibility along those lines, though I should be inclined to believe the initial time would be not less than six months. I have the feeling it will take more than a month for the man, when he first comes here, to learn his way around, and in addition, I would like to handle this problem in as nominal a way as possible. Normal employment policies are such that we would not like to employ a man for less than six months. There would be exceptions, and individual cases will have to be handled on an individual basis. If we had to bring him here, send him back, etc., it

413

would be difficult to explain to the general accounting office if he stayed less than six months; therefore, we set up this period of time as expectation of employment. It might be feasible to arrange a six months' employment period with possibly subsequent and shorter periods. We would do it on leave of absence basis.

Dr. Buchta (Minnesota): I presume it will be a shorter period for staff members. Have you any idea what proportion of work done by students would be declassified?

Dr. Bradbury: There will be a tendency to find problems that can be declassified. It simplifies the problem in many ways, and I expect that 50 to 75% of the work will be declassified. This is not necessarily so, but I would gather, from the tenor of this meeting, that there should be a definite effort made to find declassifiable problems. We are interested in the progress of science, and this means getting as much material declassified as possible, and we expect there will be a large amount of work declassifiable. However, we hope to find many people who are willing to undergo temporary difficulties of classified research in view of the physical interest which the problem that they are working on presents.

Dr. Glockler (Iowa): I have been thinking about the possibility of picking these graduate students, and it occurs to me there are two ways—to let him take his course work, then his languages and qualifying examinations, and then research work. There are also quite a number of graduate students interested in chemistry who don't follow the plan. They will start out with research and languages, follow with course work, and then the thesis. In that case, of course, it is quite evident we could not enter any arrangement with you. We leave more or less up to the student which arrangement he wishes to follow. Personally, I must say, I urge him to get through with the course hurdles, his languages, and other qualifying examinations.

When will you pick the research problem? Just when will I find that this man wants to work on a certain problem I hope could be worked on in cooperation with this laboratory? It isn't so simple to know how and at what stage to pick the man. Here is an example of our procedure at Iowa. I will describe research topics to the student. Some of them state right away they want to work in physical chemistry, and questions are discussed. They will then do nothing but readings. It is quite usual that the thesis will be published with the major advice of the senior professor, and it will usually be published under these two names. If you must have

414

another advisor here, there will be three names on the paper. Dr. Bradbury, I would like to have your thoughts on these matters.

Another question: the student here might work on several machines. We have hardly enough time to get the students to work on their theses, and if you, in addition, send them all over the place I don't believe they'd ever get to their theses.

Here at Iowa the student must take his final examination on the campus, and it sometimes even happens that students leave for a job and have to come back to Iowa and take their final examination before receiving their degrees. In case of a doctor's degree, anything short of sickness or death cannot keep them from taking the test.

Dr. Bradbury: I quite agree, and we have discussed the problem here. There are two ways of proceeding. One is getting a large part of this course work done and then concentrating on the thesis work. In research, this has advantages and disadvantages. We would have to deal largely in this way because, as you pointed out, jumping back and forth is not satisfactory. We would hope by choosing men with some graduate training that a man would not find himself without a moderate amount of formal instruction. We would let him write his thesis without the necessary prior academic formalities of the classroom. I think we could probably work this out with your departmental staff.

Dr. Glockler (Iowa): Do you have any problems as far as publication is concerned?

Dr. Bradbury: We have taken the definite point of view on several theses done here that the man publish his own paper; however, I think this should be arranged by the man and his guiding professor, and whatever policies were adopted in that case would depend on relative guidance of your staff. We do feel that the man should have the majority of credit for his job.

As far as the use of different machines is concerned, this again would be a problem in which it would be arranged so that the university would be satisfied. Generally speaking, the man's activities would confine him to one special problem, or particular equipment.

The problem of final examination, I am sure, can be worked out, and from our point of view, leaves of absence, periodic or at the end, or whatever was required, could be worked out.

Dr. Kirkpatrick (Stanford): This plan, as proposed, is already going on to some extent—is that true?

415

Dr. Bradbury: We have had quite a few; I can't give you the exact number at present. Perhaps five or ten theses have already been presented and accepted at some universities. I have appointed a committee that reports to me and I report what the committee concludes. In every case, I have tried to include members on the staff of the particular university if possible.

Dr. Kirkpatrick (Stanford): The details of the plan, as you presented them, have been lenient, and from the experience to date it has been found they have worked?

Dr. Bradbury: We find there is considerable difference in philosophy between different universities, as might be expected, but in general, we believe these things can be worked out.

Dr. Kirkpatrick (Stanford): The Laboratory has been carrying on work in nuclear physics, etc. One might think that they ought to be interested in metallurgy, and perhaps in electronics. Do you anticipate any other branches of engineering?

Dr. Bradbury: Yes, to some extent I do, although it is not felt to be feasible to go into details here because most of our electrical engineering problems and chemical engineering problems do not lend themselves very well to theses. However, there are special cases, and we would be very glad to discuss them. In particular, we have problems along electronic lines that may be of the stature that could be developed by one man in a reasonable length of time.

There are also some other problems of great interest such as those concerned with equations of state, hydrodynamics, and shock waves. These are not problems that would ordinarily occur to the average institution. We experience difficulty in getting these problems out to the light of day, and to the student, frequently because of classification. But problems of that nature, as indicated this morning, are problems in which we have a great interest and which would set up a very good thesis.

Dr. Gustavson (Nebraska): Would the field also go into biological work, or is it restricted mostly to physical sciences?

Dr. Bradbury: We are undertaking, with the active cooperation of Washington University, a very extensive program in biophysical research, associated with particular problems that the material of this project brings to us. Actually, Washington University is taking the responsibility of carrying out that work and Professor Hughes might want to comment more on this arrangement.

416

Dr. Hughes (Washington): So far as the medical work and associated biological research, conversations have been conducted by representatives of Los Alamos and Washington University—plans whereby the university will take the responsibility for supplying men of the right caliber, subject to acceptance by Los Alamos; the advantage seems to be mutual. As far as I know, the actual preliminaries seem to be satisfactory to both sides.

Dr. Jacobs (Iowa): Does that mean that any medical work, biochemical and biophysical work, will be conducted by the University of Washington?

Dr. Bradbury: Yes, in a general way, although the responsibility of the project will, of course, remain.

Dr. Worchester (Colorado): Concerning theoretical physics, it has been mentioned that a year's time might be required for a Ph.D. thesis and part of that time to learn the laboratory procedure, etc. Could that be done, perhaps, largely at our own university, perhaps with consultant work here, making use of experimental data? Would that be available in case of theoretical problems?

Dr. Bradbury: When we spoke of a year, we were thinking of experimental problems. I think this could be worked out, depending upon the nature of the problem and extent to which the problem could be set up, but this involves considerably more discretion. An outline would probably have to be set up and I'm afraid frequent visits would be required of the man.

Dr. Worcester (Colorado): I have a man working on a theoretical problem that might involve use of machines, that's why I ask.

Dr. Bradbury: That could be worked out.

Dr. Gustavson (Nebraska): I think our institutions of higher learning have a tremendous responsibility to keep America strong in the fields we have been discussing today, and we have a marvelous opportunity in this. The project has taught a great lesson of cooperation, and institutions through the country are recognizing it. In connection with the Argonne Laboratory, there is a very definite plan of cooperation in research, in connection with university facilities. There is a very definite program being considered of accepting work carried on at other universities, which, it seems to me, is a very fine thing and would indicate America growing up, academically. We do have some precedent for that sort of thing where, e.g., Woods Hole work has been carried out in a fashion—outside of the fact that no problem of security is involved in that research. I was very glad to get the report on biological research, and I have a few questions I would like to ask. Undoubtedly, as men come to work at Los Alamos they will be covered by insurance, I presume, against accidents and health hazards. What is the legal responsibility of Los Alamos to the university relative to injuries on the project?

Question: In connection with classified research, or partially classified research, how are we going to avoid problems of duplication if there is established in the northeast a regional laboratory, one at Argonne, one here at Los Alamos, and in California? Wouldn't it be wise to have some sort of a clearing-house on problems to assure us secrecy is not leading us into duplication?

Question: With respect to the size of the stipend—seems the logical thing would be to have a salary range of $2400 to perhaps $3500. In that case, what about the living costs at Los Alamos?

Question: Couldn't we meet some of the problems of interchange between universities? After all, we are getting a tremendous gift from Los Alamos in having the opportunity to use the facilities of this laboratory.

Dr. Bradbury: An employee of this project may be covered by several types of insurance. Workmen's Compensation under the laws of the State of New Mexico is applicable. Conventional group accident and health insurance is available to the employee at his own expense on very reasonable terms. In addition, there exists a fund that may be available as compensation in the event of accidents arising out of some of the special hazards of the project. I do not believe that a University would have any liability, but I would have to have legal advice on that point.

The problem of classified science is not a new one. Since the close of the war, I think the situation in the Northeast, Argonne, and Clinton has been quite similar to that here. We in the Manhattan District have instituted a system of information interchange that at present consists in periodically listing document titles pertaining to all fields of research in chemistry and physics, except those fields which actually pertain to the construction of the weapon. In the latter case, no one is particularly interested, except ourselves. These title lists are available within the confines of the Manhattan District. The problem of the dissemination of information is handled by that method and partly by recently instituted periodic research meetings. The Manhattan District Laboratory is set up so we can know what is going on within the District even though it can't all be published now. I think that we can say that the necessity for access to the work that is being done in other parts of the Manhattan District has been recognized and has resulted in such a mechanism for the internal publication of documents.

Living costs are extremely modest. The housing basis we use is that set up for government housing by the FHA or Civil Service. Apartments will run from $33 to $67 a month. Dormitory rooms will run from $15 a month up. At the present time, we have a commissary at which all employees are entitled to purchase supplies. One can get meat, butter, etc., and I think living costs, in general, are less. There will certainly be no objection to universities bearing part of costs of travel if it turns out that students will have to make more trips than one. We would have to govern our interpretation of whether we could allow consultation travel expense. If you are going back to consult with a man who is working on a problem, we would be able to support from this end; if taking an examination, this could not be done.

Dr. Nielsen (Oklahoma): I would like to make a couple of remarks from a point of view different from that of Dean Stewart. I'm not a dean, although I happen to be on a committee such as Dr. Buchta mentioned. I'm from the University of Oklahoma where, up until now, no work has been done in nuclear physics. We're going to have two faculty members who have done work in nuclear physics, trained at the University of California, and I think I'm right when I say that we, at the University of Oklahoma, which is a rather young university, look upon this possibility of cooperation as an opportunity. We hope that through this cooperation it will be possible for young men from Oklahoma to get a better chance in this field than we can give them. We also hope that this project will help us to develop nuclear physics and chemistry faster than we could do without this cooperation.

Dean Stewart emphasized the point of differentiation between degrees done under faculty supervision from honorary degrees. Frankly, I think that these fears are more or less imaginary. I think the only thing we need to consider is whether or not the job that the student has done, the course work, and research is worthy of a doctor's degree; whether it is done under supervision here or in our own university. We have, as all universities have, a large amount of red tape, more than I think we ought to have, or need to have. But most faculties are perfectly ready, whenever the occasion arises, to cut through the red tape. I don't think these formal

419

difficulties will be serious. I think the realities will guide us rather than the formalities.

I'd like to point out what seems to me to be the greatest difficulty in such a program. This question was touched upon by Professor Glockler. This scheme of the student's first doing all course work and general examination, and then spending a year on research here, is somewhat difficult. It is customary, as you know, for the student to do research and the course work more or less at the same time, and work for these two years on research, and if he holds an appointment as graduate assistant, sometimes work three years. I'm wondering if it would be possible to make some such arrangement as this? If a university has a student who is capable, and has an interest in the type of work being done here, to approach this laboratory and go through some of the first procedures of clearance and plan the work several months or a year before the student actually comes here. In that way it would be possible for the student to plan his work in such a way that he will be better prepared for his work here and the concentration of research here in one year would not be quite as difficult under other circumstances.

Dr. Stewart (California): Maybe I didn't express myself clearly about this matter of degrees, earned and unearned, correspondence and honorary, and regularly earned degrees. I think I am speaking for most of the graduate deans in the country. We have discussed comparable things at the Association of American Universities at deans' meetings. I agree there is a lot of red tape and I assure you that I believe in cutting it, but there are some problems that do count up. The point I had in mind is a matter of precedence: the larger university, the more you have to guard against establishing undesirable precedents. If we set up a cooperative project, such as that under discussion, we must recognize the probability that there will be comparable suggestions presented to our various universities. So far as this matter of doing work, good work that is deserving of a degree, is concerned, there is no argument against it, and yet, as considered above, it is full of danger. I don't want to guess the number of requests I get from students who have been at the University of California doing some graduate work, possibly a good many years ago, and since that time have been working for an oil company, or some other organization, where there has been enough to do in research. These people write in regularly asking to submit that work as partial fulfillment, the remaining fulfillment, for the degree. We consistently and regularly refuse such a man. A man cannot go more than four or five years, after taking qualifying examinations, and continue towards his degree. In nine out of ten cases, he has forgotten a number of things he should know. I probably did not make my point clear. I think it is a real issue, administratively, and in drawing up any plans here, I think it is absolutely necessary to take into consideration this danger and protect ourselves against it. There is a distinct difference between an honorary degree and an earned degree, and certainly your universities are going to insist upon keeping these distinctions. I know the University of California is.

Dr. Bradbury: The remarks of Dean Stewart are pertinent. We, of course, are interested in maintaining the standards of the Ph.D. degree. We also feel that the university must maintain the academic control of the student; that, I think, is the point you made. We have been very careful to insist that this is a relation between the university and student that is satisfactory to each. Locally, I think both your point of view and that of Dr. Nielsen are completely reconcilable. Universities may differ in the extent in which they may have a hand in the direction of the student. This may depend on the staff, but I think there is no argument that the university must be satisfied that it is their student doing the work. This is what we insist upon. If this cannot be done, we have no interest in it.

With regard to the second question, the procedure whereby the student may consider a problem in advance: to tie down a specific problem six months in advance would be ill advised. The man could work in a general field in preparation for his research, but to settle on a given problem would mean to tie it up for six months and so delay its solution and run the risk that it will be done elsewhere by someone else. In addition, if it's a good problem it's hot, and we would like to see it done with dispatch.

Dr. Weniger (Oregon): I am an administrator of graduate work and would like to suggest that when the graduate writes in to present a thesis from an oil company, etc., that it be considered as work toward an engineering degree. We just made an engineering degree and got rid of all the people very nicely. Now, of the things that have been mentioned here, and the various rules, I haven't noticed any that couldn't be satisfied. If there is no objection, we could have an arrangement with the student to come back to the original campus for one year, or the necessary time after he completes work at Los Alamos.

Dr. Bradbury: We would have no concern as to when or where he satisfied his academic requirements. It may even be possible for the actual writing of theses to be done away from Los Alamos. It may be necessary to make return trips here periodically; that, again, could be handled in one way or another. The man himself would probably have to handle it.

Question: Is it necessary that the appointment go up to the end of a certain fiscal year?

Dr. Bradbury: No, we have no tie-in with the fiscal year.

Dr. Van Atta (Southern California): This brings out the question of the permanency of the proposition.

Dr. Bradbury: We felt we would not begin to see any results before the beginning of the academic year in the fall. The project is permanent, at least until June 1947. Funds already exist, so I see the proposition at least as far as that. This will mean, I think, that since students will show up at the universities first, we will see students coming here in October, November, or December; otherwise, there is no particular time.

Question: Will you be able to make any estimate of this: suppose a student suddenly decides he would like to do that at a particular time. What would be the time in which he could be cleared before we would start here?

Dr. Bradbury: Security clearance with the governmental agencies takes four to six weeks: sometimes it is shorter, and sometimes longer—it depends entirely on the individual, where he has lived, etc. I would say something like four weeks would elapse from our acceptance and the beginning of the investigation before he would go on the payroll.

Question: Would a man advising students have a choice in problems? Would he be able to choose from a list?

Dr. Bradbury? We would feel it our responsibility to suggest problems that had occurred to us as jobs falling very closely in connection with our work. We would, of course, first discuss with the student the merits of doing this job as opposed to some other job.

Dr. Jacobs (Iowa): I'm particularly interested in this question of classified information. I don't know what the general opinion here is, but quite a few of us feel that now that the war is over, we would be very reluctant to engage in work that must be classified. I think we would have a great deal of difficulty if a thesis could not be written and then submitted to the university.

Dr. Bradbury: I know you represent a general feeling on this matter. I can only say this is a circumstance that is completely out of the control of the Manhattan District. Permission to declassify material has to be obtained from the President. We feel these things are the general background of physics and chemistry and should be public knowledge,

422

but the actual plain fact is that there is no governmental authority to do this.

Dr. Jacobs (U. of Iowa): If general opinion feels that this is desirable, there might be pressure placed on governmental authorities. What is the area in which the work could be unclassified?

Dr. Bradbury: The character of the legislation that is now pending before the House implies there will be a greater release of this information than has heretofore occurred. There is declassification of a great deal of basic work, and I am inclined to think that the whole feeling of the scientists in this country is along the same line. There is a real possibility that by the time the whole program is in effect, this will be a solved problem; therefore, by considering only declassifiable material as of July 1946, we may fail to consider many problems that will be declassified later.

Dr. Gustavson (U. of Nebraska): I think that the history of the feeling of the group in Chicago will be of interest here. After the cessation of hostilities, they were not willing to engage in any work of a classified nature. Yet this has changed, and the feeling of the Chicago group is now much the same as you have expressed. They feel that this is a transition, and that a certain amount of playing along with this has to been engaged in. You also have to remember that, to start out with, although the subject seems to be clearly of a declassified nature, it may develop into something that is highly classified. Another question—can this laboratory be interested in offering training apprenticeship in certain areas; e.g., suppose someone wants to learn the techniques of making measurements on carbon, is there any plan on which he might come here and learn the techniques, and go back to his own laboratory?

Dr. Bradbury: There is no plan of that nature: There will be no objection in principle to that arrangement, but he would have to accept the fact that we would expect in advance a certain duration of employment—six months to one year, so that our normal personnel policies would apply. We have not gone into that particular phase of things. If we did, he would be on the regular employee basis. Here is a man who wants to come to us—we find he can do a certain job; we would expect, before we put him on the payroll, that he would stay a certain length of time.

Dr. Jette (Los Alamos): Our main objective here is to protect the student against having the direction of a thesis change from something that was apparently unclassified to one that might be classified. We would not like to see a student start in on a job in good faith and have it turn out at the end of his job that he can't get published. We believe if anyone takes a beating on this it is the university that should, by making arrangements to accept classified material and thesis submitted to it. I am going to suggest one way of satisfying the library requirements, whereby a certain number of copies of theses must be deposited in the university library. Namely, you have a file, a locked file in the university, the combination of which is known only to the local Army security officers.

Dr. Jacobs (Iowa): I am not sure that I approve of the spirit behind this. The rule should not be broken—that act of placing a thesis in the library to see that some of this information that has been accumulated is available to the rest of the world is in the true spirit of a free science. If it is a patriotic duty to work on secret gadgets in time of peace, it seems to me fundamental that the progress of science and secrecy aren't compatible.

I believe that in peace time most universities will not be able to say that they will be able to give blanket agreement accepting the work regardless of whether it turns out to be classified or not. It seems your responsibility should be that when a man starts on a document in good faith, someone at Los Alamos should see that it is not classified.

Dr. Bradbury: If we knew the answers in advance, we could do this. You will not find something classified occurring in most of the problems. If it does occur, I am sure that the student who has done it would be delighted to have accomplished this particular work.

Dr. Smythe (C.I.T.): Are you going to classify things that were not classified before the war? A good many things would be discovered here. Even in the field of research on nuclear physics you may easily discover things that should be classified.

Dr. Bradbury: That is true. These are problems that exist and that are being worked on. These are the difficulties characteristic of security in peace time. These things occur now, but I think next year will see a great many difficulties solved.

Question: Don't you think it is likely that the university that hadn't considered this point could find a way out when the time arrived?

Dr. Bradbury: I am sure we could.

Dr. Stewart: (California): I think we are making a more serious problem than actually exists. Someone stated that it would be the duty of the Laboratory and university not to penalize the student, and I believe that if a university is willing to cooperate and recognize as, I think, each

424

university must recognize, that each case be an individual case, it would be a very simple matter. I can see no objection why you would not easily agree in advance that it might be classified work and make arrangements in advance. If we do that, we are not going to penalize students.

In respect to the library—there is a responsibility in the matter of filing doctors' theses, something that is sometimes rather difficult, and with considerable frequency people write into the library to borrow a thesis. Students can't be reached, and we feel we cannot release that thesis because it is always dangerous that it will be published as original work, without giving credit to the student. The idea of locking it up isn't such a bad idea.

Dr. Worcester (Colorado): It does seem to me we are making a mountain of a molehill as far as classified and unclassified thesis material is concerned. If we are willing to go into this project at all, we are going into it with our eyes open. You said you might take from 15 to 25 graduate students a year. There are fifteen universities represented here now. This would make an average of one student from each university in difficulties with respect to publication—one in four or five years. Perhaps an institution might be confronted with suppressing the publication of a thesis. I don't think it serious. If we believe in taking advantage of the opportunities we have, we would believe it worthwhile not to publish a thesis every four or five years.

I wanted to ask a question. I feel that many university administrators are not going to encourage leaves of absence for members of its staff. We have all been through a long period of war—many have been teaching long hours, and we need to get away. It seems to me rather than carrying the idea to this meeting that leave of absence will be hard to get that we believe that they should be easy to get. We ought to do everything to encourage it. It may be, however, that a year's leave of absence is impossible. What about three to six months for the members of the faculty? Couldn't they do something worthwhile in this length of time?

Dr. Bradbury: We have discussed this problem and have agreed that this is possible. Three months represents a lower limit, but it would be worthwhile for a man to come for that length of time. Three months is a short time to get results unless the man is interested in a going concern, a problem in process; but between three and six months is a feasible arrangement and a man might make a definite contribution in that time.

Question: What about the possibility of exchange? Do you feel that

425

you have men here at the Laboratory that would be interested in going to a university for a year in exchange, or would it be a one-sided exchange?

Dr. Bradbury: This would be a proposition that would interest some of our people. I mentioned that the background of the project is highly academic—many people have been missing academic association. To be able to return to the classroom means a great deal to people, and while I can't say for certain, I have the feeling that this would be an attractive opportunity for some of our people. That is why we mentioned courses. People like to instruct, if there isn't too much of it.

Question: It seems to me that is a very important matter. It is extremely difficult to get competent teachers, and if we are going to be able to send men down here to the Laboratory, we must get replacements; otherwise it isn't going to work out.

Dr. Bradbury: I can certainly concur that leaves of absence are hard to get. I would like to say that I feel this problem of exchange is one we would be interested in working on. I think it is an established fact that the best research seems to be done in academic institutions in an academic atmosphere. A project of this sort should endeavor to obtain academic associations.

Question: We don't know about the effects on the bodies of individuals. This matter occurred to me when Dr. Gustavson asked about the insurance. What about long-range effects of radiation on humans?

Dr. Bradbury: This is the reason why this particular fund that I mentioned previously was set up—to take care of things we don't understand. All our knowledge is based on present experiments on animals. This is a new field, and we are prepared to take care of people over a long period of time; that is why I mentioned that we plan to study the biophysical aspect of radiation.

Dr. Smythe (C.I.T.): Special requirements in different departments at Cal Tech are entirely different, so anything I have to say has nothing to do with chemistry. I have listened to various discussions and I think there are several ways Cal Tech could use this laboratory. We have, I think, fifteen or twenty-two starting their Ph.D.'s in nuclear physics, and such students, I think, are certain to turn up problems—some theoretical and experimental—which, in the course of development, reach the limit of our facilities. When that happens, it seems to me the logical thing to do, rather than for these students to come here. Most of them will have done a considerable amount of graduate work. I am quite sure the cosmic ray

426

group will be turning up problems that will definitely indicate they must go somewhere else where the equipment is. Arrangements would be desirable for such men. One thing it was suggested I ask about, in particular, was the possibility of postdoctoral work. We have lots of men who work in the field, have theoretical, and some experimental background, but really have only a limited knowledge of nuclear physics. Is it possible for men—competent experimental men—who have a pretty good knowledge of the field to obtain a postdoctoral fellowship at Los Alamos? Could these men come and work two years, or permanently? They would want to get experience with the facilities here for research in nuclear physics.

I think that, unless we can get around the red tape, there is not a very good chance that our men will work on problems originating here for Ph.D. theses. We are under pressure to accept, as theses, work done by former students, some who have quit for some reason or another, but who have subsequently done work outside that is perfectly suitable for a Ph.D. If we ever started accepting such work for Ph.D. theses, we would be running ourselves in for something. No one can certify, for example, that the work was done by that man alone. Furthermore, he may have been working in some narrow field and forgotten all the training he has had. When you give a man a Ph.D., he is supposed to be qualified to do anything that a physicist (or chemist) is supposed to do, and he may want that degree to get a job. The problems will have to originate at Cal Tech.

About the security question. We already have locked up tight a number of theses, and it is not very satisfactory, but in every case we had a promise that the material would be released. We had certificates from faculty members cleared on that particular topic that a thesis was of suitable quality, and the graduate school was willing to accept on their recommendation.

Now, in connection with the Navy. We have been working on some subjects ourselves, and we have come up to the authorities with work about which we didn't know the security status. We were informed that it was highly classified. The fact that we had communicated with them on this subject, in their opinion, seems to make us responsible for maintenance of security on that topic. I wonder if that is likely to happen here.

Dr. Bradbury: I think I can answer the particular question regarding the authority of this project to classify experimental work done outside and not under contract with us. We would not have any authority to classify such work.

Dr. Smythe (C.I.T.): If we have communicated with you and signed a statement that we have discussed the subject with you, we are not put on the spot, then?

Dr. Bradbury: These are very delicate questions. There has been a considerable amount of voluntary classification. If the research is carried out, then as far as I know there is no applicable law other than the Espionage Act. However, research making use of government funds falls under the classification policy. All we can tell you is, if that work were being done here, we would have to classify it. If it were continued here and worked on, it would be classified. The patent feature is also related to this question of research done using government funds. The patent obtained on work done with government funds would belong to the government. There are questions as to the original patent belonging to the university and the improvements belonging to the government that will take years to straighten out.

Question: What about foreign students? Can they be cleared?

Dr. Bradbury: It is desirable that our employees be American citizens by birth or naturalization.

About the question of employing men for postdoctoral work; as far as we are concerned they would be regular employees. They would be employed in accordance with our accepted personnel policies and, of course, be paid more than a man with less experience. I think it would be possible to make very adequate salary arrangements.

Dr. Froman (Los Alamos): It should be noted that if a graduate student came to this group and invented a gadget, it belongs to the United States.

Dr. Bradbury: When a person is employed by government, any patent he has belongs to the government.

Dr. Van Atta (Southern California): Regarding arrangements between the university and the Los Alamos Laboratory: it occurs to me that for an arrangement to work out satisfactorily it should be of permanent standing. A record should be kept of the participation in the work by members of the university staff, through consultant capacity, for members of the university staff. It seems to me that for the work to progress satisfactorily, it would be desirable for that contract to be made for a definite period and continue so that the consultant would have a very good prospective for carrying on the work done here; so he could

428

direct students into some definite channels. Then I believe an arrangement would be satisfactory.

Generally, a student gets into his research job by degrees. If an arrangement were worked out by which students could come here for a period of three months, they might assume the status of helper in Laboratory work. They would learn a great deal that would mean a great deal in their future studies. There is one disadvantage. The interval spent on that basis would be short—three months—equivalent to a summer vacation. However, I would believe such an experience would be extremely valuable. Even from the point of view of contributions to the Laboratory, for he would eventually come on a thesis job and be more effective when he does.

Dr. Bradbury: It would not be possible for the project to pay transportation, as I have indicated. That prerogative is only available for a reasonably long-term employment. However, if a man wanted to get himself here it might be possible to allow this. It might be necessary to stagger such requests, but I quite agree that such primary training would be desirable and would help the man.

Question: Some of the men who come here have children of school age. Is there a school?

Dr. Bradbury: Yes, also adequate hospital facilities.

Dr. Smythe (C.I.T.): About the postdoctoral arrangement? I think these men would not wish to be quite regularly employed, they would not wish to be primarily permanent regular staff members, although it is quite possible they might consider this. They wish to become familiar with the different techniques in nuclear physics. They would not be requesting a very high salary—they would merely want to break even.

Question: How do you intend to proceed from this point to implement the plan of university cooperation we have been discussing today?

Dr. Bradbury: I think we have underscored the fact that there would have to be almost individual arrangements with each institution. We are now ready, or will be next September and thereafter, to undertake the sort of program we have indicated here, to receive students subject to our approval, with whatever degree of consultation or supervision the university in question finds it desirable to suggest and that is acceptable to us. I do not believe that, with the problems of different universities, it is possible to write down a set of rules. We can probably find our way to cooperate on all problems. What I think we can do, and should do at this time, is to prepare, at the Project's expense, an abstract of this meeting.

We will arrange to send you several copies so you can distribute them where you think best.

I think it would be appropriate for you to write me a letter stating the situation as it appears to others at your institution. In that letter they may wish to ask questions that apply to your university.

APPENDIX E

Nuclear Physics Conference Program
August 19-24, 1946
Los Alamos, New Mexico

Monday Morning—Chairman, J. M. B. Kellogg
Opening Address	N. E. Bradbury
Accelerating Equipment at Los Alamos	John Manley
Chain Reactors at Los Alamos	Philip Morrison

Monday Afternoon—Chairman, R. F. Taschek
High-Temperature Pile	Oliver Simpson
Reactor at Oak Ridge	Harry Soodak
Particle Detection	Hana Staub

Tuesday Morning—Chairman, Egon Bretscher
High-Energy Accelerators	Robert Serber
Linear Accelerator	Luis Alvarez
Synchro Cyclotron	Robert Thornton
Synchrotron	Edwin McMillan

Tuesday Afternoon—Chairman, R. R. Wilson
Fission Process	David Inglis
Some Heavy Isotopes	Glenn Seaborg
Fission Process, Chain Lengths	Katherine Way

430

Energy Spectrum of Spontaneous
Fission Fragments   Emilio Segrè
Eniacalculations on Liquid Drop   Nicholas Metropolis

Wednesday Morning—Chairman, R. P. Feynmann
Light Particle Scattering   Julian Schwinger
Resonance in Particle Reactions   E. P. Wigner

(Possible Short Subjects)
Resonances in Disintegration of
Fluorine by Protons   T. W. Bonner
n-d Scattering   C. L. Critchfield

Wednesday Afternoon
NO MEETING

Thursday Morning—Chairman, L. D. P. King
Neutrons as Waves   Enrico Fermi
Scattering by Micro Crystals   R. G. Sachs

Thursday Afternoon—Chairman, Rolf Landshoff
Fast Neutron Processes   Victor Weisskopf

(Possible Short Subjects)
Potential Well Calculation   Frederick Reines
Scattering of Fast Neutrons   A. L. Hughes

Friday Morning—Chairman, Edward Teller
Nuclear Induction   Felix Bloch
Quadrupole Moments   I. I. Rabi
Very Short-Lived Isomer   S. De Benedetti

The Technical Series

Note, 1961. This series has not been made available for outside distribution except for Vol. I, published by McGraw-Hill Book Company, Inc., in the National Nuclear Energy Series; Vol. IV, published by the Government Printing Office; and Vol. VII, part of which was issued as LA-2000 (Unclassified) available from the Office of Technical Services, Department of Commerce.

Volume No.	Title and Description	Editor
0	"Relation Between the Various Activities of the Laboratory"	S. K. Allison

(A general survey of the work of the Los Alamos Laboratory during the war years, with particular emphasis upon the problems of the critical mass and of the efficiency. In addition to a discussion of the gun- and implosion-type bombs, the volume contains a section dealing with other methods of attaining the explosive release of nuclear energy.)

| I | "Experimental Techniques" | Darol K. Froman |

(A description of the experimental physics equipment used by the Los Alamos Laboratory. The volume has three parts: the first dealing with electronics; the second with ionization chambers and counters; and the third with miscellaneous techniques used in obtaining physical measurements.)

| II | "Numerical Methods" | Eldred C. Nelson |

(A survey of the methods used in performing numerical calculations of various types of equations by hand computation and with the use of International Business Machines.)

III   "Nuclear Physics"   R. R. Wilson

(A comprehensive report of nuclear physics measurements made by the Los Alamos Laboratory, together with theoretical evaluations of results and a detailed discussion of the fission process.)

IV   "Neutron Diffusion Theory"   George Placzek

(The theory of diffusion with and without a change in velocity, including a discussion of statistical fluctuations.)

V   "Critical Assemblies"   O. R. Frisch

(A report of critical mass experiments made at Los Alamos with uranium-238 and plutonium assemblies for various tampers. A theoretical discussion is included.)

VI   "Efficiency"   V. F. Weisskopf

(A theoretical method for calculating the energy release of a nuclear explosion.)

VII   "Blast Wave"   Hans A. Bethe

(A study of blast wave phenomena, both from a theoretical and an experimental point of view. Particular emphasis is placed upon the behavior of the blast wave in large explosions, and an effort has been made to interpret blast data from studies made at Trinity, Hiroshima, and Nagasaki.)

VIII   "Chemistry of Uranium and   Joseph Kennedy
                    Plutonium"

(A survey of the problems concerned with the chemical purification and recovery of uranium and plutonium, together with a discussion of the preparation of their various compounds and of the analytical methods used in their study.)

IX   Not assigned.

X   "Metallurgy"   Cyril S. Smith

(A report on the metallurgy of uranium, plutonium, and all other metals fabricated by the CMR Division.)

433

XI   "Explosives"   G. B. Kistiakowsky

(A survey of the experimental work done by the Los Alamos Laboratory on the behavior of explosives and on the techniques of explosive casting.)

XII   "Implosion"   R. F. Bacher

(A report on the experimental implosion program from the early tests to the development of the Trinity bomb. The volume covers work done on polonium, radiobarium, and radiolanthanum.)

XIII   "Theory of Implosion"   R. E. Peierls

(A theoretical survey of the implosion process. The volume contains discussions of shock hydrodynamics, equations of state, and various implosion designs.)

XXI   "The Gun"   F. Birch

(A survey of the experimental gun program from the early tests to the development of the Hiroshima bomb. This volume includes design specifications and a discussion of the interior ballistics of the gun.)

XXII   "Fuzes"   R. B. Brode

(A study of work done by the Los Alamos Laboratory in designing detonating fuze assemblies for the implosion and gun-type bombs.)

XXIII   "Engineering and Delivery"   N. F. Ramsey

(The history of Project "A," together with a discussion of engineering problems encountered in the delivery program. Particular attention has been given to the mechanical design and assembly of the Model 1561 implosion bomb.)

XIV   "Trinity"   K. R. Bainbridge

(A complete report on the 100-ton TNT calibration and rehearsal shot and the July 16, 1945, atomic bomb test at the Alamogordo Air Base. The volume includes both experimental and theoretical discussions of the various phases of the test. A large appendix contains all pertinent Trinity memoranda and all LA and LAMS reports concerning the Trinity explosion.)

434

APPENDIX G

Bradbury's Letter to the Atomic Energy Commission
November 14, 1946

Atomic Energy Commission
Washington, D. C.

Gentlemen:

1. Of the many problems facing your Commission, that presented by the Los Alamos Laboratory may well not be the least. For this reason, the senior technical personnel of the Laboratory have given much thought to their proper role in peacetime under the existing legislation. Although no single statement can completely reflect all the varied opinions of many individuals, there is, nevertheless, sufficient unanimity of thought to warrant its presentation to you in this manner.

2. The Los Alamos Laboratory was established under war conditions and with the greatest secrecy in 1943. The aim of the Laboratory could be stated simply: to devise and produce an atomic bomb, and, if necessary, continue to produce more bombs. Since adequate knowledge of many basic nuclear constants was lacking, an elaborate laboratory for experimental physics was set up with all the relevant equipment for research of this character. To interpret these results, to design a weapon around them, and to predict its behavior, mathematicians and theorists of the highest caliber were active in a Theoretical Division. The chemistry, chemical engineering, and metallurgy of the active materials, except in microscopic quantities being almost unknown, heavily staffed divisions were set up to contend with these new and complicated problems. When it became apparent that high explosives would play a prominent role in a heretofore unknown way in the assembly of active material into a super-critical assembly, appropriate experts attacked the problems and established a research pilot plant for the study and production of high-explosive charges of the necessary quality and character. Other divisions

435

attacked the problem of the experimental study of the assembly techniques in order that the probability of success of the weapon might be high on its initial attempt. Finally, due to the urgency and secrecy of the whole problem, another large group of scientists and engineers formulated and solved the entire problem of ordnance engineering including the method of delivery of the weapon and its ballistic behavior. In many of these diverse responsibilities, the Laboratory drew upon other organizations and subcontractors and was responsible for the direction and coordination of their efforts.

3. The success of all these efforts is known to the world. With the close of the war, the original directive of the Laboratory was completed and no new general directives carrying national approval or acceptance were to be found. The confusion of the nation and the world with respect to this new weapon was reflected in Los Alamos. The majority of the most senior technical personnel returned to their old or to new academic institutions; many of the younger personnel hastened to accept appointments, positions in industry, or returned to colleges and universities to complete work for academic degrees.

4. It soon became clear that atomic energy legislation would not immediately be forthcoming and that the Manhattan Engineer District would be directed to maintain essentially the status quo until such legislation was available. From a variety of personal motives, but all having in common the belief that to abandon work on atomic weapons and the fundamental processes involved therein was contrary to the best interests of the nation at this time, a considerable number of individuals elected to remain with the Los Alamos project until there was clarification of national policy in this field. During the course of the following year, other scientists, many of whom had not previously been connected with the Manhattan District, joined the Los Alamos Laboratory. Despite the departure of most of the nationally and internationally known scientists associated with the project during the war, many of the men who had worked with them remained and took over technical and administrative responsibilities. While not at its former extraordinary level, the technical status of the Laboratory remained high.

5. The philosophy and directives of the Laboratory during the last year have been developed partly internally and partly externally. The Manhattan District took a definite interest in the stockpiling of weapons of the type in existence at the close of the war. Toward the end of the period, an interest was also taken in establishing purely military units to take over

436

the ordnance assembly, delivery, and logistic problems associated with the weapon. The participation of the Laboratory in the Crossroads Operation was the result of a directive from the Joint Chiefs of Staff through the Manhattan Engineer District.

6. For itself, the Laboratory proceeded upon a general philosophy that it would endeavor to set up and maintain a program that represented the best approach to its own conception of the nation's need for a laboratory concerned with atomic weapons, both present and future. Believing that the present weapon was primarily the fruit of fundamental research extending over a period of years, it was considered that such research should play a definite role in the life of the Laboratory. Such research has been conducted in all the technical fields that bore upon the development of the weapon. This includes not only nuclear physics and chemistry, but high explosives, equations of state, radiation, hydrodynamics, and phenomena of solids. Since the development of many portions of the weapons had proceeded on an almost entirely empirical basis, attempts were now to be made to increase the understanding of the processes involved.

7. Because of this interest in basic research and our conviction that such research alone can provide the foundation for a strong nation, we have undertaken two major developments, which, while they will bear upon fundamental weapon research, will contribute equally strongly to basic scientific advance and to the peaceful uses of atomic energy. These developments are: first, the construction of a large pressurized Van de Graaff accelerator to replace the one returned from this project to the University of Wisconsin, but capable of reaching voltages up to eight or ten million volts; and second, the construction of a nuclear reactor of approximately 10-kW output operating on fast neutrons and utilizing plutonium as active material. Design work has been completed and some procurement started upon the accelerator, and the reactor is expected to be in preliminary operation early in 1947.

8. Since the close of the war found the present type of weapon in a stage of engineering development such that the probability of successful firing and functioning was unknown, a program of establishing this reliability was set up. The weapon itself was aptly termed during its development a "gadget." Some effort was to be spent in improving the engineering of the whole model. The demands of the Crossroads Operation interfered seriously with these objectives, which cannot yet be said to have been obtained.

437

9. The possibility of a slightly different type of weapon . . . model was known at the close of the war. The potential improvement in performance of such a weapon. . . under these circumstances indicated this to be one of the problems upon which effort should be expended, and these efforts have been, in the main, successful.

10. Again, at the close of the war, it was known that a possibility existed of employing elements of low atomic weight in a "Super" weapon, which, if capable of development, would be thousands of times as effective as the present weapon. Since the program for such a weapon as then conceived would involve a laboratory fully as extensive as Los Alamos at the peak of its activity, and would require, as well, large developments in other portions of the Manhattan District, the interest of the Laboratory was restricted to determining the feasibility of the weapon and to research and theoretical calculations bearing toward this end. These investigations, in which we have had the advice and consultation of previous experts in this field, have led to no decrease in our expectations that such a weapon could be constructed were the necessary effort to be expended thereupon. Furthermore, there has appeared a somewhat different suggestion as the result of these considerations that indicates the definite possibility of a weapon many times superior to the present one but lying reasonably within the capabilities of the Laboratory.

11. The explosives research has concentrated upon gaining a better understanding of the unusual techniques involved in the weapon. Conspicuous advances have been made in the development of certain types of materials, which have an explosive-like behavior in lenses and have certain special advantages for the weapon, but which are by themselves largely inert.

12. The production aspect of active material chemistry and metallurgy has required extensive development as, at the close of the war, a new plant for the production of active material had not yet been put into actual operation. Preliminary efforts to place this plant in operation indicated the necessity of major design changes and the desirability of extensive research on a radically modified process. Recently, such a process has been developed and put into operation with extremely gratifying results and enormous increases in the safety and efficiency of the operation.

13. The declassification of documents concerned with basic scientific data and techniques that could be released by the project under the

provisions of the recommendations of the Tolman Committee has resulted in the release of about two hundred documents and a corresponding reentry of Los Alamos personnel into the scientific world. Approximately an equal number of documents are in process of release. Arrangements for cooperation with universities of the region with respect to the training of graduate students have been undertaken with satisfactory preliminary results and good expectations for the future.

14. The Manhattan Engineer District has realized the inadvisability, both from the point of view of our restricted facilities and from the standpoint of their possible loss, of concentrating production activities relating to atomic weapons at this site. During the past year, the routine production of "standard" high-explosive charges has been taken over by Salt Wells Plant of the Naval Ordnance Test Station at Inyokern, California. Plans are in progress to transfer the routine production of current-type nuclear initiators to the Monsanto Chemical Company at Dayton, Ohio, and personnel to take over this work are currently being trained at Los Alamos. The production of electric detonators is still continued at this site in view of the research aspects of the problem that still remain and that require pilot plant production facilities to solve. The production of active material into the required shapes has been maintained here since there are, at present, no other facilities in the country that can participate. We have also continued, for the time being, the production of normal uranium and aluminum parts associated with the active material in the weapon. The electronic components and mechanical components were largely stockpiled shortly after the close of the war and have represented only an inspection and modification burden upon the project personnel, with development being carried on by a subcontractor.

15. The Sandia Laboratory of this project, with increasing military participation, has taken cognizance of the stockpile storage problem, of the weapon assembly problem, the modification of stockpile parts, and the test and acceptance of components, other than nuclear or high explosive for the stockpile. In addition, this base has furnished the facilities for the limited amount of ordnance engineering development and tests of reliability carried out by the project.

16. In retrospect, it is believed that the project has functioned reasonably well in fields that involve basic physics, chemistry, metallurgy, and high explosives. It has made progress in the design of nuclear components for weapons. It has not made satisfactory progress

439

in the ordnance engineering of weapons. Administratively, some progress has been made in employment practices for personnel; the payment of return travel expenses has been discontinued except for special cases; and property procedures have been modified to meet more nearly the requirements of peacetime practice. Insufficient research has been carried out directed towards, establishing a basis for more satisfactory health safety practices for the special hazards affecting this type of work, although progress has been made in the routine administration of known tests and precautions. The practice of making critical assemblies has been discontinued until a remote control technique is available early in 1947. There has been inadequate integration of the community with respect to the relationships between the Technical Area and the Military Post. In spite of continued efforts by senior technical and senior military personnel, there remains an incompatibility or antagonism apparently impossible to overcome, resulting in a continual concern by the civilian technical personnel that the "Army will take over." The roots of this difficulty lie far in the past, and the problem itself may not be unique to Los Alamos.

17. Your Commission now faces the problem of determining the character and future directives of Los Alamos. Unfortunately, the local project is so small that the problems of the community bear upon the character of work done by the Technical Area, and reciprocally, the existence of the Technical Laboratory determines the existence of the community. While these problems can be discussed separately, their simultaneous successful solution is required for the success of either.

18. The Los Alamos Laboratory does not presume to indicate to the Commission what the policy of that body should be with respect to the national need for atomic weapon development. Nor should the Laboratory as such express its views on the relationship of such a national program to the international scene. The discussion that follows is based upon the assumption that the United States will require, for an unknown time to come, a program in atomic weapon development and research. Such a program should be directed not only at maintaining an immediate superiority for the United States in this field, but also toward maintaining general scientific progress and a concern for basic and long-range developments, which will make for strength in the future. It is also assumed that the government of the United States must know what weapons might be arrayed against it for the proper formulation of its own national and international policies. The ensuing discussion is based, in

addition, upon an assumption, which the Laboratory can only suggest, that the Commission shares with the established armed forces of the United States a responsibility for the security and defense of the country; that the atomic weapon plays a fundamental role in any security program set up at this time; and that, therefore, the Commission and the Army and Navy are jointly concerned with this problem.

19. It has been noted that, up to the present time, the Los Alamos Laboratory has been responsible for the atomic weapon in its entirety. The atomic bomb has been employed by the armed forces exactly as received from Los Alamos and assembled with only Los Alamos personnel. There has remained, ever since the close of the war, concern as to the engineering reliability of the weapon, as well as a conviction that engineering improvements were not only possible but desirable. The skepticism of the armed forces with respect to the ballistic determinations of Los Alamos personnel has already been apparent, and it may be anticipated that this feeling will grow to include the fusing and firing mechanisms and the complexity of weapon assembly. It is further noted that a demand is already apparent for weapons of somewhat different engineering properties—e.g., a weapon that will penetrate the surface of water and detonate at a predetermined depth. Other requests from the armed forces, including the guided missile investigators, may be expected to appear shortly.

20. It is the belief of the senior technical personnel at Los Alamos that this laboratory should not attempt to carry out these purely ordnance engineering aspects of atomic weapon development. Conversely, it is strongly suggested that these problems should be handled using the Sandia Laboratory and the existing ordnance facilities of the Army and Navy, as well as additional laboratories that may have to be set up.

21. It is suggested to your Commission that the Los Alamos Laboratory may be most effective if its concern is limited to the nuclear components of atomic weapons including, naturally, the technique of supercritical assembly of active material. The Laboratory would then be expected to carry out research on both long-range and short-range modifications in the nuclear structure of atomic weapons, but would not be expected to present to ordnance engineering laboratories more than a functional design for a weapon with the exception of those parts intimately concerned with the nuclear reaction.

22. Such a division of responsibility will clearly call for the most active liaison between this laboratory and such other laboratories as are

441

carrying out the engineering development. While such liaison will present problems, they are not believed to be insurmountable. To maintain the present philosophy and localized Los Alamos responsibility for complete weapon development will not only result in a practical strangulation of effort devoted to long-range research, but will curtail the responsibility of the armed forces in a problem in which they are presumably able and anxious to participate.

23. It is further suggested that Los Alamos retain the responsibility for testing the nuclear reactions for new atomic weapons, but that such tests as have a purely military significance be carried out by the armed forces. The distinction intended is that of separating a test of the "Alamogordo" type from a test of the "Crossroads" type. In view of the limited facilities of this laboratory, however, the most active assistance of the armed forces would be required in subsequent "Alamogordo"-type tests, but the directive responsibility would come from this laboratory.

24. There is attached herewith a statement of the Laboratory program for 1947. Following the suggestion above, this project would be relieved of essentially all the program of the Weapon Engineering and Development Division (Paragraph "F"). In its place would be formed a much smaller engineering group whose responsibility to the weapon program would be to represent this laboratory in its relation with the Sandia Laboratory or other agencies carrying out atomic ordnance engineering, and to prepare for such agencies the preliminary suggestions for the design of weapons having a different ordnance character. The Sandia satellite of Los Alamos would become an independent entity and would probably have to be considerably enlarged.

25. The problem of the "production" of atomic weapons has been considered above. It is believed that no immediate change can be made in the extent of the limited actual "production" carried out by Los Alamos. However, if the philosophy of maintaining Los Alamos as an atomic weapon research laboratory is carried out, it is suggested that plans be made to remove as much as possible of this routine activity from this site. This has the additional advantage of disseminating the knowledge of the necessary techniques, as well as decreasing the seriousness to the nation of a major accident or catastrophe at Los Alamos.

26. A program of training regular military personnel in the assembly and component part testing of the current atomic weapons has already been instituted with personnel based at Sandia, but receiving the nuclear and high-explosive phases of their indoctrination at Los Alamos. The Los

Alamos Laboratory is no longer adequately staffed with personnel whose primary responsibility is the assembly of atomic weapons, although such assembly could probably be done in a grave national emergency.

27. The stockpiling and stockpile storage of atomic weapons is becoming a purely military responsibility for which the current headquarters are located at the Sandia Base.

28. It is probably true that the above activities in weapon research and development are by no means sufficient if this country is to engage actively in an atomic armament race with any hope of ultimate success. At the most, such a program can hope to achieve only a temporary security; at the least, it preserves a framework from which an expansion can occur if this becomes inevitable for the nation.

29. Up to the present time, this laboratory has not concerned itself with the application of nuclear energy to problems of military propulsion. Although the preliminary phases of such development fall within the experimental and theoretical background of Los Alamos, it is dubious if an adequate program could be carried out at this time without an expansion of the facilities of both the Laboratory and community. However, the character of the experimental research required by such a program, as well as its intermediate nature between a reactor and a nuclear explosion, suggests the possibility of ultimate participation in this field by the Los Alamos Laboratoy.

30. The intimate relationship between the Los Alamos community and its technical activities has been mentioned above. During the war and the postwar year, this relationship has had a hybrid civilian-military character with the community considered as a rather unusual variety of Army Post. The exact character of the community status has varied with the dispositions and directives of the various Commanding Officers and the Director of the Laboratory. It cannot be said that there has ever been unanimity or complete satisfaction on this subject.

31. For many reasons, it is desired to suggest most earnestly to the Commission that they give the strongest consideration to operating Los Alamos—if it is to be operated with approximately the above philosophy—as a one-contractor civilian operation under the jurisdiction of the Commission and one or more of its Directors. We state with reluctance, but with conviction, that we do not believe that a continued Army operation of Los Alamos as a research laboratory and attached community will be successful. Nor do we believe that under Army operation it will retain or attract personnel adequate for the tasks facing

it. Even the purely military guarding function will probably be better done ultimately with civilian guards.

32. The operation of Los Alamos as a "company town" in itself is characterized by a number of complications of which not the least arises from a general desire of personnel to own their homes and to be responsible through election and community taxation for the conditions under which they live. At the present time it is believed that these problems permit of solution, the accomplishment of which may be difficult and slow, but not beyond the powers of enlightened management.

33. Whether or not Los Alamos should be continued over a long period of time is doubtless a problem that will be considered by the Commission. This question has naturally received consideration here, and having received a tentative affirmative answer, has resulted in extensive programs of permanent construction. Many, but not all, of the activities proposed for this laboratory should not be conducted near populated areas. The isolation of the site represents certain community problems which are largely if not entirely balanced for personnel now here by the attractions of the climate and of the present mountainous location. The isolation of the technical community is more easily handled by a policy of encouraging attendance at national and regional scientific meetings, both of regular scientific societies and within the Manhattan District. The absence of railroad connections has contributed to a somewhat higher cost of transportation of materials to the project. Not a negligible factor involved in a proposed change of location is the fact that a large number of technical personnel have remained with the project because they and their families enjoy this location more than urban communities. It is hoped, should a new location be considered, that its advantages will be conspicuous.

34. Should the international situation develop to the point at which the United States may cease to have any concern for further weapon development or production, the Los Alamos Laboratory program would require careful reconsideration. Since, presumably, this is not a point at issue at the present time, it need not be considered here except to state that the operations involving plutonium, the basic chemistry and physics, the fast reactor, the large Van de Graaff accelerator, studies of materials at high temperatures, pressures, and radiation densities are all activities that will undoubtedly play a role in the peaceful applications of atomic

energy no less important than the role they play in a program whose objective is weapon research.

35. The above discussion in no sense deals with the programs or problems of the Laboratory, but it is believed that the major ones have been presented. The staff of the Laboratory is, of course, deeply concerned with the attitude of the Commission with respect to these matters, and is, of course, available for any more detailed discussion that the Commission may desire.

Respectfully submitted,

N. E. Bradbury
Director

NEB/b

Glossary

(α,n) reaction   Any nuclear reaction in which an alpha particle (helium nucleus) is absorbed by a nucleus, with subsequent emission of a neutron

autocatalytic assembly   Any method of assembling supercritical amounts of nuclear explosive in which the initial stages of the explosion assist further explosive assembly; for example, by expulsion or compression of neutron absorbers placed in the active material

baratol   A castable explosive mixture of barium nitrate and TNT

Baronal   Castable explosive mixture of barium nitrate, TNT, and aluminum

betatron   Induction electron accelerator for generating electron beams of very great energies

branching ratio   The ratio of the capture cross section to the fission cross section

Cockcroft-Walton   An accelerator using voltage multiplication of the rectified output of a high-voltage transformer to obtain a high potential

Composition B   A castable explosive mixture containing RDX and TNT, $C_{6.85}H_{8.75}N_{7.65}O_{9.3}$

critical mass   That amount of fissionable material that, under the particular conditions, will produce fission neutrons at a rate just equal to the rate at which they are lost by absorption (without fission) or diffusion out of the mass

tamped critical mass   When the active material is surrounded by a tamper

critical radius   The radius of spherical fissionable material equal to one critical mass under existing conditions

447

cross section   A quantitative measure of the probability per particle of the occurrence of a given nuclear reaction; the number of nuclear reactions of a given type that occur, divided by the number of target nuclei/cm^2 and by the number of incident particles

absorption cross section   Where a neutron is absorbed by a given nucleus

capture cross section   In the (n,γ) reaction, when a nucleus absorbs a neutron with subsequent emission of gamma radiation

fission cross section   Where a neutron is absorbed, followed by fission

scattering cross section   The cross section for neutron scattering by the nuclei of some target material; because scattering is quantitative, the definition is incomplete; the differential scattering cross section scatters at an angle between θ and $\theta + d\theta$; the transport cross section is an average or integral scattering cross section, which gives the average scattering in the forward direction,

$$\sigma_T = 2\pi \int_0^{\pi} (1 - \sin \theta) \, \sigma_s (\theta) \sin \theta \, d\theta \, ,$$

where $\sigma_s(\theta)$ is the differential scattering cross section defined above

cyclotron   Magnetic resonance accelerator, used in investigating atomic structures

D(d,n) reaction   The nuclear reaction produced by bombarding deuterons with deuterons, producing high-energy neutrons

D-D source   The above reaction used as a source of high-energy neutrons; at Los Alamos, the Cockcroft-Walton accelerator was used principally

deuterium   Heavy hydrogen, D_2 or $_2^2H$, the hydrogen isotope of mass two

deuteron   A nucleus of deuterium or heavy hydrogen

448

electron volt The energy acquired by an electron falling through a potential of 1 volt; about 1.6×10^{-12} ergs; in thermodynamic units, 1 electron volt corresponds to about 12 000 K; thus 1/40 V per particle corresponds to "room temperature;" energies of this range are called "thermal;" 1 MeV corresponds to a temperature of 1.2×10^{10} K

fission spectrum Energy distribution of neutrons emitted in the fission process

inelastic scattering Neutron scattering in which energy is lost to excitation of target nuclei

Li(p,n) reaction Neutrons are produced by lithium bombardment by protons

neutron number The average number per fission; statistically variable

(n,γ) reaction A nuclear reaction in which a neutron is captured by a nucleus, with subsequent emission of gamma radiation

PETN Pentaerythritol tetranitrate, $C_5H_8N_4O_{12}$

RDX Cyclotrimethylenetrinitramine, $C_3H_6N_6O_6$

thermonuclear reaction Induced by thermal agitation of the reactant nuclei; self-sustaining if the energy release counterbalances the energy losses that may be involved

tamper A neutron reflector placed around a fissionable mass to decrease the neutron loss rate

Taylor instability A hydrodynamical principle stating that when a light material pushes against a heavy one, the interface between them is unstable, and that when a heavy material pushes against a light one, the interface is stable

tritium The hydrogen isotope of mass three discovered in the Cavendish Laboratory by Oliphant in 1934, where it was produced by deuterium-deuterium bombardment; radioactive gas with a half-life of about twenty years

triton A nucleus of tritium

thermal neutrons Neutrons of thermal energy; see **electron volt**

T-D reaction The nuclear reaction of tritons with deuterons

449

Torpex   A castable explosive mixture of RDX, TNT, and aluminum, $C_{1.8}H_{2.0}N_{1.7}O_{2.2}Al_{0.7}$

Van de Graaff   An accelerator using the electrostatic charge collected on a mechanically driven belt to obtain a high potential

Bibliography

Amrine, Michael
GREAT DECISION: the secret history of the atomic bomb
New York, Putnam, 1959.

Baker, Paul R, ed.
ATOMIC BOMB: the great decision. 2nd rev. ed.
Hinsdale, Illinois, Dryden Press, 1976.

Batchelder, R. C.
IRREVERSIBLE DECISION, 1939-1950
New York, Macmillan, 1965.

Blackett, Patrick M. S.
FEAR, WAR, AND THE BOMB: military and political consequences of
atomic energy
New York, Whittlesey House, 1949.

Blow, Michael
HISTORY OF THE ATOMIC BOMB
New York, American Heritage, 1968.

Brown, Anthony Cave and Charles B. McDonald, eds.
SECRET HISTORY OF THE ATOMIC BOMB
New York, Dial Press/James Wade, 1977.

Chambers, Marjorie Bell
TECHNICALLY SWEET LOS ALAMOS: the development of a federally
sponsored scientific community
Doctoral dissertation, University of New Mexico, Albuquerque, 1974.
University Microfilms, 1975.

Chevalier, Haakon
OPPENHEIMER: the story of a friendship
New York, Braziller, 1965.

451

Church, Peggy Pond
HOUSE AT OTOWI BRIDGE
Albuquerque, University of New Mexico Press, 1959.

Clark, Ronald W.
BIRTH OF THE BOMB
London, Phoenix House, 1961.

Compton, Arthur H.
ATOMIC QUEST
Oxford, England, Oxford University Press, 1956.

Curtis, Charles P.
THE OPPENHEIMER CASE
New York, Simon and Schuster, 1955.

Dean, Gordon
REPORT ON THE ATOM: what you should know about the atomic energy
program of the United States, 2nd ed.
New York, Knopf, 1957.

Feis, Herbert
JAPAN SUBDUED: the atomic bomb and the end of the war in the Pacific
Princeton, N. J., Princeton University Press, 1961.

Fermi, Laura
ATOMS IN THE FAMILY
Chicago, University of Chicago Press, 1954.

Fogelman, E.
HIROSHIMA: the decision to use the A-bomb
New York, Scribner, 1964.

Gellhorn, Walter
SECURITY, LOYALTY AND SCIENCE
Ithaca, N. Y., Cornell University Press, 1950.

Gilpin, Robert
AMERICAN SCIENTISTS AND NUCLEAR WEAPONS
POLICY
Princeton, N. J., Princeton University Press, 1962.

Gilpin, Robert and Christopher Wright, eds.
SCIENTISTS AND NATIONAL POLICYMAKING
New York, Columbia University Press, 1964.

Giovannitti, Len and Fred Freed
DECISION TO DROP THE BOMB
New York, Coward-McCann, 1965.

Goldschmidt, Bertrand
ATOMIC ADVENTURE: its political and technical aspects (translated from
the 2nd French edition)
Oxford, England, Pergamon, 1964.

Goudsmit, Samuel
ALSOS
New York, Schuman, 1947.

Gowing, Margaret
BRITAIN AND ATOMIC ENERGY, 1939-1945
New York, St. Martin's, 1964.

Grodzins, Morton and Eugene Rabinowitch, Eds.
ATOMIC AGE: scientists in national and world affairs
New York, Basic Books, 1963.

Groueff, Stephane
MANHATTAN PROJECT: the untold story of the making of the atomic
bomb
Boston, Little, Brown and Co., 1967.

Groves, Leslie R.
NOW IT CAN BE TOLD: the story of the Manhattan Project
New York, Harper, 1962.

Hewlett, Richard G. and Oscar E. Anderson, Jr.
HISTORY OF THE UNITED STATES ATOMIC ENERGY
COMMISSION. Volume 1. The New World, 1939-1946.
University Park, Pa., Pennsylvania State University Press, 1962.

Johnsen, Julia Emily, compiler
ATOMIC BOMB
New York, H. W. Wilson, 1946.

453

Jungk, Robert
BRIGHTER THAN A THOUSAND SUNS
New York, Harcourt, Brace, 1958.

Knebel, Fletcher and Charles W. Bailey II
NO HIGH GROUND
New York, Harper, 1960.

Kunetka, James W.
CITY OF FIRE: Los Alamos and the atomic age, 1943-1945. 2nd ed.
Albuquerque, University of New Mexico Press, 1979.

Lamont, Lansing
DAY OF TRINITY
New York, Atheneum Publishers, 1965.

Libby, Leona Marshall
URANIUM PEOPLE
New York, Crane, Russak and Co., 1979.

Mcphee, John
CURVE OF BINDING ENERGY
New York, Farrar, Straus and Giroux, 1974.

PURCELL, John
BEST-KEPT SECRET
New York, Vanguard, 1963.

Rouze, Michel
ROBERT OPPENHEIMER: the man and his theories
New York, Eriksson, 1965.

Savage, John and Barbara Storms
REACH TO THE UNKNOWN
(The Atom, Volume 2, Number 8 (Special Issue, January 16, 1965)
Los Alamos, NM, Los Alamos Scientific Laboratory, 1965.

Smith, Alice Kimball
PERIL AND A HOPE; the scientists' movement in America: 1945-47
Chicago, University of Chicago Press, 1965.

Smyth, Henry De Wolf
ATOMIC ENERGY FOR MILITARY PURPOSES: the official report on
the development of the atomic bomb under the auspices of the United States
Government, 1940-1945
Princeton, N. J., Princeton University Press, 1945.

Strauss, Lewis L.
MEN AND DECISIONS
Garden City, N. Y., Doubleday, 1962.

Strout, Cushing
CONSCIENCE, SCIENCE AND SECURITY: the case of Dr. J. Robert
Oppenheimer
Chicago, Rand McNally, 1963.

Teller, Edward and Allen Brown
LEGACY OF HIROSHIMA
Garden City, N. Y., Doubleday, 1962.

U. S. Atomic Energy Commission
IN THE MATTER OF J. ROBERT OPPENHEIMER. Transcript of hearing
before Personnel Security Board, Washington, D. C., April 12, 1954 through
May 6, 1954
Washington, D. C., U. S. Government Printing Office, 1954.

Name Index

457

458

461

Subject Index

A Division (see Administrative Division)

"Able Test" 274, 327

accelerating equipment, original 316

accidents
  compensable 300-302
  critical materials 164, 199, 270, 271, 325, 326
  fire 230, 339
  plutonium 55
  prevention activities 116, 289, 294
  radiation 291, 293, 300, 301
  rates 290

accident insurance 46, 161

Accounting Office, Los Angeles, 43, 46

active material receipt 240

administration
  of Laboratory 29
  recommendations of Reviewing Committee 23
  reorganization 42-43, 155
  Trinity 234-236

Administration and Services (A and S) Division 278-303

Administrative Assistant 278

Administrative Board 156

Administrative (A) Division 34, 155, 272

Administrative Group 279

Administrative Officer 32

Advanced Designs Group, T Division 306

Aero Insurance Underwriters 46

age distribution, civilian personnel Graph 1

"age theory," Fermi's 79

Air Blast and Earth Shock Group, Trinity 236

Airborne Measurements Group, Trinity 235

airborne tests, gun 194, 195, 247, 249

air burst test 274

Aircraft Ordnance Team, Alberta Project 252

aircraft release mechanism 248

Air Force, 20th 250, 253

Air Transport Command 253

Alamogordo
  Air Base 238, 280, 294
  Bombing Range 233
  Test (see Trinity Test)

Albemarle 276, 286

Alberta Project 158, 159, 166, 167, 171, 172, 191, 195, 247-258

Albuquerque, New Mexico 160, 273

Albuquerque District Office, U.S. Army Corps of Engineers 9, 36, 59, 60, 250

allocation of new cars 283

alloys (see plutonium alloys; uranium alloys)

alpha particles
  investigation 107, 183, 200

altitude chamber 345

American Federation of Labor (AFL) 59

americium 178

Ames biscuit metal 345, 335

Ames, Iowa (see Iowa State College of Agriculture and Mechanical Arts)

"Amos" unit 195

amplifiers 207, 314

Analysis Group, CM Division 134, 151, 217, 227-228

Analytical Chemistry Group, CMR Division 329-332

analyzer, electronic 314

Anchor Ranch Proving Ground 111, 115, 117, 124, 127, 128, 194, 212

antimony-beryllium neutron source 99

Appalachian 348

April conference (see conferences)

APS/13 radio altimeter (Archie) 120, 121, 195

463

464

bomb (cont)

reduction technique 71, 219
stockpiling 344
testing 131, 132, 253, 273-277, 343
bombing tables 191, 195, 196
bonding agents 139
boron
absorption measurements 98, 101
"bubble" autocatalysis 188
compacts 224
fabrication techniques 71, 147
impurities in calcium, uranium, and plutonium mixtures 152
neutron absorbers 71, 146, 187
-paraffin screen 95, 187
separation 109
trifluoride counters 109, 149, 227
"branching ratio" 100-102, 180, 181, 446
bridge wire detonator 206 (see also detonators)
British mission 16, 26-28
airplane (Lancaster) 131
early work 19, 96
high-explosive bomb data 176
lens development work 129
personnel 26-28, 280
Brookhaven 269
Brown University 40, 136
refractories 136
Bruceton (see Explosives Research Laboratory)
Bruns General Hospital, Santa Fe 302
"bucket chamber," neutron counter 149
Buell, T. H. Co. 34
Bureau of Ordnance 250
Bureau of Standards 8, 131
Bureau of Yards and Docks, Navy Depart-
ment 250
Business Manager 43, 280, 303
Business Men's Assurance Company 303
(see also insurance)
Business Office 34, 43-46, 61, 160, 161, 280,
294-303
Business Officer, University of California 6
Buying Group, Procurement Section 50

C-54 transport 251, 253, 255, 276
cadmium-paraffin-covered counter 320, 322
cadmium plate, control valve 106
Cadwalder, Wickersham, and Taft 34
calcium fluoride in RaLa program 332
calcium oxide impurities 228
Calculations Group
Ordnance Division 191
California (see University of California)

California Institute of Technology 61, 137,
159, 160, 163, 210 (see also Camel Project)
California State Employees Retirement System (S.E.R.S.) 44, 296
calorimeter 224
Camel Project 159, 160, 196, 210, 216, 248,
250, 273
cameras (see also Photographic Group)
armored 207
color 245
Crossroads 327, 328
drum 117
Fairchild aero view, Trinity 245
Fastax 116, 233, 245
gamma ray 245
high-explosive flash light 126, 127
oscilloscope, 207 (see also oscillograph)
rotating mirror 206, 207, 210, 213
rotating prism 127, 207, 210
stereoscopic 207
x-ray 127, 213
Canadian Project (see Montreal Project)
"canning" rods 313
car allocation plan 283
carbon
microdetermination, gasometric analysis 153
tamper 103, 104
Carnegie Institution, Washington, D.C. 4, 8
work on fission cross sections 19
cast explosives (see high explosives)
casting plant 127, 128, 211-214, 273
Cavendish Laboratory 28
censorship of mail 38
Central Mail and Records Group, A
Division 280, 284
centrifuge reduction method 145
cerium production 136, 141, 145
cerium sulfide 136, 144, 146, 147, 224
chain reaction 10, 177, 198
chain reactor, controlled (see Water Boiler)
Charge Inspection Section, X Division 208,
209
"Charlie" Test 274
check cashing facilities 44, 45, 297
Chemical Research and Development Group,
CMR Division 329, 330, 332
Chemistry and Metallurgy (CM, later called
CMR) Division 134-154, 167, 217-231,
271, 272, 328-337
control of plutonium hazard 54, 55, 170,
293
early program 56, 134, 155, 156
projectile and target fabrication 167, 170,
324
Safety 289

Chicago (see University of Chicago)
Christy implosion, solid plutonium spheres 223
chronograph, Potter 116
chronotron development 320, 321
civilian personnel
age distribution Graph 1
employed Graphs 2-6
Tinian 252, 253
cladding techniques 219, 220
clearance of personnel 37
Clinton Laboratories (see also Oak Ridge)
betatron recording 128, 174, 202, 207
cloud chamber data 96, 108, 178
consultants 306
plutonium 72, 74, 138, 142, 167
plutonium spontaneous fission rate 95, 155, 178
polonium 73
radiobarium and radiolanthanum 137, 226
tritium 88
CM Division (see Chemistry and Metallurgy Division)
cobalt 104
Cockcroft-Walton accelerator 8, 48, 91, 102, 104, 188, 207, 311, 314, 316, 317, 446
Cockcroft-Walton Accelerator Group, P Division 310, 311
College of the City of New York 306
Colloquium 30, 33, 272
Columbia University 6, 8, 47, 286, 306
isotopic analysis, mass spectrographic method 107
Combined Policy Committee, Great Britain-USA 26, 28
Commandant, Navy Yard, Mare Island 250
Commandant, 12th Naval District 250
Commander, Western Sea Frontier 250
Commanding Officer, 509th Group 247
Commanding Officer, Special Engineering Detachment 41, 42
Community Council 35
composite core (see implosion bomb)
Composite Weapon Group, T Division 172, 304
Composition B 211-213, 341, 446
compression studies 202, 203, 210, 215
computations Group, T Division 172, 304, 305, 306
condenser blast gauges, Trinity 232
condenser microphone method of investigating implosion 204, 244
conferences
Berkeley, June 1942 86
Los Alamos, April 1943 10, 124
Los Alamos, January 1944 39

conferences (cont)
Los Alamos, February 1945
Nuclear Physics Conference, August 1946 270, 430, 431
"Super" Conference, April 1946 269
Technical and Scheduling, December 1944 159
University of Chicago, summer 1942
University Affiliations Conference, July 1946 269, 270, 380-430
Consolidated Steel Company 320
construction
and maintenance 58-60, 288, 289, 304
betatron 202
CM 135
DP Site 218, 230
Pajarito 326
RaLa 203
S Site 128
Tinian 251, 253
Trinity 234, 237-239
Van de Graaff 334
Consulting Engineering-Physics Group, M Division 323, 328
Consultants Program 269, 270, 296, 306
Contracting Officer 49
procurement 47
salary policy 39, 43, 162
Contractor's representative (see Business Office; University of California)
contractors, construction 9, 33, 59, 60
controlled nuclear reaction
supercritical with prompt neutrons 199
Coordinating Council 30, 266, 272, 356-368
at Trinity 243
copper crusher gauges 116
copper spark analysis 136, 151, 152
cordite propellant 117
Cornell University 4, 305, 306, 313
delayed neutron experiments 20
velocity selector equipment 99
Corrosion Protection Group, CMR Division 329
Cosmic Rays Group, P Division 311, 321
counters 320, 322, 337
"Cowpuncher Committee" 158, 160, 169
Trinity 239
crater survey, Trinity 246
critical assemblies
enriched uranium hexafluoride 188
health hazard of 164, 270, 271, 293
investigation 177, 181, 197, 313, 324
plutonium-239 199, 309, 326
power producers 308, 309, 313
Critical Assemblies Group, G Division (later M Division) 166, 170, 197-199, 323, 325

466

467

Detroit Office 111, 113, 122, 131, 163, 194
  centerline plant 341
deuterium 447
  equation of state 154
  ignition difficulties 87
  liquefaction plant 21, 87, 154
  production and storage of liquid 88
  thermonuclear reaction 15, 86
  tritium reaction cross sections 86, 169, 188
deuterium bomb 185 (see also Super bomb)
Development, Engineering, and Tests Group,
  X Division 208, 232, 234, 338
Development of Substitute Materials (DSM)
  Project 3, 20, 64
Dickie granules in blood 292
Diffusion Problems Group 172, 304, 305
diffusion theory 76, 80
Diffusion Theory Group, T Division 172,
  173, 304-306
Director 31, 33, 34, 38, 49, 160, 232, 266,
  270, 272, 278 284, 303, 347, 352
Directorate of Tube Alloys 26 (see also Brit-
  ish mission)
discriminators 207
Division Leader
  Administrative (A) 34, 155
  Administration and Services (A and
  S) 278, 279
  Bikini Planning (B) 274, 345
  Chemistry and Metallurgy (CM, later
  CMR) 134, 217
  Engineering (E) 232
  Experimental Physics (R and F) 90
  Explosives (X) 155, 271, 272, 338
  F 155, 271
  M 323-328
  Ordnance (O) 155
  Ordnance Engineering (E, later Z) 271,
  272, 342, 345, 346
  Physics (P) 271, 310, 311
  Research (R) 271
  Theoretical (T) 76, 271, 305, 306
  Weapon Physics (G) 155, 271
Document Room 34, 52, 348
Documentary (D) Division 279, 347-352
DP Site (see sites)
DP Site Group, CM Division 217, 230
draft deferment policy 39, 40
Drafting Section, Shops Group 290, 350
"Dragon" 188, 198, 305, 308
drop tests 216
  Camel program 160, 196, 248, 250, 273
dry purification 138
  plutonium 142, 222
  uranium 138, 141, 218
dust-borne product survey 248

DuPont Company 88, 137

Editor 34, 161
Editorial Group, A Division (later,
  D Division) 279, 348, 350
Editorial Section 52
E Division (see Ordnance Engineering
  Division)
efficiency 12, 13
  blast 175, 274, 276
  early calculations of 18, 82-86, 175
  nuclear 175, 274, 276
  of implosion 174
Efficiency Theory Group, T Division 172,
  304, 305
Electric Detonators Group, G Division 191,
  197, 215, 247
electric
  circuits 207
  detonators (see detonators)
  fusing (see fusing)
  method 204
Electric Method Group, G Division (later, M
  Division) 197, 323
Electronics Group, G Division 53, 197, 207
Electronics Group, P Division 90, 99, 109,
  110, 310, 311, 314, 321
Electronic Engineering Group, Z Division 342
Electronics Mechanics 320
electroplated metal coating on plutonium 223
electrostatic generator (see Van de Graaff)
Electrostatic Generator Group 90, 93, 177
  boron absorption measurements 98
  capture cross-section measurements 182
  fission cross sections 179, 181
  fission spectrum measurement 96, 103,
  173, 178
  long counter 109
  mass spectrographic analysis 183
  measurement of branching ratio, ^{235}U 101,
  102
  multiplication experiment 180, 190
  neutron spectroscopy 97, 100
  scattering studies 103, 104
emission time after fission, neutron measurement 92
energy release 12
  estimates from expansion velocity of ball of
  fire 307
  nuclear fission 10
  nuclear measurements 343
  predictions 175
  Super 185
  test 171, 275
Engineering Group, O Division 191, 192, 216

468

469

470

High-Explosives Section, X Division 208, 247

High-Explosives Development Group, E Division 112, 191, 208, 232

High-Power Water Boiler 189, 224

High-Vacuum Research Group, CM Division 134, 153, 217

Hiroshima 254, 255, 260, 261, 268, 304, 307
and Nagasaki damage compared with Super 186, 187
bomb 182, 194, 220, 246
teletype from bomber plane 274, Fig. 4

History Group, A Division 279

Hospital 37, 53, 278, 283, 297, 299

hot chemistry laboratory 225, 226

hot plugs 325

hot pressing 147

housing 33
guest ranches 45
shop personnel 162
shortages 36, 41, 162, 282, 287
temporary, financing 45

Housing Office 35, 283, 298

hydride (uranium hydride) 18, 23, 65, 66, 138, 139, 177
bomb 97, 98, 103, 104
critical assemblies 107, 167, 170, 198, 308
gun 118
-plastic cubes 198, 218
implosion 69
program, abandonment 70, 118

Hydrodynamics Group, T Division 305

hydrodynamics of implosion 68, 76, 172, 304, 305, 353
contribution of Taylor 28, 85, 175, 448

hydroxide-oxalate process (see RaLa)

IBM calculations 81, 174 (see also efficiency; critical mass)
hydrodynamics of implosion 69, 76, 168

IBM Computations Group, T Division 172, 304-306

Illinois (see University of Illinois)

immersion testing (see underground or underwater testing)

implosion bomb (see also bomb; Fat Man)
assembly 67-70, 249, 343
asymmetries in 82
composite core 172, 173, 198, 259
design 28, 58, 122, 158, 159, 199, 273
dynamics 69, 129, 130, 177, 200, 210, 224, 304
efficiency calculations 166, 174-176
experiment (see Trinity)
investigation 157, 197, 199-205, 210, 211, 234, 259

implosion bomb (cont)

jets (see jets)
lens program 82, 168, 169
magnetic method 89
mockups, "pumpkins" 160
modulated neutron initiator 174
photography 130, 200
predetonation (see predetonation)
solid-core lens with modulated nuclear initiator 168
stability of convergent shock waves 174
temperature effects 174

Implosion Group 76, 125

Implosion Dynamics Group, T Division 172, 304

Implosion Experimentation Group 111, 191

implosion program
at Camel 159, 160, 196, 210, 216, 248, 250, 273
expansion and reorganization 74
growth and development 22, 124-130

Implosion Research Group, X Division 208, 210, 211

Implosion Steering Committee 112

implosion studies
betatron 128, 174, 200-202, 207, 213, 320, 321, 446
magnetic method 129, 200
RaLa method 129
x-ray studies 129

Implosion Studies Group, X Division 200

impulse gauges, Trinity 245

Indemnity Insurance Company of North America 161, 302

Indianapolis 253

indium cross sections 98, 107

Informer Group, Z Division 345, 346

"informers" 119, 120, 240, 314

initiator
chemistry 148, 149, 333
design, contributions of Bohr and Taylor 27, 28
development 67, 134, 148, 149, 169, 170, 174, 192, 197, 200, 205, 211, 324
modulated 74
neutron background measurement 183
polonium 73
radon-beryllium 73

Initiator Chemistry Group, CMR Division 330

Initiator Group, G Division (later, M Division) 158, 197, 204, 205, 224, 323

inspection (see testing)

instability, Taylor 28, 85, 175, 448

471

472

magnesium oxide impurities 228
magnetic coil velocity measurements 116
"Magnetic Method" 127, 129, 168, 200, 201, 213
Magnetic Method Group, G Division (later, M Division) 197, 323, 328
"magnetic model" 127
mail service, 38, 160, 283
Maintenance and Service Section, X Division 209
Maintenance Group, Tech Area 34, 60, 161, 288, 289
manganese cross sections 101, 104
Manhattan Engineer District
  auditing 287
  construction 312
  contract 6
  draft deferment 39
  Joint Task Force, Operation Crossroads 273, 274
  Laboratories 53, 270, 324
Manhattan Project Editorial Advisory Board 351, 352
maps 233, 234, Figs. 2 and 3
Mark IV primers 117
Marshall Islands 274
Martin, Nebraska plant 247
mass spectrographic method, isotopic analysis 107, 108, 182, 183
Massachusetts Institute of Technology (M.I.T.)
  crucibles 141, 224
  Radiation Laboratory 7, 10, 159, 249, 305, 306, 319
  research on refractories 136
Master Policies 46, 161
  postwar policy 265, 278, 353
Mathematical Methods Group, T Division 306
McDonald's ranch house, Trinity 240
McKee, R. E., construction 60, 162
McNierney cattle case 298
M Division 323-328
Measurement Group, Trinity 235
"mechanical chemist" 73, 150
Mechanical Engineering Group, Z Division 342, 346
Mechanical Laboratory Group, Z Division 346
Medical Group, Trinity 236
medical officer, Trinity 241
medical research 270, 292
Medical Section 37
  salary proposals 38
  site selection 4-6

specifications for original buildings 58-60
Memorial Hospital, New York 8
Messenger Service
Metal Fabrication Group, CMR Division 329, 330, 334, 335
Metal Physics Group, CMR Division 329, 330, 335
Metal Production Group, CMR Division 329, 330, 332, 335
Metallurgical Laboratory (see University of Chicago)
metallurgy (see also micrometallurgy)
  plutonium 71
  uranium 70
Metallurgy Group, CM Division 118, 224
Meteorology Group, Trinity 233, 235, 236, 239, 246
Michigan (see University of Michigan)
microflash photography 116
microtorsion blanche 224
microvolumetric assay 228
microwaves 116
military personnel 6, 40-42, 265-268, 285-289, 325, Graphs 2 and 3
  Alberta Project 247-258
  509th Composite Group, USAAF 247, 252
  Joint Task Force Operation, Crossroads 273-277, 294
  Maintenance group 288
  Navy 67, 111, 112, 115-117, 119, 120, 131, 194, 210, 250, 273, 340
  property records 287
  Special Battalion, U. S. Army 345
  Special Engineering Detachment 40-42, 128, 130, 281, 282, 285
Minnesota (see University of Minnesota)
Miscellaneous Metallurgy Group, CM Division 56, 134, 146, 217, 219, 224
Mitchell cameras, Trinity 245
mock bombs 160, 196
mock-fission source
  design 97
  multiplication experiment 180
Model Design, Engineering Service, and Consulting Group, X Division 209
Moffett Wind Tunnel 132
Mold Design Section 123
mold development 122, 123, 209
Molds Committee 123
molybdenum, blue 152
monitoring equipment 53, 55, 293, 294
monitoring personnel safety (see Health Groups)
Monroe jets 125

473

475

plutonium (cont)

discovery of ^{240}Pu 73, 74, 108, 136, 145, 146, 227

fission cross sections 98, 179, 315

foils 73

gallium oxide pyroelectric method 227

gun 67, 192, 220, 247

metallurgy and micrometallurgy 20, 31, 144-147, 222-224

multiplication experiments 182, 224

near-critical systems 312

neutron number measurements 64, 73, 167, 188, 190

oxidation and valence states 135

physical properties 223, 331

processing 3, 65, 218, 230, 331

production 336

projectile specifications 114

purification 24, 71, 72, 138, 141-144, 152, 170, 182, 183, 220-222, 229, 330, 336

rates of corrosion and diffusion 331

radioassays 142, 293, 332

reactor 312

recovery 72, 142

reduction 136, 223

self-diffusion studies 335

solubility tests 332

specific heat 335

spontaneous fission rates 73, 167, 178

sulfates and sulfides 228, 229

thermal conductivity 335, 336

toxicity 54, 55, 72, 163, 170, 220, 230, 293

uranium as stand-in 72, 141, 145

Plutonium Chemistry Group, CM Division 226

Plutonium Metallurgy Group, CM Division 217

Plutonium Production Group, CMR Division 329, 330

Plutonium Purification Group, CM Division 217

Plutonium Recovery Group, CM Division 226

polonium

chemical and metallurgical properties 331

extraction of 135

hazard 163, 293

initiators 67, 73, 95, 148, 149, 218

Joliot effect 73, 95

mock-fission sources 179

processing 205, 336

toxicity 170, 293

used in mock-fission sources 150

Polonium Chemistry Group, CMR Division 329

Polonium Group, CM Division 217, 230

postdetonation 14

Post Operations Office 60

Post Safety Engineer 289

Post Safety Section 289

Post Supply Section 48

powder metallurgy 71

Powder Metallurgy Group, Water Boiler Specifications 105

power boiler 164

power-producing devices 308, 309

predetonation 13, 67, 166, 167, 174

Prescott microgas analyzer 153

President's Interim Committee on Atomic Energy 259

pressure gauges 116, 244

Primacord systems 123, 168, 203, 205, 245

Prime Contractor, University of California 6

Princeton University 7, 8, 306, 323

PRM radar device 195

Process Development Group, CMR Division 330

procurement early difficulties 6, 33, 47-50

Procurement Group, A Division 279, 283, 285-287

Procurement Group, O Division 121, 191

Procurement Office 6, 34, 47, 160, 162, 163, 286, Graph 7

Procurement Section 50, 286

Production Group, Z Division 346

Production Section, X Division 209

Project A (see Alberta Project)

Project Technical Committee, Alberta Project 253

Project Trinity (TR) 234 (see also Trinity)

Project Y, selection of site 4

Projectile, Target, and Source Group, E Division 111, 118, 191

Prompt Measurements Group, Trinity 240

prompt neutrons 198, 199

propellants 114, 115, 117

property audit 287

Property Group (see Supply and Property Group)

property inventory 163, 287

Property Inventory Section, Procurement 50, 163

protactinium 98, 147, 178

Proving Ground Group 111, 115, 118, 191

proximity fuses (see fuses)

public liability 300-302

pulse-height analyzer, 10-channel 207

"pumpkin" program 160, 196

Purchasing Offices

Chicago 48, 49

local 48, 296-303

Los Angeles 6, 25, 43, 46-51, 300

476

477

478

481

War Production Board (WPB) 48
Warren Grove, New Jersey 121
Washington Liaison Office
  draft deferment 40
  overseas communications 257
Washington State University 321
Washington University, St. Louis, Missouri,
  34, 35, 270, 293
water baffle recovery 215
water column effects 276
Water Boiler 104-107
  chemistry 149, 150, 218, 333
  critical mass calculation 79, 104
  development 56, 57, 155, 189, 313
  early discussion of 20
  first successful operation 74, 75, 91
  health hazards 164
  problems 226
  tamper 147
  thermal neutron calculations 79, 80
Water Boiler Group, P Division 90, 169, 177,
  181, 184, 190, 198, 226, 294, 310, 311,
  315
Water Delivery and Exterior Ballistics Group,
  O Division 191, 196
water shortage 271, 317, 318
Weapon Engineering Program 272-273
Weapon Physics (G) Division 107, 123, 130,
  152, 155, 156, 158, 167, 173, 177, 181,
  183, 191, 197-207, 210, 247, 271, 272,
  338
Weapons Committee 122, 157, 158, 192,
  248-250
Weapons Panel 272
Weather Division, USAAF (see also
  Meteorology Group, Trinity)

Welfare Fund 297, 301, 302
Wendover Field, Utah 159, 160, 194, 195,
  216, 247-250, 273
West Area, DP Site 230
Westinghouse Research Laboratories 8, 33,
  139
wet purification
  plutonium 142-144, 220, 221
  uranium 138, 141, 218, 219
wind tunnel testing 131, 132
Wisconsin (see University of Wisconsin)
Women's Army Corps 40, 41 (see also military personnel)
Workmen's Compensation, New Mexico 44,
  46, 300-302 (see also Welfare Fund;
  public liability)

X Division (see Explosives` Division)
x rays
  charge examination 199, 214, 249
  flash photography 129, 200-203, 205, 210,
    211, 213
X-Ray Method Group, G Division 197, 201
X units (see Fat Man)

Y-12 Plant 89
yield (see energy release)
Yorktown Naval Munitions Depot 210, 250

Z Division (see Ordnance Engineering
  Division)
Zia Company 283, 284, 287, 288, 298, 299
zirconium
  analysis 141, 152
  cupferron precipitable refractories 152
  nitride 147

482

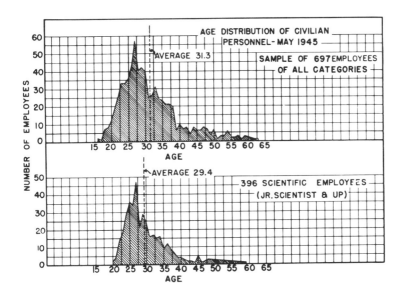

Graph 1. Age distribution of civilian personnel in May 1945.

Two curves are shown, one a sample of all employees, the other of scientific employees only. The averages for both are low—29.4 for the scientists and 31.3 for the others—with 27 the most probable age for both. Actually, there is only one man over 58 among the scientific employees. These figures emphasize the importance of the draft deferment problem. Information was obtained from the active card file of the Personnel Division in June 1945.

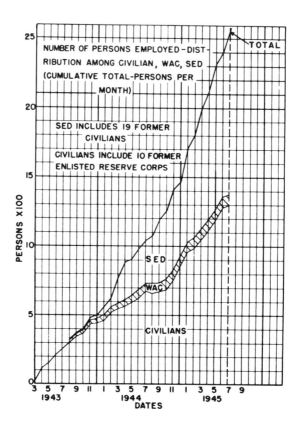

Graph 2. Number of persons employed and distribution among civilians, WAC, and SED.

Shows sharp and continuous increase of personnel from beginning of project. Civilians increase at a steady rate, WAC contingent remains about the same, and SED contingent increases very rapidly. Information was obtained from records of Technical Area and SED personnel offices.

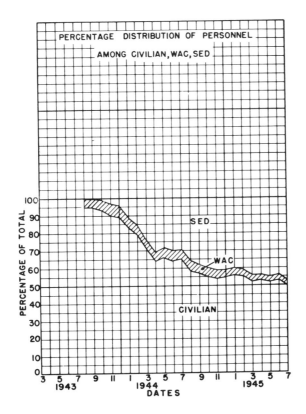

Graph 3. Percentage distribution of personnel among civilians, WAC, and SED.

Data of previous graph replotted on percentage basis. Project changed from being 100% civilian during first five months to 50% civilian in July 1945.

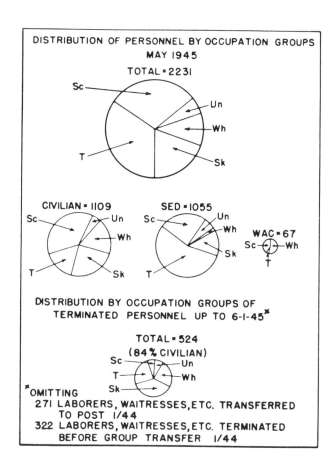

Graph 4. Distribution of personnel by occupation groups.

Classification of personnel into five large categories, according to occupation, as of May 1945. In the chart for the total number, one sees the preponderance of scientific and technical personnel; in the civilian chart, the preponderance of scientific personnel; in the SED chart, the preponderance of technical personnel. The chart of terminations shows the very small proportion of scientific personnel terminating and the relatively large proportion of skilled labor terminating. The latter fact reflects some of the difficulties encountered by the shops in retaining personnel, as well as a difference in motivation. Information was obtained from card files in Tech area and SED personnel offices.

Un - Unskilled (Laborer, Messenger, or Warehouse Assistant)

Wh - White Collar (Clerk, Secretary, Nurse, or Teacher)

Sk - Skilled (Machinist, Toolmaker, or Glassblower)

T - Technical (Technician, Draftsman, or Scientific Assistant)

Sc - Scientific & Administrative (Junior Scientist and up)

486

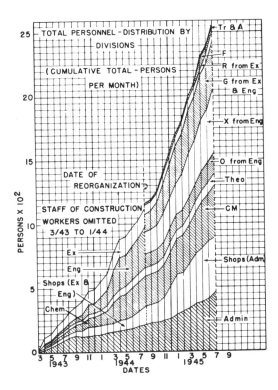

Graph 5. Total personnel distribution by divisions.

Shows growth of various divisions, reflects change in emphasis from research to engineering, especially after reorganization in August 1944. Engineering divisions G, X, and O assume large proportions while research divisions R and T remain small. Information was obtained from group assignment records of Tech Area and SED personnel offices. Abbreviations and letters refer to various divisions.

Exp	Shops	R	O	Chem
Eng	Admin	G	T	F
Theo	Tr & A	X	CM	

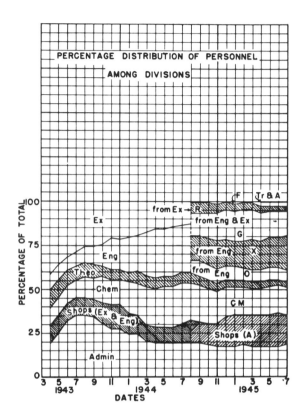

Graph 6. Percentage distribution of personnel among divisions.

Data of previous graph replotted on percentage basis. Shops, G, X and O account for more than half of total personnel. Abbreviations and letters refer to various divisions.

Exp	Shops	R	O	Chem
Eng	Admin	G	T	F
Theo	Tr & A	X	CM	

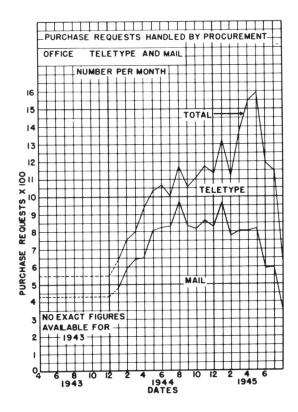

PURCHASE REQUESTS HANDLED BY PROCUREMENT OFFICE TELETYPE AND MAIL

NUMBER PER MONTH

TOTAL

TELETYPE

MAIL

NO EXACT FIGURES AVAILABLE FOR 1943

PURCHASE REQUESTS X 100

DATES

Graph 7. Purchase requests handled by procurement office by teletype and mail.

Total number of requests handled each month by Procurement during 1944 and part of 1945. Mail requests represent bulk of routine business; teletype requests those items needed with special urgency. Peak month, especially for teletype requests, was May 1945, in preparation for Trinity. Note the sharp slump which follows. Each request involves at least 60 pieces of paper, according to Procurement records. Information was obtained from a monthly record of purchase requests kept in the request file section.

489

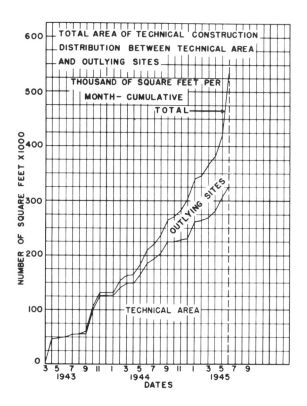

Graph 8. Total area of technical construction in Technical Area and outlying sites.

Shows steady growth of construction both in Technical Area and outlying sites. Sharp rate of increase in outlying site construction in June 1945 represents completion of first buildings at DP site. For more information see Site map, Fig. 3. Information was obtained from files of D. Dow in Director's Office, files of Post Construction Officer, and files of W. C. Kruger, Project Architect.

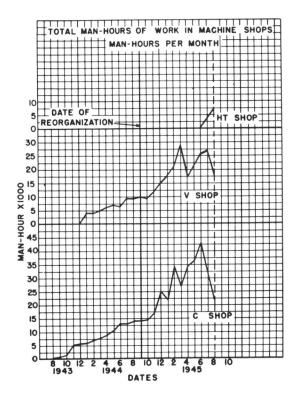

Graph 9. Total man-hours of work in machine shops.

Shows rapid shops expansion after reorganization. Slump in January and sharp rise in February indicate results of fire. Peak of activity in C Shop in June preparatory to Trinity, followed by sharp decrease in activity; one-month lag in peak for V Shop, but same sharp decrease follows. Information was obtained from weekly records kept in office of machine shops.

491

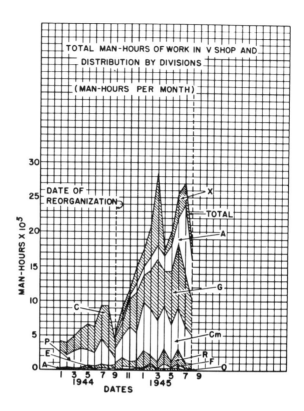

Graph 10. Total man-hours of work in V Shop and distribution by divisions.

Shows largest proportion of work done in V Shop for G and CM Divisions. Work done for A Division represents work done for shops themselves. Decrease in activity for all divisions except A after June 1945. Information was obtained from weekly records kept in machine shop office. Letters refer to various divisions.

C	E	A	G	F
P	R	X	CM	O

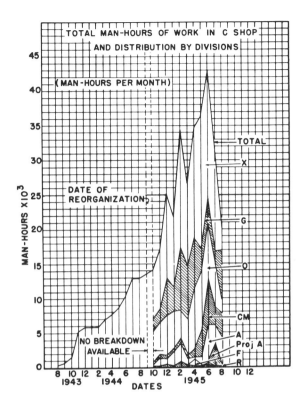

Graph 11. Total man-hours of work in C Shop and distribution by divisions.

Shows largest proportion of work done in C Shop for X Division. Fire accounts for slump in activity in January; no apparent reason for subsequent slump in March. Information was obtained from weekly records kept in machine shop office. Letters refer to various divisions.

X	CM	F	O
G	A	R	Project A

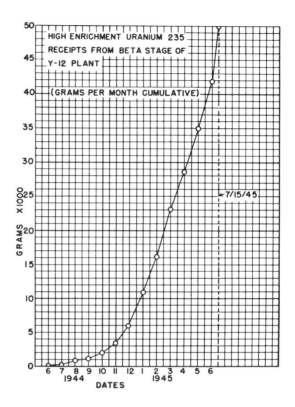

Graph 12. High-enrichment ²³⁵U receipts from beta stage of Y-12 plant.

Cumulative total of ²³⁵U received up to date of Trinity test. This represents all highly enriched ²³⁵U produced by the District. Such material was shipped after the final processing done in the beta stage of the Y-12 plant at Oak Ridge. Enrichment of ²³⁵U in tuballoy increased from 63% to 89%. Information was obtained from records of receipts in Director's Office, now filed with the Quantity Control Section of the Chemistry and Metallurgy Division.

OFFICE OF SCIENTIFIC RESEARCH AND DEVELOPMENT

1530 P STREET NW.

WASHINGTON, D. C.

ANNEVAR BUSH
Director

February 25, 1943

Dr. J. R. Oppenheimer
University of California
Berkeley, California

Dear Dr. Oppenheimer:

We are addressing this letter to you as the Scientific
Director of the special laboratory in New Mexico in order to confirm our many conversations on the matters of organization and
responsibility. You are at liberty to show this letter to those
with whom you are discussing the desirability of their joining
the project with you; they of course realizing their responsibility
as to secrecy, including the details of organization and personnel.

I. The laboratory will be concerned with the development
and final manufacture of an instrument of war, which we may desig-
nate as Projectile S-1-T. To this end, the laboratory will be
concerned with:

A. Certain experimental studies in science,
engineering and ordnance; and

B. At a later date large-scale experiments
involving difficult ordnance procedures
and the handling of highly dangerous
material.

The work of the laboratory will be divided into two periods in
time: one, corresponding to the work mentioned in section A; the
other, that mentioned in section B. During the first period, the
laboratory will be on a strictly civilian basis, the personnel,
procurement and other arrangements being carried on under a contract arranged between the War Department and the University of
California. The conditions of this contract will be essentially
similar to that of the usual OSRD contract. In such matters as
draft deferment, the policy of the War Department and OSRD in regard
to the personnel working under this contract will be practically
identical. When the second division of the work is entered upon
(mentioned in B), which will not be earlier than January 1, 1944,
the scientific and engineering staff will be composed of commissioned
officers. This is necessary because of the dangerous nature of the

Fig. 1. Groves-Conant letter. This is the original directive of the Los Alamos Laboratory.

worκ and tne need for special conditions of security. It is expected that many of those employed as civilians during the first
period (A) will be offered commissions and become members of the
commissioned staff during the second period (B), but there is no
obligation on the part of anyone employed during period A to accept
a commission at the end of that time.

    II. The laboratory is part of a larger project which has
been placed in a special category and assigned tne highest priority
by tne President of tne United States. By his order, the Secretary
of War and certain other high officials have arranged that the
control of this project shall be in the hands of a Military Policy
Committee, composed of Dr. Vannevar Bush, Director of OSRD, as
Chairman, Major General W. D. Styer, Chief of Staff, SOS, Rear
Admiral W. R. Purnell, Assistant Chief of Staff to Admiral King;
Dr. James B. Conant serves as Dr. Bush's deputy and alternate on
this Committee, but attends all meetings and enters into all discussions. Brigadier General L. R. Groves of the Corps of Engineers
has been given over-all executive responsibility for this project,
working under the direction of the Military Policy Committee. He
works in close cooperation with Dr. Conant, who is Chairman of the
group of scientists who were in charge of the earlier phases of
some aspects of the investigation.

    III. Responsibilities of tne Scientific Director.

    1. He will be responsible for:

    a. The conduct of the scientific work so that
the desired goals as outlined by tne Military Policy
Committee are achieved at the earliest possible dates.

    b. The maintenance of secrecy by tne civilian
personnel under his control as well as their families.

    2. He will of course be guided in his determination of
policies and courses of action by tne advice of his scientific
staff.

    3. He will keep Dr. James B. Conant and General Groves
informed to sucn extent as is necessary for them to carry on the
work which falls in their respective spheres. Dr. Conant will be
available at any time for consultation on general scientific
problems as well as to assist in tne determination of definite
scientific policies and research programs. Through Dr. Conant
complete access to tne scientific world is guaranteed.

Fig. 1. (cont).

Dr. J. R. Oppenheimer - 3 - February 25, 1943

IV. Responsibilities of the Commanding Officer.

1. The Commanding Officer will report directly to General Groves.

2. He will be responsible for:

a. The work and conduct of all military personnel.

b. The maintenance of suitable living conditions for civilian personnel.

c. The prevention of trespassing on the site.

d. The performance of duty by such guards as may be established within the reservation for the purpose of maintaining the secrecy precautions deemed necessary by the Scientific Director.

V. Cooperation.

The closest cooperation is of course necessary between the Commanding Officer and the Scientific Director if each is to perform his function to the maximum benefit of the work. Such a cooperative attitude now exists on the part of Dr. Conant and General Groves and has so existed since General Groves first entered the project.

Very sincerely yours,

James B. Conant

Leslie R. Groves

Fig. 1 (cont).

TECHNICAL AREA AS OF DECEMBER 1942

Building No.	Designation
7	Infirmary
16	Gatehouse
25	T - Main Tech Building
26	U - Chem. and Phys. Labs
27	V - Shop (Machine)
28	W - Van de Graaff
29	Y - Cryogenics Lab
30	X - Cyclotron
31	Z - Cockcroft-Walton
32	Covered walk
33-36	Ranch houses
37	Chem. Stock
41	Warehouse
42	Icehouse
44	Boiler
47	Guard tower
48	Ranch house - PX
56	Cooling towers

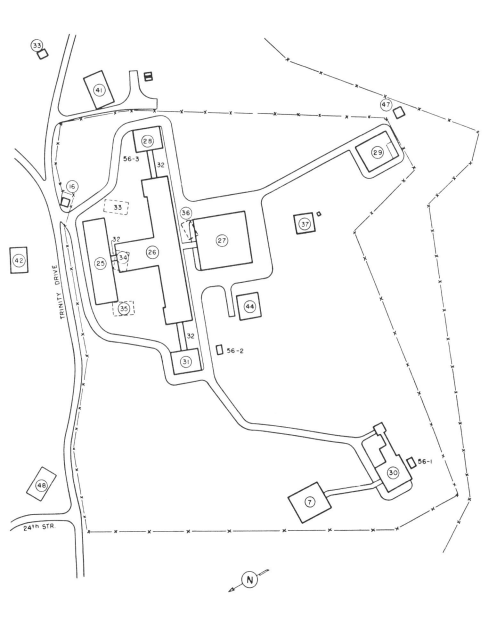

Fig. 2. Technical Area plot map. Map showing building layout of the Technical Area, as drafted in December 1942. Technical Buildings T, U, V, W, X, Y, and Z were constructed as map indicates.

499

Scale 1.8 in. = squares are 1/2 mile by 1/2 mile

Hard-surfaced roads   _____
Foot trails   . . . .
Site and designation   ▼ · VI
Water supply main   —< —< —< —<
Power line   .—.—.—.—.—.—
Firing sites   ★
DP Site   ⌐__⌐__⌐__⌐__⌐

Number	Site	Division	N/S Coordinate	E/W Coordinate
I	Post Technical Area		100	135
II	Omega	G	93	121
III	South Mesa	G	89	158
IV	Alpha	G	68	108
V	Beta	G	69	94
VI	2-Mile Mesa (upper)	X	74	171
VII	2-Mile Mesa (lower)	Q	69	147
VIII	Anchor Gun Site	O	65	184
IX	Anchor HE	X	65	183
X	Bayo	G	107	71
XI	K	G	38	157
XII	L	X	59	139
XIII	P	G	47	171
XIV	Q	X	52	152
XV	R	X	49	138
XVI	S	X	46	187
XVII	X	G	72	192
XVIII	Pajarito	O-X	45	91
XIX	East Gate Lab	R	93	72
XX	Sandia	G	77	82

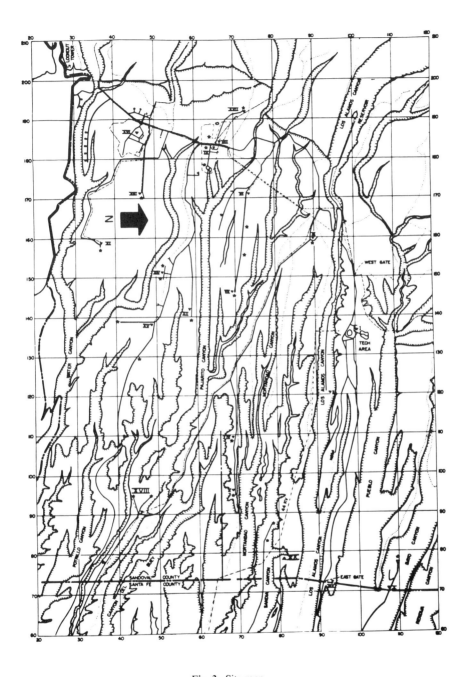

Fig. 3. Site map.

WA 137

FROM WASH LIAISON OFC WASH DC AUG 450520402

TO COMMANDING OFFICER CLEAR CREEK

FIVE PARTS- TART ONE

SW

KC

FLASHED FROM THE PLANE BY PARSONS ONE FIVE MINUTES AFTER RELEASE

AND RELAYED HERE WAS THIS INFORMATION QUOTE PAREN REF EIDM WL

TO OPPENHEIMER FROM GENERAL GROVES THIS RESUME OF MSSSAGES PREPARED

BY DOCTOR MANLEY PAREN CLEAJ CUT RESULTS COMMA IN ALL RESPECTS SUCCES

FUL PD EXCEEDED TR TEST IN VISIBLE EFFECTS PD NORMAL CONDITINXXXXX

CONDITIONS OBTAINED IN AIRCRAFT AFTER DELIVERY WAS ACCOMPLISHED PD

VISUAL ATTACK ON HIROSHIMA AT ZERO FIVE TWO THREE ONE FIVE Z WITH

ONLY ONE TENTH CLOUD COVER PD FLACK AND FIGHTERS ABSENT UNQUOTE AFTER

RTXXXXX RETURN TO BASE AND GENERAL INTERROGATION FARRELL SENT THE

FOLLOWIGXXXX FOLLOWING INFORMATION QUOTE ALARGE OPENING IN CLOUD

COVER DIRECTLY OVER TARGET MADE BOMBING FAVORABLE PD EXCELLENT RECORD

REPORTED FROM FASTAX PD FILMS NOT YET PROCESSED BUT OTHER OBSERVING

XXXXXXSOALSO ANTICIPATE GOOD TREXXXX RECORDS NXX PD NO APPRE

QQXD QCFA

R NIL

K HOW MANY LINES DID U GET

R 12 LINWA

PLANES ALSO ANTICIPATE GOOD RCXXX RECORDS PD NO APPRECIABLE NOTICE OF

SOUND PD BRIGHT DAYLIGHT CAUSED FLASH TO BE LESS BLINDING THAN TRPXXX

TR PD A BALL OF FIRE CHANGED IN A FEW RECORDS TO PURPLE CLOUDS AND

BOILING AND UPWARD SWIRLING FLAMES PD TURN JUST COMPLETED WHEN FLASH

WAS AXXX OBSERVED PD INTENSLY BRIGHT LIGHT CONCEALED BY ALL AND RATE

OF RISE OF WHITE CLOUD FASTER THAN AT TR PD IT WAS ONE THIRD GREATER

IN DIAMETER REACHING THIRTY THOUSAND FEET IN THREE MINUTES PD MAXIMUM

ALTITUDE AT LEAST FORTY THOUSAND FEET WITH FLATTENED TOP AT THIS

LEVEL PD COMBAT AIRPLANE THREE HUNDRED SIXTY THREE MILES AWAY AT

TWENTYXXVXXXXXXXTHOUSAND FEET OBSERVEDIT PD D

NIL AGN

.. OK OPR WELL JUST HAVE TO KEEP TRING AS THESE MESSAGES AR IMP

MIN PLS

OPR U STARTED THIS MSG AS PART TWO ISNT IT PART OF PART ONE

M MIN OPR I TOLD U I WD START PART TWO WHERE PART ONE NILED

IS THAT CLEAR

BUT OPR I DIDNT GET PART ONE COMPLETE

AND THE I TOLD TO U TO SA START WITH 12 LINE

AND THE 12 LINE U L O WELL I THOT U MEANT U GOT 12 OK

M THIS IS A AWFUL MESS ISNT IT IT SH SURE IS DOU THINMI WNGEFG

MIN PLS

TRY ANOTHER MACHINE MAYBE IT WILL DO VETTER

OPR IT ISNT UG MACH AND I KNOW IT ITS MINE AND THERE ISNT

A THING CAN BE DONE AS THE REPAIR MAN SAYS THERE ISNT ANYTHING WRONG

WITH IT HES BEEN HERE ALL DAY AND THIS IS AS GOOD AS IT WAILL RUN

I HAVE LOADS TO GO UXX TO U TONIGHT BUT WELL HAVE TO DO IT THIS WAY

A FEW LINES AT A TIME   MIN I WANT TO TALK TO THE LT A MIN

OK

OPR ILL CALL U BACK IN A BT 10 MINUTES

Fig. 4. Hiroshima teletype announcing success of Hiroshima mission received at Los Alamos from Washington office, prepared by Manley. Note comments by teletype operators at end. They were Technician Third Class Flora L. Little of Jackson, Mississippi, in the Washington office and Technician Third Class Mildred Weiss of New Orleans, Louisiana, in the Los Alamos office.

TRINITY PROJECT DETAIL LOCATION PLAN

Station	Group Leader	Symbol
Piezo Gauge	Walker	×
Sentinel (Type A)	Moon	⊕
Sentinel (Type B)	Moon	✳
Geophone	Houghton	△
Paper Box Gauge	Hoogterp	□
Flash Bomb	Mack	◪
R 4 Ground Station	Segrè	⊠
R 4 Balloon Winch	Segrè	♀
E. D. G.	Moon	‡
Mack Slit Camera	Mack	ᚱ
Impulse Meter	Jorgensen	◓
Condenser Gauge	Bright	⯇
Excess Velocity Gauge	Barschall	⊕
Tank Range Poles	Anderson	△
Tank Flag Poles	Anderson	�የ
Primacord Station	Mack	-δ
Metal Stake (Earth Disp)	Penney	○
Piezo Gauge Amplifier	Walker	◎
Balloon	Richards	♀
Balloon Winch	Richards	⊖
Ground Station	Richards	⊕

Roads ══ ══ ══ ══ ══ ══

Buried Wires or Cables — — — — — — — — — -

Center Lines - ———— - ———— - ———— - —

Tank Right of Way ⌐__⊥__⊥__⊥__⊥__⊥__⊥__⊥__⌐

Note: Angles are Azimuths on "OA" Line
Distances thus (800) are Radial Yards from "O"
Distances thus (75') are Offsets from L of Roads and Center Lines.

Scale: 1500 Yard circle - 1" = 300 Yards. - Sheet 1
10,000 Yards - 1" = 2750 Yards. - Sheet A

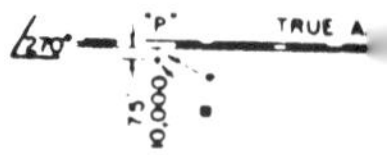

504

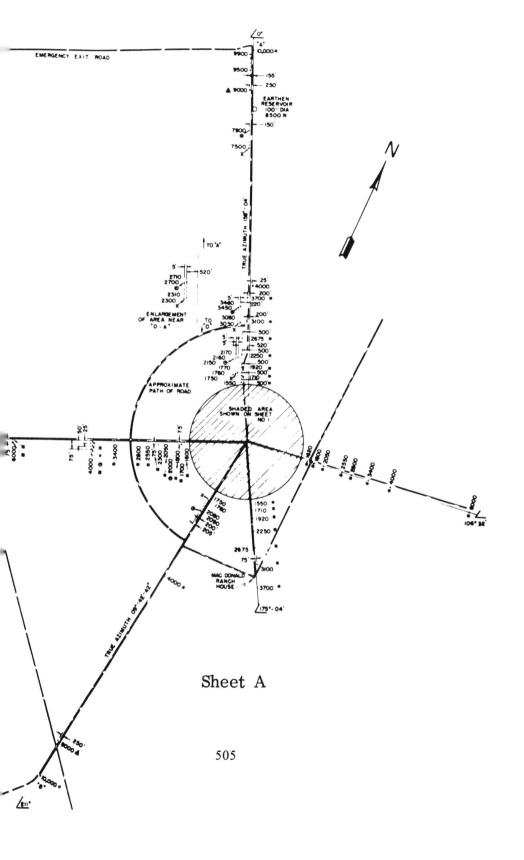

Sheet A

505

Sheet 1

TRINITY PROJECT DETAIL LOCATION PLAN

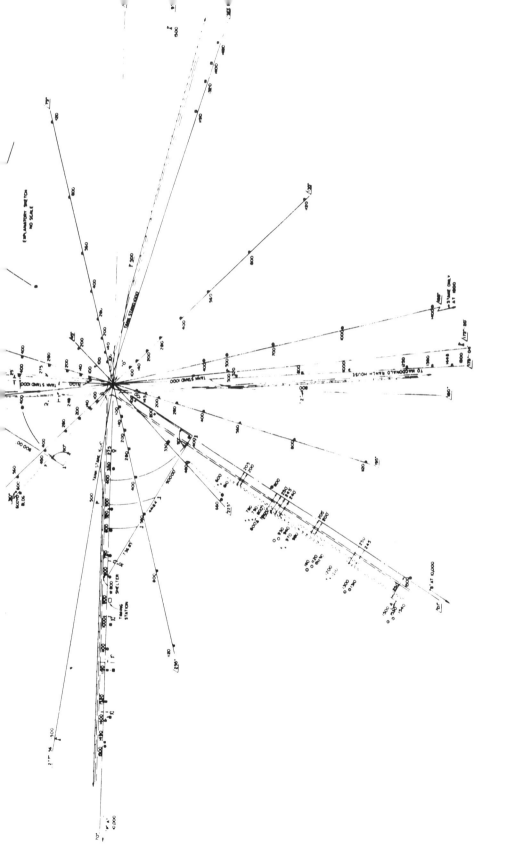

*This limited edition,
produced with the cooperation of the
Los Alamos Laboratory,
is printed on Howard Permalife Olde White Text,
bound in full buckram-type blue cloth,
with blue Elephant Hide matching slipcase.*

—

Design and production by Joseph Simon